Innovative Applications in Smart Cities

Editors

Alberto Ochoa-Zezzatti
Universidad Autónoma de Ciudad Juárez

Genoveva Vargas-Solar
French Council of Scientific Research (CNRS)
Laboratory of Informatics on Images and Information Systems
France

Javier Alfonso Espinosa Oviedo
University of Lyon, ERIC Research lab
France

CRC Press
Taylor & Francis Group
Boca Raton London New York

CRC Press is an imprint of the
Taylor & Francis Group, an informa business
A SCIENCE PUBLISHERS BOOK

First edition published 2021
by CRC Press
6000 Broken Sound Parkway NW, Suite 300, Boca Raton, FL 33487-2742

and by CRC Press
2 Park Square, Milton Park, Abingdon, Oxon, OX14 4RN

Library of Congress Cataloging-in-Publication Data

Names: Ochoa Ortiz Zezzatti, Carlos Alberto, 1974- editor. | Vargas-Solar, Genoveva, 1971- editor. | Espinosa Oviedo, Javier Alfonso, 1983- editor.
Title: Innovative applications in smart cities / editors, Alberto Ochoa-Zezzatti, Universidad Ciudad Juárez, México, Genovera Vargas-Solar, French Council of Scientific Research (CNRS), Laboratory of Informatics on Images and Information Systems, Cedex, France, Javier Alfonso Espinosa Oviedo, University of Lyon, ERIC Research Lab, Cedex, France.
Description: First edition. | Boca Raton : CRC Press, Taylor & Francis Group, 2021. | "A science publishers book." | Includes bibliographical references and index. | Summary: "This research book is a novel, innovative and adequate reference that compiles interdisciplinary perspectives about diverse issues related with Industry 4.0 and Smart Cities on different ways about Intelligent Optimisation, Industrial Applications on the real world, Social applications and Technology applications with a different perspective about existing solutions. Chapters report research results improving Optimisation related with Smart Manufacturing, Logistics of products and services, Optimisation of different elements in the time and location, Social Applications to enjoy our life of a better way and Applications that increase Daily Life Quality. This book is organised into three scopes of knowledge: (1) applications of Industry 4.0; (2) applications to improve the life of the citizens in a Smart City; and finally (3) research associated with the welfare of the working-age population and their expectations in their jobs correlated with the welfare - work relationship"-- Provided by publisher.
Identifiers: LCCN 2021000974 | ISBN 9780367820961 (hardcover)
Subjects: LCSH: Smart cities.
Classification: LCC TD159.4 .I486 2021 | DDC 307.760285--dc23
LC record available at https://lccn.loc.gov/2021000974

ISBN: 978-0-367-82096-1 (hbk)
ISBN: 978-1-032-04256-5 (pbk)
ISBN: 978-1-003-19114-8 (ebk)

Typeset in Times New Roman
by Radiant Productions

Preface

"Innovation" is a moto in the development of current and future Smart Cities. Innovation understood by newness, improvement and spread, is often promoted by Information and Communication Technologies (ICTs) that make it possible to automate, accelerate and change the perspective of the way economy and "social good" challenges can be addressed.

In economics, innovation is generally considered to be the result of a process that brings together various novel ideas to affect society and increase competitiveness. In this sense, future Smart Cities societies' economic competitiveness is defined as increasing consumers' satisfaction given by the right products price/quality ratio. Therefore, it is necessary to design production workflows that maximise the resources used to produce the right quality products and services. Companies' competitiveness refers to their capacity to produce goods and services efficiently (decreasing prices and increasing quality), making their products attractive in global markets. Thus, it is necessary to achieve high productivity levels that increase profitability and generate revenue. Beyond the importance of stable macroeconomic environments that can promote confidence, attract capital and technology, a necessary condition to build competitive societies is to create virtuous creativity circles that can propose smart and disruptive applications and services that can spread across different social sectors strata.

Smart Cities have been willing to create technology-supported environments to make urban, social and industrial spaces friendly, competitive and productive contexts in which natural and material resources can be accessible to people, where citizens can develop their potential skills in the best conditions possible. Since countries in different geographic locations, natural, cultural and industrial ecosystems have to adapt their strategies to these conditions, Smart Cities solutions are materialised differently. This book shows samples of experiences where industrial, urban planning, health and sanitary problems are addressed with technology leading to disruptive data and artificial intelligence centred applications. Sharing applied research experiences and results mostly applied in Latin American countries, authors and editors show how they contribute to making cities and new societies smart through scientific development and innovation.

Contents

Preface iii

Prologue vii
Khalid Belhajjame

Part I: Daily Life in a Smart City

1. **Segmentation of Mammogram masses for Smart Cities Health Systems** 1
Paula Andrea Gutiérrez-Salgado, Jose Mejia, Leticia Ortega, Nelly Gordillo, Boris Mederos and *Alberto Ochoa-Zezzatti*

2. **Serious Game for Caloric Burning in Morbidly Obese Children** 10
José Díaz-Román, Alberto Ochoa-Zezzatti, Jose Mejía-Muñoz, Juan Cota-Ruiz and *Erika Severeyn*

3. **Intelligent Application for the Selection of the Best Fresh Product According to its Presentation and the Threshold of Colors Associated with its Freshness in a Comparison of Issues of a Counter in a Shop of Healthy Products in a Smart City** 22
Iván Rebollar-Xochicale, Fernando Maldonado-Azpeitia and *Alberto Ochoa-Zezzatti*

4. **Analysis of Mental Workload on Bus Drivers in the Metropolitan Area of Querétaro and its Comparison with three other Societies to Improve the Life in a Smart City** 34
Aarón Zárate, Alberto Ochoa-Zezzatti, Fernando Maldonado and *Juan Hernández*

5. **Multicriteria analysis of Mobile Clinical Dashboards for the Monitoring of Type II Diabetes in a Smart City** 47
Mariana Vázquez-Avalos, Alberto Ochoa-Zezzatti and *Mayra Elizondo-Cortés*

6. **Electronic Color Blindness Diagnosis for the Detection and Awareness of Color Blindness in Children Using Images with Modified Figures from the Ishihara Test** 75
Martín Montes, Alejandro Padilla, Julio Ponce, Juana Canul, Alberto Ochoa-Zezzatti and *Miguel Meza*

7. **An Archetype of Cognitive Innovation as Support for the Development of Cognitive Solutions in Smart Cities** 89
Jorge Rodas-Osollo, Karla Olmos-Sánchez, Enrique Portillo-Pizaña, Andrea Martínez-Pérez and *Boanerges Alemán-Meza*

Part II: Applications to Improve a Smart City

8. **From Data Harvesting to Querying for Making Urban Territories Smart** 107
Genoveva Vargas-Solar, Ana-Sagrario Castillo-Camporro, José Luis Zechinelli-Martini and *Javier A. Espinosa-Oviedo*

9. **Utilization of Detection Tools in a Human Avalanche that Occurred in a Rugby Stadium, Using Multi-Agent Systems** 117
Tomás Limones, Carmen Reaiche and *Alberto Ochoa-Zezzatti*

10. **Humanitarian Logistics and the Problem of Floods in a Smart City** 135
Aztlán Bastarrachea-Almodóvar, Quirino Estrada Barbosa, Elva Lilia Reynoso Jardón and *Javier Molina Salazar*

11. **Simulating Crowds at a College School in Juarez, Mexico: A Humanitarian Logistics Approach** 145
Dora Ivette Rivero-Caraveo, Jaqueline Ortiz-Velez and *Irving Bruno López-Santos*

12. **Perspectives of State Management in Smart Cities** 155
Zhang Jieqiong and *Jesús García-Mancha*

Part III: Industry 4.0, Logistics 4.0 and Smart Manufacturing

13. **On the Order Picking Policies in Warehouses: Algorithms and their Behavior** 165
Ricardo Arriola, Fernando Ramos, Gilberto Rivera, Rogelio Florencia, Vicente Garcia and *Patricia Sánchez-Solis*

14. **Color, Value and Type Koi Variant in Aquaculture Industry Economic Model with Tank's Measurement Underwater using ANNs** 186
Alberto Ochoa-Zezzatti, Martin Montes Rivera and *Roberto Contreras Masse*

15. **Evaluation of a Theoretical Model for the Measurement of Technological Competencies in the Industry 4.0** 203
Norma Candolfi-Arballo, Bernabé Rodríguez-Tapia, Patricia Avitia-Carlos, Yuridia Vega and *Alfredo Hualde-Alfaro*

16. **Myoelectric Systems in the Era of Artificial Intelligence and Big Data** 216
Bernabé Rodríguez-Tapia, Angel Israel Soto Marrufo, Juan Miguel Colores-Vargas and *Alberto Ochoa-Zezzatti*

17. **Implementation of an Intelligent Model based on Big Data and Decision Making using Fuzzy Logic Type-2 for the Car Assembly Industry in an Industrial Estate in Northern Mexico** 229
José Luis Peinado Portillo, Alberto Ochoa-Zezzatti, Sara Paiva and *Darwin Young*

18. **Weibull Reliability Method for Several Fields Based Only on the Modeled Quadratic Form** 235
Manuel R. Piña-Monarrez and *Paulo Sampaio*

Index **267**

Prologue

Khalid Belhajjame

Nowadays, through the democratisation of internet of things and highly connected environments, we are living in the next digitally enriched generation of social media in which communication and interaction for user-generated content are mainly focused on improving the sustainability of smart cities. Indeed, the development of digital technologies in the different disciplines in which cities operate, either directly or indirectly, alters expectations among those in charge of the local administration and of citizens. Every city is a complex ecosystem with a lot of subsystems to make it work, such as work, food, clothes, residence, offices, entertainment, transport, water, energy, etc. As they grow, there is more chaos and most decisions are politicized, there are no common standards and data is overwhelming. The intelligence is sometimes digital, often analogue, and almost inevitably human.

The smart cities initiative aims to better exploit the resources in a city to offer higher-level services to people. Smart cities are related to sensing the city's status and acting in new intelligent ways at different levels: people, government, cars, transport, communications, energy, buildings, neighbourhoods, resource storage, etc. A smart city is much more than a high-tech city, it is a city that takes advantage of the creativity and potential of new technologies to meet the challenges of urban life. A smart city also helps to solve the sensitive issues of its citizens, such as insecurity, urban mobility problems, water resources management and solid waste. They are not the instruments that make a smart city, but everything that is achieved through the implementation of those processes.

A vision of the city of the "future", or even the city of the present, rests on the integration of science and technology through information systems. This vision implies re-thinking the relationships between technology, government, city managers, business, academia and the research community. Conclusions and actions are determined by the social, cultural and economic reality of the cities and the countries in which they are located. Therefore, beyond smart cities as an object of study, it is important to think about urban spaces that can host smart cities of different types and built with different objectives, maybe those that have more priority and that can ensure the well-being of citizens.

Smart Cities and Urban Computing applications

Smart cities where the focus was on how cable, telephones and other wired media were changing our access to services [1,8,22] have been around for a long time. Today, the concept "smart city" refers to a world initiative leading to better exploit the resources in a city to offer value-added services to people [15] considering at least four components: industry, education, participation, and technical infrastructure [5].

University Paris Dauphine, LAMSADE, France.
Email: Khalid.Belhajjame@dauphine.fr

The advent of technology and the existence of the internet have helped transform traditional cities into cities that are more impressive and interactive. There are terms analogous to "smart cities", such as a digital, intelligent, virtual, or ubiquitous city. The definition and understanding of these terms determine the way challenges are addressed and projects are proposed to go towards a "smart urban complex ideal" [2,6,21].

Overall, the evolution of a city into a smart city must focus on the fact that network-based knowledge must not only improve the lives of those connected but also bring those who remain unconnected into the fold, creating public policies that truly see the problems faced by big cities and everyday citizens. Several smart cities in the most important capitals of the world and well-known touristic destinations have developed urban computing solutions to address key issues of city management, like transport, guidance to monuments, e-government, access to leisure and culture, etc. In this way, citizens of different socio-economic groups, investors and government administrators can have access to the resources of the city in an optimised and personalised manner. Thus, a more intelligent and balanced distribution of services is provided thanks to technology that can improve citizens life and opportunities. This has been more or less possible in cities where the socio-economic and technological gap is not too great. Normally, solutions assume that cities provide good quality infrastructure, including internet connection, access to services (energy, water, roads, health), housing, urban spaces, etc. Yet, not all cities are developed in these advantageous conditions, there are regions in the world where exclusion prevails in cities and urban spaces, where people have little or no access to electricity, technology and connectivity, and where services are not regulated. It is in this type of city that smart cities technology and solutions face their greatest challenges.

In Mexico, projects on Smart Cities have been willing to promote sustainable urban development through innovation and technology. The objective of the smart cities project has addressed the improvement of life quality for inhabitants. Areas promoted in Smart Cities in Mexico are quite diverse, ranging from environment, safety and urban design to tourism and leisure. This book describes solutions to problems in these areas. Chapters describing use cases are also analysed to determine the degree of improvement of citizens quality of life, human logistics within urban spaces, of the logistics strategies and the access and distribution of services like transport, health or assistance during disasters and critical events. The experiments described along the chapters of the book are willing to show the way academics, inspired in living labs promoted in other cities, have managed to study major smart cities problems and provide solutions according to the characteristics of the cities, the investment of governments and industry and the willingness of people to participate in this change of paradigm. Indeed, citizen participation is a cornerstone that must not be left aside. After all, it is citizens who are beginning transformation and who constantly evaluate the results of information integration. Citizen satisfaction is the best way to calibrate a smart city's performance.

Urban computing[1] is defined as the technology for acquisition, integration, and analysis of big and heterogeneous data generated by a diversity of sources in urban spaces, such as sensors, devices, vehicles, buildings, and human, for tackling the major issues that cities face. The study of smart cities as complex systems is addressed through this notion of urban computing [23]. Urban computing brings computational techniques to bear on urban challenges such as pollution, energy consumption, and traffic congestion. Using today's large-scale computing infrastructure and data gathered from sensing technologies, urban computing combines computer science with urban planning, transportation, environmental science, sociology, and other areas of urban studies, tackling specific problems with concrete methodologies in a data-centric computing framework.

[1] Computing and Smart Cities Applications for the Knowledge Society. Available from: https://www.researchgate.net/publication/301271847_Urban_Computing_and_Smart_Cities_Applications_for_the_Knowledge_Society [accessed Jul 21 2020].

Table 1: Urban computing applications.

Urban planning	• Gleaning Underlying problem in Transportation Networks • Discover Functional Regions • Detecting a City's Boundary
Transportation	• Improving Driving Experiences • Improving taxi Services: dispatching, recommendation, ride sharing • Improving Public Transportation Systems: bus, subway, bike
Environment	• Air quality • Noise pollution
Social & Entertainment	• Estimate user similarity • Finding local experts in a region • Location recommendation • Itinerary planning • Life patterns and styles understanding
Energy	• Gas consumption • Electricity consumption
Economy	• Finding trends of city economy • Business placement
Safety & Security	• Detecting traffic anomalies: distance based, statistics based • Disaster detection and evacuation

Table 1 presents a summary of the families of applications that can be developed in the context of urban computing: urban planning, transportation, environment, social and entertainment, energy, economy and safety and security. Often, these applications can be organised on top of a general urban computer framework reference architecture, enabling platforms that provide the technical underlying infrastructure [16] necessary for these applications to work and be useful for the different actors populating and managing urban territories.

In urban computing, it is vital to be able to predict the impact of change in a smart city's setting. For instance, how will a region's traffic change if a new road is built there? To what extent will air pollution be reduced if we remove a factory from a city? How will people's travel patterns be affected if a new subway line is launched? Being able to answer these kinds of questions with automated and unobtrusive technologies will be tremendously helpful to inform governmental officials' and city planners' decision making. Unfortunately, the intervention-based analysis and prediction technology that can estimate the impact of change in advance by plugging in and out some factors in a computing framework is not well studied yet. The objective would be to use this technology to reduce exclusion and make citizens' life more equal. How to guide people through urban spaces with little or no land registry? How to compute peoples' commute from home to work when transport is not completely regulated? How to give access to services through applications that can be accessible for all? For example, Latin American cities that appear as first in the Smart Cities rankings (Buenos Aires, Santiago de Chile, São Paulo, Mexico City) are megacities of more than 10 million inhabitants. For many Smart Cities ideologies, the big urban "spots" are the antithesis of the ideas and values of a truly smart city. Thus, there is room for scientific and technological innovation to design smart cities solutions in these types of cities and thereby tackle requirements that will make citizens' lives better. This book is original in this sense because it describes smart cities solutions for problems in this type of city. It provides use case examples of prediction solutions for addressing not only smart cities issues, but urban computing as a whole.

Smart Cities as Living Laboratories

At the beginning of 2013, there were approximately 143 ongoing or completed self-designated smart city projects: North America had 35 projects, Europe 47, Asia 50, South America 10, and the Middle East and Africa 10. In Canada, Ottawa's "Smart Capital" project involves enhancing businesses, local government, and communities using Internet resources. Quebec was a city highly dependent upon its provincial government because of its weak industry until the early 1990s when the city government kicked off a public-private partnership to support a growing multimedia sector and high-tech entrepreneurship. In the United States, Riverside (California) has been improving traffic flow and replacing ageing water, sewer and electric infrastructure through a tech-based transformation. In San Diego and San Francisco, ICT have been major factors in allowing these cities to claim to be a "City of the Future" for the last 15 years. Concerning Latin America, the Smart Cities council recognizes the eight smartest cities in Latin America: Santiago (Chile), Mexico City (Mexico), Bogota (Colombia), Buenos Aires (Argentina), Rio de Janeiro (Brazil), Curitiba (Brazil), Medellin (Colombia) and Montevideo (Uruguay). Each city focusses on different aspects, including automating pricing depending on traffic, smart and eco-buildings, electrical and eco-subway, public Wi-Fi and public tech job programs, weather, crime, emergency monitoring, university and educational programs.

In Mexico, the first successful Smart City project "Ciudad Maderas" (2013–2020) was developed in Querétaro in the central part of the country. This project included the construction of technology companies, hotels, schools, shopping centres, residential areas, churches and huge urban spaces dedicated as a natural reserve in El Marques district. The purpose has been to integrate technological developments into the daily lives of Queretaro's inhabitants. Concerning e-governance, the State Government has launched the Querétaro Ciudad Digital Application. The purpose of this application is to narrow the gap between the citizens and the government. The application is regarded worldwide as second-to-none technology. Cities like Mexico City have focused on key services, such as transportation. A wide range of applications is readily available to residents to accomplish their daily journeys from A to B: Shared Travel Services, Uber, Easy, Cabify. Since 2014, the city of Guadalajara has been working on the Creative Digital City project to promote the digital and creative industry in the region. The city of Tequila, also in the state of Jalisco, promotes the project "Intelligent Tequila" for attracting tourism to the region. One of the smart technologies already in use are heat sensors which help to measure massive concentrations in public places. In Puebla, the Smart Quarter project develops solutions for improving mobility, safety and life quality for the inhabitants. For example, proposing free Wifi in public areas and bike tracks equipped with video – monitoring and alarm systems.

The European Union has put in place smart city actions in several cities, including Barcelona, Amsterdam, Berlin, Manchester, Edinburgh, and Bath. In the United Kingdom, almost 15 years ago, Southampton claimed to be the country's first smart city after the development of its multi-application smartcard for public transportation, recreation, and leisure-related transactions. Similarly, Tallinn has developed a large-scale digital skills training program, extensive e-government, and an award-winning smart ID card. This city is the centre of economic development for all of Estonia, harnessing ICT by fostering high-tech parks. The European Commission has introduced smart cities in line 5 of the Seventh Framework Program for Research and Technological Development. This program provides financial support to facilitate the implementation of a Strategic Energy Technology plan [4] through schemes related to "Smart cities and communities".

Statistics of the Chinese Smart Cities Forum report six provinces and 51 cities have included Smart Cities in their government work reports in China [14,17]; of these, 36 are under new concentrated construction. Chinese smart cities are distributed densely over the Pearl and Yangtze River Deltas, Bohai Rim, and the Midwest area. Moreover, smart cities initiatives are spread in all first-tier cities, such as Beijing, Shanghai, and Shenzhen. The general approach followed in this city is to introduce some ICT during the construction of new infrastructure, with some attention to environmental issues but limited attention to social aspects. A modern hi-tech park in Wuhan is considered an urban complex that is multi-functional and ecological. Wuhan is a city that is high-tech and self-sufficient. It is an eco-smart city designed for exploring the future of the city. It is a natural and healthy environment and incubator of high culture and expands the meaning of the modem hi-tech park [18]. Taipei City clarified that the government must provide an integrated infrastructure with ICT application and service [10]. China has encouraged the transition to urbanism by improving public services and improving efficiency for transformation to the government model, enhancing the economic development of the city. In 2008, a "digital plateau" was proposed; in 2009, more than ten provinces set goals to build a smart city. China improved the construction of the city, industrial structure, and social development. To start the implementation strategy, a good plan is necessary, as well as knowledge of the importance of smart city construction.

Several Southeast Asian cities, such as Singapore [11], Taiwan, and Hong Kong, are following a similar approach, promoting economic growth through smart city programs. Singapore's IT2000 plan was designed to create an "intelligent island," with information technology transforming work, life, and play. More recently, Singapore has extensively been dedicated to implementing its Master Plan in 2015 and has already completed the Wireless@SG goal of providing free mobile Internet access anywhere in the city [7]. Taoyuan in Taiwan is supporting its economy to improve the quality of living through a series of government projects such as E-Taoyuan and U-Taoyuan for creating e-governance and ubiquitous possibilities.

Korea is building the largest smart city initiative in Korea, Songdo, a new town built from the ground in the last decade and which plans to house 75,000 inhabitants [13]. The Songdo project aims at developing the most wired smart city in the world. The project is also focused on buildings and has collaborated with developers to build its networking technology into new buildings. These buildings will include telepresence capabilities and many new technologies. The plan includes installing telepresence in every apartment to create an urban space in which every resident can transmit information using various devices [12], whereas a city central brain should manage the huge amount of information [19]. This domestic offering is only the first step; Cisco aims to link energy, communications, traffic, and security systems into one smart network [20]. At present, there are 13 projects in progress towards the smart city initiatives of New Songdo [9].

Despite an increase in projects and research to create smart cities, it is still difficult to provide cities with all the features, applications, and services required. Future smart cities will require a re-thinking of the relationships between technology, government, city managers, business, academia and the research community. This book shows some interesting results of use cases where different communities and sectors interact to find alternative solutions for cities that are willing to become smart and address their problems innovatively and effectively.

Driving Urban Phenomena Foresight

Data science and urban computing have been developing and applying a great number of analytics pipelines to model and predict the behaviour of urban spaces as complex systems. For example, there are projects devoted to addressing adaptive street lighting to modulate public lighting, i.e., to locally adjust the luminous intensity of each lamp post according to parameters, and to take into account maintenance of the equipment as precisely as possible (anticipate failures). This regulation is done according to external conditions: luminosity, but also humidity level or even with presence sensors (pedestrians coming, car traffic). Systems for industrial predictive maintenance and prioritisation of interventions can be implemented applying data science pipelines that can use operational research/ multi-criteria optimization applied to light modulation. The expected benefits are a reduction in consumption and maintenance costs and an improvement in the quality of service (feeling of security for citizens, immediate replacement of defective streetlamps). Finally, by adjusting light intensity, cities reduce the level of light pollution, thus improving their aesthetics and their impact on the immediate environment.

Multi-channel urban transport projects also introduce phenomena and situations foresight requirements. Urban transport can exploit collected data sharing in order to offer the passenger a global offer based on all the means of transport in a city, i.e., multi-channel transport. An intermodal platform can consolidate information on the use and operation of all means of transport at the local authority level (bus, tramway, bicycles, car, transport conditions). Thanks to this consolidated and interpreted information, the city can offer its citizens the most appropriate solution, taking into account the context and requirements of the traveller. In terms of Data Science, just as for parking in the city, intermodality is first and foremost a subject of optimisation of resources under constraints. More broadly, the addition of new data (video, traffic conditions) and the identification of non-linear patterns (formation of traffic jams, congestion measurement) makes the subject rich and complex. Finally, scoring and customer knowledge algorithms are exploited to take into account user preferences to improve the recommendation. Perhaps it is not necessary to systematically propose bicycle travel times to a daily bus user?

Providing fluid displacement of people within urban spaces is also an important problem that can be solved through data science. Focussing on traffic management that is a daily problem in cities, the coordination of traffic lights can be an effective means of regulating road traffic and limiting congestion situations. By smoothing the flow and reducing the number of vehicles passing through bottlenecks, it is possible to increase the daily flow and reduce the level of traffic congestion. For instance, the reduction in speed on the ring road has had the effect of reducing traffic jams and congestion. Despite the counter-intuitive side of this effect, the physical explanation comes from the fact that by reducing the maximum speed, the amplitude of speed variations has been reduced (fluid mechanics enthusiasts already know that it is better to be in laminar rather than chaotic flow situations). Another contribution of the knowledge of traffic conditions is the use of models to test the impact of road works or the construction of new infrastructures. In terms of data science, the aim is to identify forms of congestion and to detect and use the most effective means of contrasting them. Among them, one can generally act on the maximum speed allowed or on the regulation of the timing of traffic lights according to traffic conditions (detected via cameras or GPS signals), visibility conditions and, in general, the weather, the presence of pedestrians, the time of day or other parameters that emerge as significant. Perhaps the most direct benefit expected is the reduction of traffic jams and slowdowns and, indeed, the time required for a trip. Other collateral benefits are the reduction of air and noise pollution and a reduction in the number of accidents caused by traffic problems.

This book provides examples of solutions that can be proposed through data science solutions that apply machine learning, data mining and artificial intelligence methods to data intended to make people live better in their daily lives as citizens of cities with unfair distribution of services. It also shows how data science can contribute to human logistics problems, particularly in the presence of critical events happening in cities with few infrastructures or with a huge population. The use cases are issued from the Mexican context, but they can be present in Latin American cities and, thus, solutions can be reproduced in cities with similar characteristics. Having systems and solutions that can promote foresight and planning are key in these kinds of urban places.

Book content and organisation

The book consists of eighteen chapters organized into three parts, namely, Daily Life in a Smart City (part I), Applications to Improve a Smart City (Part II) and Industry 4.0, Logistics 4.0 and Smart Manufacturing (Part III). These general topics address problems regarding smart cities as environments where citizens behaviour, health, commercial preferences and use of services (e.g., transport) can be observed. As mentioned before, the originality of the chapters is that they address topics regarding cities in Latin American countries, in particular Mexican cities, where citizens behaviour models change given the socio-cultural diversity and the unequal access to services.

- *Part I: Daily Life in a Smart City* consists of seven chapters that focus on important problems in Latin American smart cities. In these smart cities, solutions must deal with massive protocols that can use technology to develop and implement strategies for dealing with diseases common in the Latin American population, like obesity, breast cancer, colour-blindness and mental workload.

- *Part II: Applications to Improve a Smart City*. The way people move around in cities and urban spaces gives clues as to when many events of interest come up and in which hotspots. Part II of the book consists of five chapters that describe technology-based techniques and approaches to observing civilians' behaviour and their quality of life. This part starts with an initial chapter that surveys techniques for dealing with smart cities data, including collection strategies, indexing and exploiting them through applications. Then, the four remaining chapters address the analysis of citizens' behaviour with the aim of proposing strategies for dealing with human logistics in the presence of critical events (human avalanches, floods) and the way services distribution to the population can be improved (distribution of shelter and evacuation routes).
- *Part III: Industry 4.0, Logistics 4.0 and Smart Manufacturing*. Mexico is the Latin American country with the highest number of Smart Cities that offer economic advantages for they represent niches enabling potential economic activities that have not been considered yet; for example, the 4.0 technology sustainable promotion, the energy and agricultural sectors. A wide range of growth opportunities lie ahead for companies falling in these categories. In the long run, Smart Cities push forward towards Mexican economy diversification. This part of the book consists of six that address the important services that activate the economy of smart cities. Indeed, smart manufacturing is an important activator of industry smart cities, and it is activated through techniques that are being developed in Industry and Logistics 4.0 ecosystems. Along with its six, this part addresses algorithms for managing orders in warehouses and supply chains in sectors like automotive industry and retail; and the impact of using technology and data analytics methods in the aquaculture industry.

In conclusion, this book is important because it shows that it is possible and important to show how key problems in non-ideal urban contexts, like the ones that characterize some cities in Latin America and particularly in Mexico, can find solutions under the perspective of smart cities. Being smart is maybe a good opportunity for academia, government and industry to work with society and find alternatives to reduce unequal access to services, as well as sustainability and exclusion in the complex urban megapolis.

Bibliography

[1] Abdulrahman, A., Meshal, A. and Imad, F.T.A. 2012. Smart Cities: Survey. Journal of Advanced Computer Science and Technology Research, 2(2): 79–90.
[2] Deakin, M. and Al Waer, H. 2011. From intelligent to smart cities. Intelligent Buildings International, 3: 3, 140–152.
[3] Douglas, D. and Peucker. T. 1973. Algorithms for the reduction of the number of points required to represent a line or its caricature. Canadian Cartographer, 10(2): 112–122.
[4] Duravkin, E. 2010. Using SOA for development of information system; Smart city International Conference on Modern Problems of Radio Engineering. Telecommunications and Computer Science (TCSET).
[5] Giffinger, R., Fertner, C., Kramar, H., Kalasek, R., Pichler-Milanović, N. and Meijers, E. 2007. Smart Cities: Ranking of European Medium-sized Cities (Vienna: Centre of Regional Science, 2007).
[6] Gil-Castineira, F., Costa-Montenegro, E., Gonzalez-Castano, F.J., Lopez-Bravo, C., Ojala, T. and Bose, R. 2011. Experiences inside the Ubiquitous Oulu Smart City. Computer, 44(6): 48–55.
[7] IDA Singapore, "iN2015 Masterplan", 2012, http://www.ida.gov.sg/~/media/Files/Infocomm%20Landscape/iN2015/Reports/realisingthevisionin2015.pdf.
[8] Ishida, Toru. 1999. Understanding digital cities. Kyoto Workshop on Digital Cities. Springer, Berlin, Heidelberg.
[9] Ishida, T. 2002. Digital City Kyoto. Communications of the ACM, 45: 7, 78–81.
[10] Jin Goo, K., Ju Wook, J., Chang Ho, Y. and Yong Woo, L. 2011. A network management system for u-city. 13th International Conference on Advanced Communication Technology (ICACT).
[11] Kloeckl, Kristian, Oliver Senn and Carlo Ratti. 2012. Enabling the real-time city: LIVE Singapore! Journal of Urban Technology, 19.2: 89–112
[12] Kuikkaniemi, K., Jacucci, G., Turpeinen, M., Hoggan, E., Mu, X. et al. 2011. From Space to Stage: How Interactive Screens Will Change Urban Life. Computer, 44(6): 40–47,

[13] Lee, J.H., Hancock, M.G. and Hu, M. 2014. Towards an effective framework for building smart cities: Lessons from Seoul and San Francisco. Technological Forecasting and Social Change.
[14] Liu, P. and Peng, Z. 2013. Smart Cities in China. IEEE Computer Society Digital Library, http:// doi.ieeecomputersociety.org/10.1109/MC.2013.149.
[15] Pan, Yunhe et al. 2016. Urban big data and the development of city intelligence. Engineering, 2.2: 171–178.
[16] Psyllidis, Achilleas et al. 2015. A platform for urban analytics and semantic data integration in city planning. International conference on computer-aided architectural design futures. Springer, Berlin, Heidelberg.
[17] Shi, L. 2011. The Smart City's systematic application and implementation in China. International Conference on Business Management and Electronic Information (BMEI).
[18] Shidan, C. and Siqi, X. 2011. Making Eco-Smart City in the future. International Conference on Consumer Electronics, Communications and Networks (CECNet),
[19] Shwayri, S.T. 2013. A Model korean ubiquitous eco-city? The politics of making songdo. Journal of Urban Technology, 20: 1, 39–5.
[20] Strickland, E. 2011. Cisco bets on South Korean smart city. Spectrum, IEEE, 48(8): 11- Kaufmann.
[21] Townsend, A.M. 2013. Smart Cities: Big Data, Civic Hackers, and the Quest for a New Utopia, New York: W.W. Norton & Company.
[22] Vanolo, A. 2014. Smart mentality: The smart city as disciplinary strategy. Urban Studies, 51: 5, 883–898.
[23] Zheng, Yu et al. 2014. Urban computing: concepts, methodologies, and applications. ACM Transactions on Intelligent Systems and Technology (TIST), 5.3: 1–55.

PART I

Daily Life in a Smart City

CHAPTER-1

Segmentation of Mammogram masses for Smart Cities Health Systems

Paula Andrea Gutiérrez-Salgado, Jose Mejia, Leticia Ortega, Nelly Gordillo, Boris Mederos* and *Alberto Ochoa-Zezzatti*

One of the fundamental aspects of smart cities is an improvement in the health sector, by providing its citizens with better care and prevention and detection of diseases. Breast cancer is one of the most common diseases and the one with the highest incidence in women worldwide. In smart cities, to improve the quality of life of its citizens, especially for women, is to diagnose breast tumors in shorter periods with simpler and automated methods. In this chapter, a new deep learning architecture is proposed to segment breast cancer tumors.

1. Introduction

According to the World Health Organization (WHO), breast cancer is one of the most common diseases with the highest incidence in women worldwide, about 522 thousand deaths are estimated annually with data collected in 2012 (OMS, 2019). In smart cities, as regards to the health sector, it seeks to improve the quality of life of its citizens (Kashif et al., 2020; Abdelaziz et al., 2019; Rathee et al., 2019), thus there is a need to diagnose breast tumors in shorter periods with simpler and automated methods that can produce accurate results. The most common method to an early diagnostic is through mammographic images, however, these images usually have noise and low contrast which can cause the doctor to have difficulty classifying different tissues. In some mammogram images, malignant tissues and normal dense tissues are presented, but it is difficult to contrast between them by applying simple thresholds when automatic methods are used (Villalba, 2016). Because of these problems, it is necessary to develop various approaches that can correctly identify the malignant tissues, which represent higher intensity values compared to background information and other regions of the breast. Also, regions where some normal dense tissues have intensities similar to the tumor region have to be excluded (Singh et al., 2015).

Universidad Autónoma de Ciudad Juárez, Av. Hermanos Escobar, Omega, 32410 Cd Juárez, Chihuahua, México.
* Corresponding author: jose.mejia@uacj.mx

The interpretation of a mammogram is usually difficult, sometimes it depends on the experience of medical staff. In approximately 9% of the cancers detected, tumors were visible on mammograms obtained from two years earlier (Gutiérrez et al., 2015). The key factor for early detection is the use of computerized systems. The segmentation of tumors takes a very important role in the diagnosis and timely treatment of breast cancer. Currently, there are methods to delimit tumors using artificial neuronal networks (Karianakis et al., 2015; Rafegas et al., 2018) and deep learning networks (Hamidinekoo et al., 2018; Goodfellow et al., 2016), but there is the possibility of improving them.

In this chapter, a new architecture with aims to segment mammary tumors in mammograms using deep neural networks is proposed.

2. Literature Review

There exist several diagnostic methods to perform timely detection, the use of mammography being the method most used by medical staff because of the effective and safe results of the method. The examination is carried out by firm compression of the breast between two plates, using ionizing radiation to obtain images of breast tissue, which can be interpreted as benign or malignant (Marinovich et al., 2018). Here, we review some methods for an automatic segmentation/detection of malignant masses by processing the mammography image.

In (Hanmandlu et al., 2010), a comparison of two different semi-automated methods was performed, using level sets method and watershed controlled by markers. Although both methods are not very accurate, they were found to have a short processing time.

In the work of (Lempka et al., 2013), two automated methods were presented based on the improvement of region growing and the segmentation with cellular neural networks. In the first stage, the segmentation was carried out through an automated region growing whose threshold is obtained through an artificial neural network. In the second method, segmentation is performed by cellular neural networks, whose parameters are determined by a genetic algorithm (GA). Intensity, texture and shape characteristics are extracted from segmented tumors. The GA is used to select appropriate functions from the set of extracted functions. In the next stage, ANNs are used to classify mammograms as benign or malignant. Finally, they evaluate the performance of different classifiers with the proposed methods, such as multilayer perceptron (MLP), vector support machines (SVM) and K-nearest neighbors (KNN). Among these methods, the MLP produced better diagnostic performance in both methods. The sensitivity, specificity and accuracy indices obtained are 96.87%, 95.94%, and 96.47%, respectively (Lempka et al., 2013).

In (Wang et al., 2014), a breast tumor detection algorithm was proposed in digital mammography based on extreme machine learning (ELM). First, they use a median filter for noise reduction and contrast improvement as data pre-processing. Next, wavelet transforms, morphological operations, and the region growing are used for the segmentation of the edge of the breast tumor. Then, they extract five textural features and five morphological features. Finally, they use ELM classifier to detect breast tumors. In the comparison of the detection of breast tumors based on SVM, with the detection of breast tumors based on ELM, not only does ELM have a better classification accuracy than the SVM, but also a much-improved training speed. Also, the efficiency of classification, training and performance testing of SVM and ELM were compared and the total number of errors for ELM was 84 while the total number of errors for SVM was 96, showing that ELM has better abilities than SVM (Wang et al., 2014).

In (Pereira et al., 2014), a computational method is presented as an aid to segmentation and mass detection in mammographic images. First, a pre-processing method based on Wavelets transformation and Wiener filtering was applied for image noise removal and enhancement. Subsequently, a method was used for mass detection and segmentation through the use of thresholds, the Wavelet transform, and a genetic algorithm. The method was quantitatively evaluated using the area overlay metric (AOM). The mean standard deviation value of AOM for the proposed method was 79.2% ± 8%. The method they propose presented a great potential to be used as a basis for the massive segmentation of mammograms in the craniocaudal and mid-lateral oblique views.

The work of (Pandey et al., 2018) presents an automatic segmentation approach that is carried out in three steps. First, they used adaptive Wiener filtering and media clustering to minimize the influence of noise, preserve edges and eliminate unwanted artifacts. In the second step, they excluded the heart area using a set of levels based on the active contour, where the initial contour points were determined by the maximum entropy threshold and the convolution method. Finally, the pectoral muscle is removed through the use of morphological operations and local adaptive thresholds in the images. The proposed method was validated using 1350 breast images of 15 women, showing excellent segmentation results compared to semi-automated methods drawn manually.

In (Chougrad et al., 2018) a computerized diagnostic system was developed based on deep convolutional neural networks (CNN) that use transfer learning, which is ideal for handling small data sets, such as medical images. After training some CNN architectures, they used precision and AUC parameters to evaluate images from different databases, such as DDSM, INbreast, and BCDR. The CNN model, named Inception v3, obtained the best results with an accuracy of 98.94%, and so it was used as a basis to build the Breast Cancer Screening Framework. To evaluate the proposed CAD system and its efficiency to classify new images, they tested it in a database different from those used previously (MIAS) and obtained an accuracy of 98.23% and 0.99 AUC (Chougrad et al., 2018).

3. Methodology

In this section, we present the methodology used for the development of an architecture based on deep learning neural networks, for segmenting tumors in digital mammography.

The schematic of the methodology is presented in Figure 1.

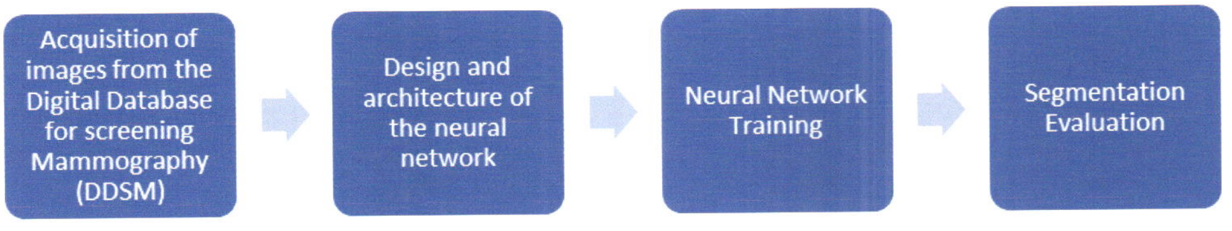

Figure 1: Diagram of the methodology.

The images used in this chapter are from the CBIS-DDSM which is a subset of the Database for Screening Mammography (DDSM) which is a database with 2620 mammography studies. It contains normal, benign and malignant cases with verified pathological information. This database is a useful tool in the testing and development of decision support systems. The CBIS-DDSM collection includes a subset of the DDSM data selected by a trained medical doctor. The images were decompressed and converted to the DICOM format. The database also includes updated ROI segmentation and delimitation tables, and pathological diagnosis for training data (Lee et al., 2017). In ROI annotations for anomalies in the CBIS-DDSM data subset, they provide the exact position of the lesions and their coordinates to generate the segmentation mask. In Figure 2, three images obtained from the CBIS-DDSM database are shown, with tumors.

3.1 Preprocessing of the images

The original size of the images obtained from the database was 6511 × 4801 pixels. A mask for the tumor contained in each image was obtained from the database. Then, the region that contained the tumor on both image and mask were trimmed, and the clipping was reduced to a 60 × 60-pixel image, this procedure was done to make the network faster and save memory since the

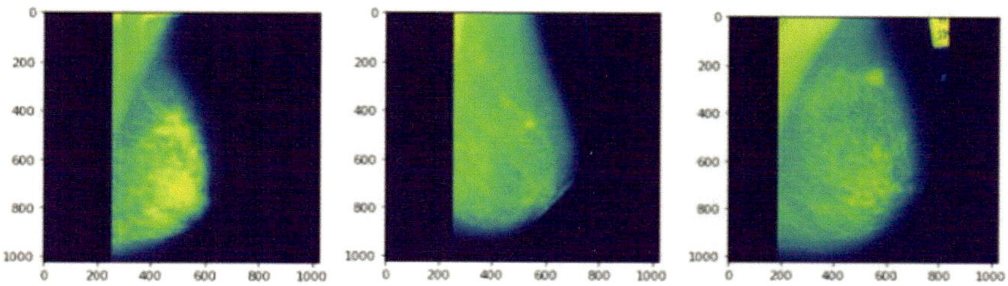

Figure 2: Examples of images obtained from the CBIS-DDSM database; images are shown with false color.

original size was too large to process. The procedure was repeated for each image in the database. In Figure 3, it is shown an image from the database, and the clipping of the region of the tumor and its mask.

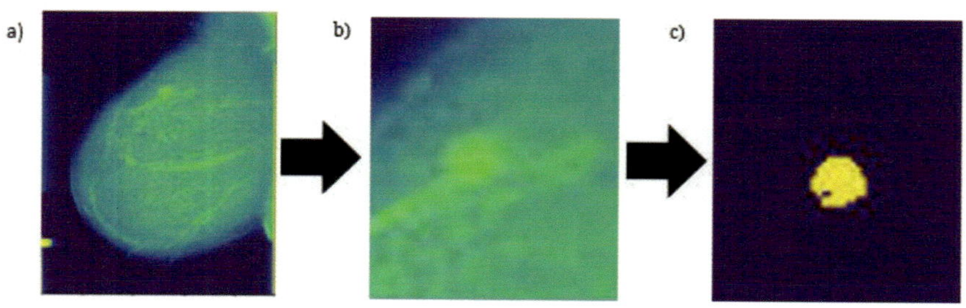

Figure 3: Pre-processing of acquired images. (a) Original mammographic image. (b) Trim delimiting the tumour area. (c) Tumour mask located by the given coordinates.

3.2 Deep learning architecture design

The network architecture is a modification of the architecture of (Guzman et al., 2018). In our modified architecture, we add a channel, using each channel for a different purpose. Thus, our architecture consists of three channels (X1, X2, X3). The input image is the mammogram, and it is directed to the three channels, each containing kernels of different size. The idea is that each channel extracts features of different sizes that help the network with its task.

- Channel X1 (for larger size features) consists of a convolutional layer with 25 filters of size 9×9.
- The second channel X2 (for medium size features) consists of a convolutional layer with 40 4×4 filters, followed by a 2×2 Maxpooling layer, then another 3×3 convolutional layer with 35 filters and finally a 2×2 size UpSampling layer.
- In the third channel X3 (for small size features), begins with a convolutional layer of 35 filters with a size of 2×2, a MaxPooling layer of 2×2, a convolutional layer of 50 filters of 2×2, a MaxPooling layer equal to the previous one, another convolutional layer with 35 filters of 3×3 and a 4×4 UpSampling.

The three channels are concatenated and then the output goes through three convolutional layers, the first two with a size of 7×7 with five and seven filters. The last convolutional layer consists of 1 filter with a size of 1×1. The network has as output, a mask over the tumor. Figure 4 shows the architecture and an example of network input and output.

Note that several other state-of-the-art architectures (Karianakis et al., 2015; Noh et al., 2015) were tested without having a favorable result. For the training of the network, a batch of 330 images was used, and we train the network with 3500 epochs. Cross entropy was used as a loss function. The optimization algorithm used was Adam. The architecture was implemented using the Keras library (Chollet, 2018; Gulli et al., 2018; Cortez, 2017).

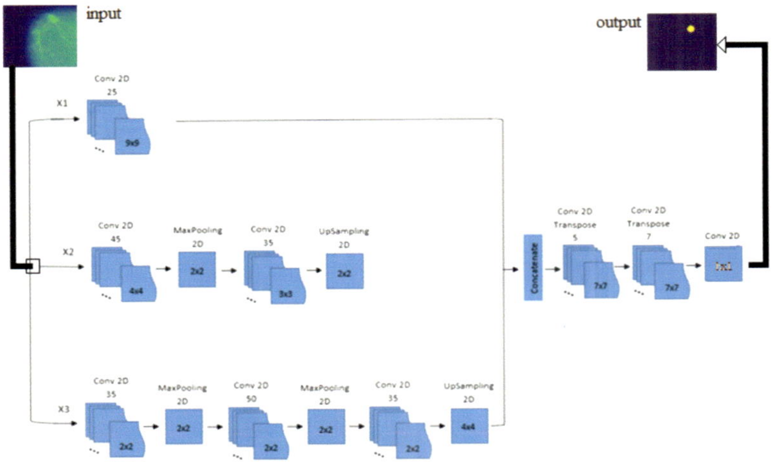

Figure 4: Architecture for the segmentation of masses in mammograms.

For segmentation evaluation, the Intersection over the Union Metric (IoU) was used. This is an evaluation metric commonly used to measure the accuracy of an object detector in a particular data set, calculating this metric is as simple as dividing the area of overlap between bounding boxes by the area of the joint (Palomino, 2010; Rahman et al., 2016). The metric is also frequently used to assess the performance of convolutional neural network segmentation (Rezatofighi et al., 2019; Rosebrock, 2019).

The formula to calculate the IoU is

$$IoU=(Area\ of\ Overlap)/(Area\ of\ union) \tag{1}$$

Another metric used is the true positive value (PPV) as (Hay, 1988; Styner, 2008).

$$PPV = \frac{TP}{(TP+FP)} \tag{2}$$

Where:

- TP are the true positive pixels, i.e., pixels that are part of the mass (object) and detected as mass;
- FP are false positive pixels, i.e., pixels that are not part of the mass (background) and detected as mass.

We also used the true positive rate (TPR), which is a true positive that represents a pixel that is correctly predicted to belong to the given class (González García, 2019). Its formula is:

$$TPR = \frac{TP}{p} = \frac{TP}{(TP+FN)} \tag{3}$$

Where:

- FN are false negative pixels, i.e., pixels that are part of the mass (object) but are classified as background.

4. Results

This section presents the results obtained from the network on the test set, which consists of 150 images. In Figures 5 and 6, two images of the test set and the output of the proposed network are presented.

In Figure 7, we show three graphical examples of the evaluation of the network. The green color represents the true positives, the red color the false positives and the blue color the false negatives.

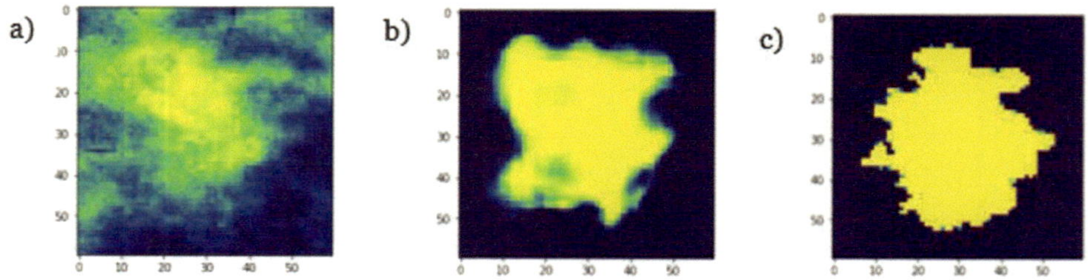

Figure 5: Image "1" of the test set. (a) Original image, (b) network output, (c) real mask delineation by a medical doctor.

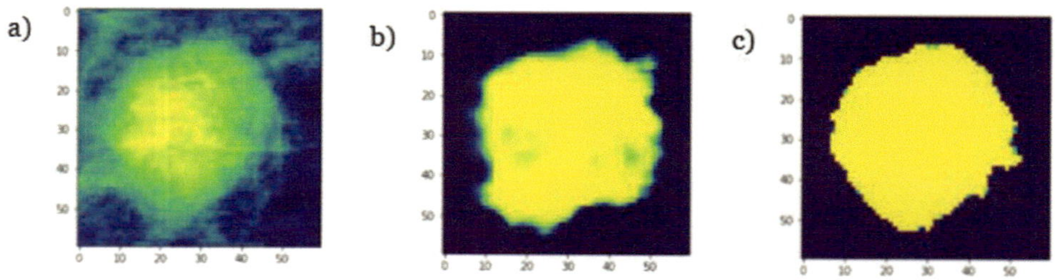

Figure 6: Image "1" of the test set. (a) Original image, (b) network output, (c) real mask delineation by medical doctor.

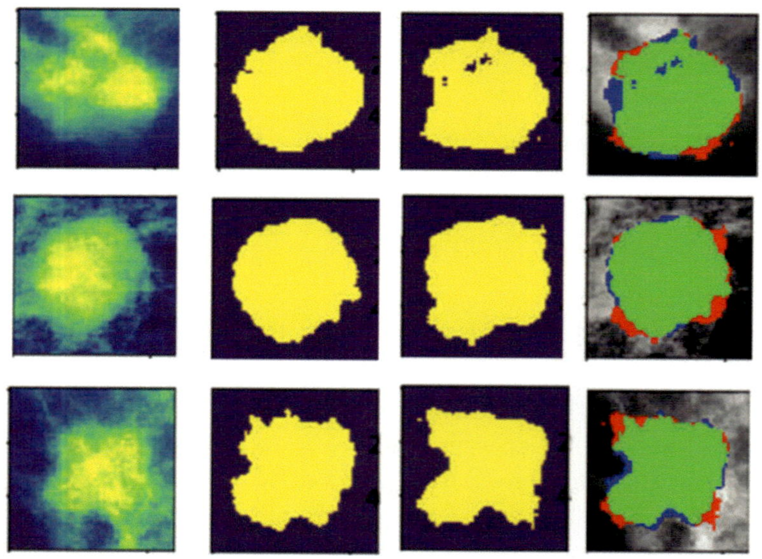

Figure 7: Each row is the input and output for the same mammogram. First column is the input mammogram, second column is the true mask, third column is the output mask from the network and finally the fourth column shows the TP, FP, and FN in green, red and blue, respectively.

Table 1 shows the IoU metric values obtained for the first eight images of the database processed by the network.

Table 2 presents the precision or positive predictive value (PPV) obtained from first eight images of the database.

Table 1: IoU metric values.

Image number	IoU value
1	0.8276077451592755
2	0.6970138383102695
3	0.7715827338129496
4	0.8355150825270348
5	0.7152858809801633
6	0.8506571087216248
7	0.6857322438717788
8	0.7023498694516971

Table 2: PPV values.

Image number	PPV
1	0.8934592043155766
2	0.8336236933797909
3	0.8148148148148148
4	0.9152119700748129
5	0.7643391521197007
6	0.8800988875154512
7	0.8203007518796992
8	0.7929255711127488

The True Positive Rate (TPR) values obtained are presented in Table 3.

Table 3: True Positive Rate (TPR) values.

Image Number	TPR Value
1	0.9182259182259183
2	0.8096446700507615
3	0.935659760087241
4	0.905613818630475
5	0.9176646706586826
6	0.9621621621621622
7	0.8069526627218935
8	0.8601119104716227

The average of the metrics for all tests was as follows: IoU as an average of 0.77, PPV averages 0.85, while TPR has an average of 0.88.

In general, the proposed architecture shows promising results; the total of TP is much greater than the sum of FP plus FN. From Figure 7, it can be seen how most of the tumor mass is correctly detected and segmented giving high TPR values. This is enough for a specialized doctor to note a possible mass in the mammography. In addition, the calculation of the total mass could serve as a quantitative measure to evaluate the response of the tumor to the treatment. On the other hand, it is also necessary to improve the network to reduce as much as possible the amount of FN and FP.

5. Conclusions and Future Work

This chapter presented the design of architecture to create a network architecture that was able to segment breast tumors. In the main structure of the architecture, three-channel convolutional neural networks with six different layers and different filters were used. The network evaluation and validation were done with the IoU metric and values PPV and TPR and indicated that the network correctly segmented the tumors with an efficiency of 88%.

As future work, it is planned to improve by the following aspects:

- Search and test more databases to obtain more variability in images.
- Improve network performance by making it more efficient by eliminating some layers.
- Evaluate the network with medical doctors.

References

Abdelaziz, A., Salama, A.S., Riad, A.M. and Mahmoud, A.N. 2019. A machine learning model for predicting of chronic kidney disease based internet of things and cloud computing in smart cities. In Security in Smart Cities: Models, Applications, and Challenges (pp. 93–114). Springer, Cham.

Adrian Rosebrock. Intersection over Union (IoU) for object detection. Machine Learning, Object Detection, Tutorials, 2016. [Online]. Available: https://www.pyimagesearch.com/2016/11/07/intersection-over-union-iou-for-object-detection/?fbclid=IwAR3ytgXlxqTNINKEgrU0JM3YeZPqFbNkdD8pSbOtGTC0c1D0bQA_LAcv0rc.

Chollet, F. 2018. Keras: The python deep learning library. Astrophysics Source Code Library.

Chougrad, H., Zouaki, H. and Alheyane, O. 2018. Deep Convolutional Neural Networks for breast cancer screening. Comput. Methods Programs Biomed., 157: 19–30.

Cortés Antona, C. 2017. Herramientas Modernas En Redes Neuronales: La Librería Keras. Univ. Autónoma Madrid, p. 60.

Dubey, R.B., Hanmandlu, M. and Gupta, S.K. 2010. A comparison of two methods for the segmentation of masses in the digital mammograms. Comput. Med. Imaging Graph., 34(3): 185–191.

Flores Gutiérrez, H., Flores, R., Benja, C. and Benoso, L. 2015. Redes Neuronales Artificiales aplicadas a la detección de Cáncer de Mama.

Gulli, Antonio, and Sujit Pal. Deep Learning with Keras. Packt Publishing Ltd, 2017.

Guzman, M., Jose Mejia, Moreno, N., Rodriguez, P. 2018. Disparity map estimation with deep learning in stereo vision. CEUR.

Hamidinekoo, A., Denton, E., Rampun, A., Honnor, K. and Zwiggelaar, R. 2018. Deep learning in mammography and breast histology, an overview and future trends. Med. Image Anal., 47: 45–67.

Hay, A.M. 1988. The derivation of global estimates from a confusion matrix. International Journal of Remote Sensing, 9(8): 1395–1398.

Ian Goodfellow and Yoshua Bengio and Aaron Courville, Deep Learning, 2016.

Juan Antonio González García. PRUEBAS DIAGNÓSTICAS (II): VALORES PREDICTIVOS. Bitácora de Fisioterapia: Noticias, comentarios, opiniones, quejas e inquietudes sobre fisioterapia, sanidad y ciencia., 2019.

Karianakis, N., Fuchs, T.J. and Soatto, S. 2015. Boosting Convolutional Features for Robust Object Proposals.

Kashif, M., Malik, K.R., Jabbar, S. and Chaudhry, J. 2020. Application of machine learning and image processing for detection of breast cancer. In Innovation in Health Informatics (pp. 145–162). Academic Press.

La, N., Palomino, S. and Concepción, L.P. 2010. Watershed: un algoritmo eficiente y flexible para segmentación de imágenes de geles 2-DE, 7(2): 35–41.

Lee, R.S., Gimenez, F., Hoogi, A., Miyake, K.K., Gorovoy, M. and Rubin, D.L. 2017. A curated mammography data set for use in computer-aided detection and diagnosis research. Scientific Data, 4: 170177.

Lempka, S.F. and McIntyre, C.C. 2013. Theoretical analysis of the local field potential in deep brain stimulation applications. PLoS One, 8(3).

Marinovich, M.L., Hunter, K.E., Macaskill, P. and Houssami, N. 2018. Breast cancer screening using tomosynthesis or mammography: a meta-analysis of cancer detection and recall. JNCI: Journal of the National Cancer Institute, 110(9): 942–949.

Noh, H., Hong, S. and Han, B. 2015. Learning deconvolution network for semantic segmentation. Proc. IEEE Int. Conf. Comput. Vis., vol. 2015 International Conference on Computer Vision, ICCV 2015, pp. 1520–1528.

OMS. "Cáncer de mama: prevención y control", Cáncer de mama: prevención y control, 2019. [Online]. Available: https://www.who.int/topics/cancer/breastcancer/es/.

Pandey, D. et al. 2018. Automatic and fast segmentation of breast region-of-interest (ROI) and density in MRIs. Heliyon, 4(12): e01042.

Pereira, D.C., Ramos, R.P. and do Nascimento, M.Z. 2014. Segmentation and detection of breast cancer in mammograms combining wavelet analysis and genetic algorithm. Comput. Methods Programs Biomed., 114(1): 88–101.

Rafegas, I. and Vanrell, M. 2018. Color encoding in biologically-inspired convolutional neural networks. Vision Res., 151(February 2017): 7–17.

Rahman, M.A. and Wang, Y. (2016, December). Optimizing intersection-over-union in deep neural networks for image segmentation. In International symposium on visual computing (pp. 234–244). Springer, Cham.

Rathee, D.S., Ahuja, K. and Hailu, T. 2019. Role of Electronics Devices for E-Health in Smart Cities. In Driving the Development, Management, and Sustainability of Cognitive Cities (pp. 212–233). IGI Global. Médicas en Cienfuegos ISSN:1727-897X Medisur 2005; 3(5) Especial", Guias buenas parcticas Clin., 3(5): 109–118, 2005.

Rezatofighi, H., Tsoi, N., Gwak, J., Reid, I. and Savarese, S. 2019. Generalized Intersection over Union : A Metric and A Loss for Bounding Box Regression, pp. 658–666.

Singh, A.K. and Gupta, B. 2015. A novel approach for breast cancer detection and segmentation in a mammogram. Procedia Comput. Sci., 54: 676–682.

Styner, Martin, et al. 2008. 3D segmentation in the clinic: A grand challenge II: MS lesion segmentation. Midas Journal 2008.

Villalba Gómez, J.A. 2016. Problemas bioéticos emergentes de la inteligencia artificial. Diversitas, 12(1): 137.

Wang, Z., Yu, G., Kang, Y., Zhao, Y. and Qu, Q. 2014. Breast tumor detection in digital mammography based on extreme learning machine. Neurocomputing, 128: 175–184.

CHAPTER-2

Serious Game for Caloric Burning in Morbidly Obese Children

José Díaz-Román,[*,1] *Alberto Ochoa-Zezzatti,*[1] *Jose Mejía-Muñoz,*[1] *Juan Cota-Ruiz*[1] and *Erika Severeyn*[2]

A new report on childhood obesity is published every so often. The bad habits of food and the increasingly sedentary life of children in a border society has caused an alarming increase in the cases of children who are overweight or obese. Formerly it seemed a problem of countries with unhealthy eating habits, such as the United States or Mexico in Latin-America, where junk food is part of the diet during childhood. However, obesity is a problem that we already have around the corner and that is not so difficult to fight in children. In the present research, the development of an application that reduces the problem of the lack of movement regarding the children of a smart city is considered a future problem. The main contribution of our research is the proposal of an improved type of Serious Game, coupled with the achievement of an innovative model to practice an Olympic sport without the complexity of moving physically in an outside space and having to invest in a space with high maintenance costs, considering the adverse weather conditions such as wind, rain and even a dust storm. We use Unity to model each Avatar associated with a set of specific sports, such as Water polo, Handball, Rhythmic Gymnastics and others.

1. Introduction

The increase in childhood obesity, a problem of great importance in a smart city, determines the challenges that must be addressed with respect to applications that involve Artificial Intelligence. Computer games to combat childhood obesity are very important to reduce future problems in our society. Children increasingly play less on the street and spend more time with video games and computer games, so they lead a more sedentary life. This, together with bad eating habits, increases the cases of obese children every year. What can parents do to avoid their children being overweight? A bet that comes to us from the University of Western Australia, Liverpool John Mores University and the University of Swansea in the United Kingdom is "exergaming", an Anglicism that comes from joining the word "exerdizze" in Turkish (exercise in English) with "gaming" (game). These are games that run on consoles such as Xbox Kinect or Nintendo Wii in which you interact through physical activity in tests in which you have to run, bike, play bowling or jump fences. The researchers tested children who performed high and low-intensity exergaming and measured their energy expenditure. The conclusion reached was that the exergaming generated an energy expenditure compared to exercise of moderate or low intensity, depending on the difficulty

[1] Universidad Autónoma de Ciudad Juárez, Av. Hermanos Escobar, Omega, 32410 Cd Juárez, Chihuahua, México.
[2] Universiad Simón Bolívar, Sartenejas, 1080 Caracas, Distrito Capital, Venezuela.
* Corresponding author: david.roman@uacj.mx

of the game. In addition, the game was satisfactory for the children, who enjoyed the activities they did. It is a tool that parents can take advantage of to prevent children from spending so many hours sitting in front of the console as it has been shown to offer long-term health benefits. In any case, it must always be one of the means we can use to encourage children to do some physical activity but not the only one. Going out the street to play, run, jump, must always be on the children's agenda, as is shown in Figure 1.

Figure 1: Intelligent application using Kinect.

The Serious game represents a practical idea of how to solve problems associated with caloric intake because they allow performing ludic aspects of a game and the regulations associated with a specific sport, which is why the research conducted took into consideration a set of sports with high mobility associated with a control group that has morbid child obesity. The remainder of this chapter is structured as follows: In Section §2, the approach of a serious game for caloric burning is presented. Methodological aspects of the implementation of serious games are presented in Section §3, where psychological and technological factors are considered to guide their development. Section §4 introduces the method for estimating caloric burning in the implementation of a serious game. Technical aspects for modeling of avatars in a serious game for caloric burning are given in Section §5. Finally, the analysis of results and the conclusions are presented in Sections §6 and §7, respectively.

2. Serious Game for Caloric Burning

After the advent (more than two decades ago) of video games that implement technologies that allow the active physical interaction of the user (active video games), there has been an increased interest regarding research into estimating the amount of energy consumed in the gaming sessions conducted by users and whether these video games promote the physical activity of players. Compared to traditional non-physically interactive video games, active video games significantly increase energy consumption to levels similar to those of moderate-intensity physical activity [1]. It has been found that a child's energy expenditure during the activity of a video game such as the Boxing and Dance Dance Revolution (level 2) of the Nintendo Wii console, is compared to the energy expenditure experienced on a treadmill at about 5.7 km/hr [2].

Studies reveal that the continuous practice of active video games generates a calorie-burning equivalent to a physical activity that is able to cover the recommendations, in terms of energy expenditure per week (1000 kcals), of the American College of Sports Medicine (ACSM) [1,3]. In 2011, Barnett et al., in a systematic review, found that the average of metabolic equivalent (MET) in young subjects during active video games was estimated at 3.2 (95% CI: 2.7, 3.7), whose value is considered to be moderate-intensity physical activity, although none of the papers reviewed in that study found that the MET reached the value of 6, which is considered to be the threshold for intense physical activity [4]. Later in 2013, Mills and co-workers found that the Kinect Sports 200 m Hurdles video game generated increases in heart rate and energy expenditure patterns consistent with intense physical activity in children, which was also related to effects that can be considered beneficial on vascular function [5]. Gao et al. conducted a study to compare the effect of physical activity performed by children in physical education classes using active video games and those who performed physical activity in regular physical education classes. The authors concluded that the positive effects of a regular physical education class can be achieved with a physical education class using active play in children with light to vigorous physical activity behavior, with similar energy expenditures in both class modalities [6]. Studies show a positive impact of active video game use on body mass index in children and adolescents [7,8]. All this reveals the potentialities of the use of video games in the control of obesity and the prevention of illnesses associated with this condition.

A remarkable aspect of active video games is the fun and entertaining nature of the video game itself, which makes it attractive to children and teenagers and represents a motivating component for physical activity. Research reveals that engaging in physical activity through an active video game is significantly more enjoyable than other traditional physical exercises, such as just walking [9] or using a treadmill [10]. On the other hand, a study showed that adolescents with obesity who were included in a physical activity program using active games reported an increase in physical activity and showed a high motivation intrinsic to the use of active video games [11].

The goal of the present research is the development of serious games based on active sports video games for increasing the burning of calories in morbidly obese children. In addition to the application of active games, the serious game incorporates metabolic equivalent analysis for the estimation of caloric burning based on the metabolic intensity indices described in the most recent compendium of physical activities for young people [12].

2.1 Components of a serious game

Let's go into detail, serious games are meant to teach. They are video games and applications with a didactic background. What does it mean? That users of serious games learn while having fun, which translates into a doubly positive experience. But what are the elements to achieve these objectives?

Narrative: A story that engages the users with a storyline will encourage them to get involved in the project, so that they do not abandon it and get more involved. Having a good argument guarantees a greater immersion and motivation, which translates into better results.

Interactivity: User participation in serious games allows communication between the tool and the user. Moreover, it is done with immediacy since the results can be seen as soon as the test is done. This provides valuable feedback, through which you can learn from mistakes and improve efficiency.

Training: The main objective. This is the virtue of serious games, to create an experience in which the user has fun, where although he plays with his eyes, in the end he learns.

With these key elements in mind, betting on serious games will provide added value in the current digital context and training in any organization that wants to build a genuine effort to help others, in this case, with the impact on promoting the exercise that helps improve the health of children in a Smart City.

3. Methodological Aspects

There are many transcendental topics when considering to establish a specific serious game associated to an aspect of the technology that allows to take the caloric control of the exercise carried out as in:

3.1 Emotions in children

The science-based child psychology emerging in the second half of the nineteenth century promised to provide a rational basis for education for the overall development of the child. All this promising area in development offers new and interesting proposals such as the study of children in their environments associated with the improvement of empathy towards exercise, not only to be used by child psychology, that is why the creation of a science focused on the child that goes beyond traditional psychology [2]. This thanks to William Preyer, who is considered the father of child psychology, but the authors of the study did so with the same rigorous exactitude as that used in their observations of their work, characterizing it as being of a very systematic type [1].

3.2 Emotional disorders in children

Mental disorders of children affect many children and their families. Children of all ages, ethnic or racial backgrounds, and from all regions of the United States have mental disorders. According to the report by the National Research Council and Institute of Medicine in the division (Prevention of mental, emotional and behavioral disorders in young people: progress and possibilities, 2009) who gathered findings from previous studies, it is estimated that 13 to 20% of children living in the United States (up to 1 in 5) have a mental disorder in a given year, and about 247,000 million dollars a year are spent on childhood mental disorders [3]. In the literature, we find several definitions to refer to emotional, mental or behavioral problems. These days, we find that they are referred to as "emotional disorders" ("emotional disturbance"). The Education Act Individuals with Disabilities Education Act ("IDEA" for short) defines emotional disorders as "a condition exhibiting one or more of the following characteristics over a long period of time and to a marked degree that adversely affects a child's educational performance" [4]. The lack of ability to learn that is inexplicable by intellectual, sensory or health reasons.

(a) A lack of ability to maintain personal relationships on good terms with their classmates or teachers.
(b) Having inconsistent behaviors or feelings under normal circumstances, including anxiety attacks.
(c) Having recurring episodes of sadness or depression.
(d) Developing physical symptoms, including hypochondriacal symptoms or fears associated with personal or school problems.

Hyperactivity: This type of behavior manifests as the child being inattentive, easily distracted and impulsive.

Assaults: When the result of the behavior ends in injury, either to themselves or their neighbors.

Withdrawal: The social life shows signs of delay or the individual shows an inability to relate to their environment. This includes excessive fears or anxiety.

Immaturity: Unwarranted crying spells and inability to adapt to changes.

Learning difficulties: Learning does not develop at the same pace as the average in their environment. It remains at a level below their peers. There are even more young children with serious emotional disturbances, i.e., distorted thought, severe anxiety, uncommon motor acts, and irritable behavior. These children are sometimes diagnosed with severe psychosis or schizophrenia [4].

3.3 Play therapy

Play therapy is defined as a therapeutic model of formal recognition of the child and has also proven its effectiveness in children with emotional stress problems that contribute to and manifest in each child during their normal development. Play therapy builds on child's play as a natural means of self-expression, experimentation, and communication. While the child plays, they learn about the world and explore relationships, emotions and social roles. It also gives the child the possibility to externalize his personal history, thus releasing negative feelings and frustrations, mitigating the effects of painful experiences and giving relief from feelings of anxiety and stress [5]. The play therapist is specialized, trained in play and uses professional technical therapeutic methods adapted to the different stages of child development. The importance of a specialized Serious Game is to capture and understand the child's emotions, as well as to get involved in the child's game to create a true relationship for the expression and management of the child's internal conflicts, an aspect of great relevance and that favors Serious Games, as well as to download and understand their emotions in order to properly manage them and not get trapped in them, making them able to recognize and explore the problems that affect their lives [6]. A serious game does not replace the play therapy of your therapist, but seeks to provide an auxiliary and also effective support tool for those who are in contact with children who at some point show signs of moodiness. It could also be used by health professionals as part of their therapeutic tools to control the emotions of their patients and to be able to give a more adequate follow-up to their therapy, especially occupational therapy.

3.4 Serious games

An old definition of the 70's that so far has remained is that serious games are those that have an educational or therapeutic purpose and are not just for fun. Within this definition is the real challenge for developers, i.e., maintaining this balance between fun and fulfilling the purpose of carefully planned learning. Once the construction of serious game is focused on learning or some specific therapeutic purpose, it often happens that the fun part of the game (so important to it) is neglected by not determining the quality and impact that serious play should have on users. The applications are varied, with the most common being education, management, policy, advocacy, and planning, among others [7]. Why games like support? The answer is simple and intuitive as these are part of the training of people who are going to use it. Nowadays most of the serious games are multiplayer, that is, collaborative, so it is very important to learn in a collaborative way. Most of these serious games are based o n cultural aspects that allow children to associate with their environment and with the people around them. The use of technology has improved the effects associated with the visual aspect of serious games [8].

Figure 2: Design of a speed skating rink in an open and arboreal space to improve the ludic aspect of the practice of this sport.

3.5 The importance of properly building the components of a Serious Game

An important aspect of Serious Games is determining the child's correct progress in it and how he adapts to changes in the stages of serious play [9]. Another aspect to consider is the environment associated with Serious Game, which should be as realistic as possible and according to the scenario where the child's learning skills are to be developed. Another great challenge is to properly organize the set of rules to follow to advance in the development of the game and achieve an intuitive understanding of it [8].

3.6 A serious game associated with the appropriate heuristics for continuous improvement

Artificial intelligence, using adequate heuristics, will allow to demonstrate the correct functioning of the avatar in the built environment for learning, achieving an adequate link with the avatar associated with the child, as can be seen in Figure 2.

4. Estimation of Caloric Burning during the Practice of Serious Game

When a person is at rest, his or her body consumes energy for the maintenance of vital functions; this energy consumed is called the resting metabolic rate (RMR). Physical activity increases energy consumption above resting levels, and the number of calories expended is related to the intensity and duration of the activity. The metabolic equivalent (MET) is an index used to measure or express the intensity of physical activity, so it can be used to estimate the amount of energy consumed during some type of exercise.

A young adult of 70 kg at rest (seated and still) normally consumes about 250 mL/min of oxygen; this is equivalent to ≈3.5 mLO2/kg·min, which represents 1 MET (standardized value), and similarly corresponds to a consumption of 1 kcal/kg·h [13]. When carrying out a physical activity, oxygen consumption increases, so this activity has a MET > 1; for example, if an activity has 4 METs, it means that 14 mLO2/kg·min are required to carry out the activity, that is to say, 4 times the consumption of energy that is presented at rest. The Physical Activity Guide for Americans classifies physical activities as light intensity with METs < 3, moderate with METs between 3 and 5.9, and intense with METs > 6 values. A walk at 2 mph represents a light physical activity of 2.5 METs; if the walk is at 3 mph it has a MET = 3, which classifies the activity as moderate intensity; and running at a speed of 10 mph corresponds to an intense physical activity with a MET = 6. Also, the guide qualifies a sedentary behavior as behavior or activity that presents low levels of energy expenditure, equivalent to MET < 1.5. It is suggested that children and adolescents aged 6 to 17 years should do 60 minutes (1 hour) or more of moderate to vigorous daily physical activity, such as Aerobic, Muscle Strengthening or Bone Strengthening [14].

Since the standardized value of 1 MET is derived from oxygen consumption for a subject with particular characteristics (healthy adult male, 40 years old, 70 kg) in resting conditions, their relationship between calorie consumption 1 MET = 1 kcal/kg·h is subject to variations depending on age, sex and body composition, which is primarily due to the fact that the energy consumed by a resting person (RMR) depends on such factors, in addition to the health condition, stress level, and others. In this sense, it should be mentioned that the RMR has higher values in men than in women, increases with height, weight and body composition of a person, and decreases with age. Research has found that the use of the standardized MET value can cause misclassification of the intensity of physical activity [15], and, in turn, inappropriately estimate the true values of oxygen consumption and energy costs during the activity [16]. This is why the compendium of physical activities (update 2011) proposes a correction factor for the calculation of the metabolic equivalent (MET corrected) based on a better estimate of the RMR, which uses the Harris-Benedict equation which takes into account the age, height, weight, and sex of the subject [17]. In [18] a correction factor that allows a more accurate MET calculation in overweight adults is proposed.

On the other hand, because children have a higher resting basal metabolism (BMR) per unit body mass than adults, adult MET values do not apply to children. In addition, as the child grows, the BMR decreases gradually. The factors that cause this decrease in BMR in children are mainly changes that occur in the mass of the organs and in the specific metabolism of some organs, and changes in muscle mass and fat mass, which in turn are linked to the sex of the child. A child's BMR may be underestimated by using the standard adult MET, since the BMR of a 6-year-old child is on average ~ 6.5 mLO_2/kg·min (1.9 kcal/kg·h) and approximately 3.5 mLO_2/kg·min for 18-year-olds [12]. Because of BMR behavior in children, calorie intake from physical activity is not constant during childhood. At the same time, for the same physical activity, a child has a higher energy expenditure per body mass than an adult or adolescent.

Thus, the most recent compendium of physical activities for youth establishes a metric for youth MET (METy) that is significantly different from that of adults, and that is age-dependent [12]. The compendium presents the METy values of 196 physical activities commonly performed by children and young people, for the following discrete age groups: 6–9, 10–12, 13–15 and 16–18 years. For the calculation of the BMR, the Schofield equations according to age groups and sex are used:

Age	BMR (kcal/min)	
	Boys	
3–10 years	[22.706 × weight (kg) + 504.3])/1440	(1)
10–18 years	[17.686 × weight (kg) + 658.2])/1440	(2)
	Girls	
3–10 years	[20.315 × weight (kg) + 485.9])/1440	(3)
10–18 years	[13.384 × weight (kg) + 692.6])/1440	(4)

For the present work, it is proposed the use of the METy values of the young compendium for the different physical activities that are implemented in the designed serious games, which belong to the category of active full body video games [12]. Table 1 shows the different METy values for the age groups 6–9, 10–12 and 13–15, for which the use of serious games is intended.

Table 1: METy values of active video games (full body) for the physical activities of the serious games [12].

Code	Specific Activity	METy by age-group (years)		
		6–9	10–12	13–15
15120X	Baseball	3.7	4.7	5.7
15140X	Boxing	3.0	4.0	4.9
15180X	Dance	2.3	3.3	4.1
15260X	Olympic games	2.6	3.6	4.5
15320X	Walking on treadmill and bowling	2.8	3.9	4.8
15400X	Wii hockey	1.4	2.4	3.2
15480X	Wii tennis	1.6	2.5	3.2

Then, knowing the BMR and METy for age group, and duration of a physical activity, the energy expenditure is calculated by:

$$EE = \text{METy} \times \text{BMR (kcal/min)} \times \text{duration (min)} \tag{5}$$

For example, if a 10-year-old girl (37 kg), with a BMI greater than the first three quartiles according to the population of her age, plays the serious game of Rhythmic Gymnastics (like Dance in table 1, METy = 3.3) for 15 minutes twice a day, her daily caloric burning due to the practice of that physical activity can be determined as follows:

- Using Schofield equation 3, BMR = [20.315 × 37 + 485.9]/1440 = 0.86 kcal/min.
- Total energy expenditure (EE) for this physical activity:

$$EE = 3.3 \times 0.86 \text{ kcal/min} \times 30 \text{ min} = 85 \text{ kcal}$$

4.1 *Using a Serious Game to determine efforts in a Virtual Sport*

As mentioned in this document, this project is currently in the construction phase. This prototype of a Serious Game has the firm intention of taking the next phase of complete and functional construction. A foundation sponsored by one of the most important children's hospitals in the city of Paso Texas in the United States has shown a genuine interest in the project, for which it mentioned providing the necessary support for its realization. In Mexico, the National System for Integral Family Development has childcare programs with interesting and very efficient strategies for reducing morbid obesity in overweight children. That is why it is intended to integrate this intelligent tool in their support programs for these children. For this, it will soon be formalized by both parties committed to supporting the project shown in this document. In future research, we try to modify a game based on collaborative work in a group—we are choosing rugby seven—with high intensity of pressure for each child and modify the importance related to the support of this type of pressure related to the responsibility of a Collective activity, an approximation will be related to what is implied for Water polo, as shown in Figure 3. A very relevant aspect is to consider that if someone asks why he likes to use our Serious Game, this user will be able to respond: because he has had a playful scope and of adequate selection with the avatar, so he could have empathy for our proposal. By analyzing in more detail the group of people who used our Serious Games, we determined that, like the role-play, it is a hobby that unites them and gives them opportunities to help each other and their videogame community.

It is a safe environment in which you can experience social interactions, something fundamental when the climate does not allow it, in Bw-type climates (according to the Köppen climate classification scale) as the place of our study. This group of users of our Serious Game says that they have witnessed the personal growth of individuals in terms of their self-esteem and the expansion of their social interactions as a result of the game. This is just one of the benefits of the game. Our research showed that it was discovered that everyone can find some hours a week to "save the universe, catch the villains or solve mysteries" while learning to practice Water polo, and that playing with the computer is as fun as any other activity in our research Playing our Serious Game can strengthen a variety of skills such as math and reading online recommendations. Increase the ability to think and speak clearly and concisely, when formulating and implementing their plans, cooperating and communicating with others, as well as increasing the ability to analyze written and verbal information. Placed on the market, our Serious Game will determine that players are cohesive members of the group in multiplayer games, and it can help people develop leadership skills and promote cooperation, teamwork, friendship, and open communication. In another study related to this kind of Kinetic Serious Game, we try to compare with our colleagues of Montenegro who propose and develop an innovative Serious Game which involves a model to Fencing practitioners, as this sport is reaching high popularity in this society. A representative model can be shown in the next Figure 4.

What would users expect from our proposal of a Serious Game in this Kinetic model to learn and practice Water polo? Improve the mood of serious game users through the components that will be used as background music, with the purpose of using music therapy techniques to lift the spirits of the players as it develops [11]. Another element of the strategy and ploy in Serious Game is the colors of the scenery. By taking into account the effect of color on mood, a better color experience can be predicted to keep the player in a positive emotional state [10]. The third element is through the sounds of the game at each stage of development. With every success and failure, the sounds in your environment can represent the stage you are playing on. And the fourth element is the recognition of achievements. Through badges, medals, trophies, scores, and appointments you want the player to have a feeling of satisfaction with the recognition of each achievement [12].

Figure 3: Our Kinetic Serious Game Model using collaborative task to play Water polo.

Figure 4: Use of Kinetic software to improve performance in Modern Pentathlon.

5. Modeling of Avatars in a Serious Game for Caloric Burning

In addition to the BDI methodology, the physiological traits of the agents intervene to improve each aspect of an avatar. The scenes structured associated with the agents cannot be reproduced in general, since they only represent a small part of the population in space and time of the different societies. These individual behaviors represent a unique and innovative form of global adaptive behavior that solves a computing problem that does not attempt to group societies only with a factor associated with their external appearance (phenotype) and therefore sports that could be practiced much more, but it tries to solve a computer problem that implies a complex change from the perspective of sport that has opted for a better empathy in relation to its practice, in order to improve competitiveness in children's health among the existing relationships with the population that practices these sports. The generated configurations can be metaphorically related to the knowledge of the behavior of the community with respect to an optimization problem (to conform to cluster social and culturally with other similar people, without being of the same sport [4]). In Table 2 is shown a sample of seven sports, describing each of the analyzed characteristics in order to determine which were the most viable to develop in a Serious Game.

$$K = [C + CS + GAL] \pm CBS \tag{6}$$

where:

C = represents if this sport can be practiced in any climate. For example, in the case of Chess, it is not an indispensable condition.

CS = represents the Symbolic Capital associated to the perspective of a society in a Smart City. The practice of Fencing is considered sophisticated and therefore has more prestige.

G = is defined as the gender in the population range. In Juarez City the population pyramid is different from the rest of Latin America because violence causes deaths and exodus from the city; that is why the values are different and their representativeness as well.

AL = Ludic aspect related to the age of the children who practice sports. For example, the trampoline is associated with a fun experience, and its playful aspect is associated with high promotion.

CBS = Social Benefits-Cost related to each sport. In the case of Rhythmic Gymnastics, it is associated with improving the health of sedentary children and helps them lose weight quickly.
We use equation 4, where K represents the performance of the practice for each of the sports in the city and their respective promotion in various collaborative spaces.

5.1 Unity implementation

To achieve everything mentioned previously, from the implementation of algorithms, we use Unity to obtain the greatest potential of the serious game. The most relevant aspect of this research is to consolidate the practice of sports in children and young people who, due to weather conditions in the practice of normal sports which is not the case of Serious Games, cannot agree to join together to practice a sport together with the self-confidence generated in the players, which allows improving their performance in other areas of their daily life through a model of emotional support in children, which entails a commitment and intrinsic complexity in their motor development. Considering that childhood is a vulnerable stage where they are also in full development and any event or occurrence may be able to cause negative effects and may leave the child permanently marked by the rest of their life [13, 15, 16, 17, 18, 19, 20], it is very important to focus more on how to obtain results associated with group performance sponsored by the individual. That is why the future of the Serious Game will require a deeper investigation that allows being of great impact by having an opportunity to help children and youth who do not have access to sports for various reasons. In the future, the Serious Games may present a natural opportunity due in large part to the acceptance that videogames have in this age and even more so with the advantage that these generations (both Gen Z and now the

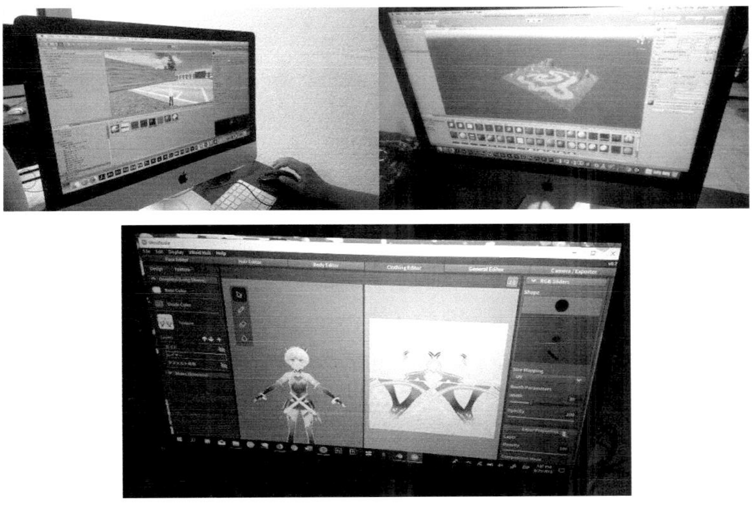

Figure 5: Representation of different virtual sports.

children of the so-called Alpha α generation) have easy access to technology. The implementation of Unity in diverse virtual sports is presented in a Collage of them, as is shown in Figure 5.

6. Analysis of results And Discussions of Our Research

As mentioned in the present investigation, this prototype wishes to continue consolidating in order to establish diverse teams associated mainly with the avatar. This prototype regarding the Serious Game to achieve the complete and functional construction phase [21, 22, 23, 24, 25, 26, 27]. At this time and due to a project with funding from the European Union and the collaboration of FINA, we want to make a Kinetic application that allows being inclusive through interesting and very efficient strategies associated with mobility. FINA is very interested in this type of application that could diversify the practice of Water polo in developing countries.

Table 2 presents the data for the multivariable analysis with the information on the number of spaces to recover for the practice of these virtual sports and their representations by gender (aspects to be covered in each sport)—specific values for both Men and Women, the social status for its practice, the fun for use of the Wii application that is being proposed, the external climate to complement with the Serious Game, the improvement of health to change the paradigms to sedentary children, and finally the relationship between the social cost/benefit associated with its practice.

Table 2: Multivariable analysis to determine the relationship between the timely improvement of some sports using our Kinect proposal and an avatar associated with its performance coupled with an intelligent system for the control of heat burn in morbidly obese children.

Sport	Virtual space	Gender	Social status	Fun	Climate	Increase Health	Cost-Benefit
Aquatic Sky	8	m-.50, f-.45	0.498	0.914	0.774	0.715	0.387
Judo	6	m- 40, f- 30	0.857	0.714	0.851	0.879	0.568
Baseball	5	m-50, f-.40	0.617	0.658	0.385	0.712	0.514
Syncronized Swimming	3	f-.47	0.275	0.637	0.416	0.748	0.885
Water polo	14	m-.45, f-.40	0.578	0.784	0.925	0.627	0.879
Bowling	3	m-.30, f-.30	0.620	0.631	0.715	0.802	0.744
BMX Bike	5	m-.40, f-.40	0.562	0.748	0.611	0.303	0. 448
Rolling Sport	4	m-.48, f-.42	0.877	0.917	0.459	0.897	0. 574
Rhythmic Gymnastics		f-.49	0.718	0.897	0.427	0.928	0. 927

7. Conclusions and Future Challenges

The main experiment consisted of detailing each one of the 47 sports, with 500 agents, and one condition of unemployment of 50 generations, this allowed us to generate different scenarios related with Time Horizons, which was obtained after comparing different cultural and social similarities in each community and to determine the existing relations between each one in relation with the Mahalanobis Distance (the number of dots indicated each sport and the size of people represents the number of people which determine the magnitude related with the society). In future research, we will try to improve the practice of a sport associated with a game and based on collaborative work in a group with high intensity of pressure for each child as professional tennis and modify the importance related to the support of this type of pressure related to the responsibility of a collective activity [14], as shown in Figure 6.

Figure 6: A serious game based on collective activities and related with the increase of social skills.

References

[1] Haddock, B.L. et al. 2012. Measurement of energy expenditure while playing exergames at a self-selected intensity. Open Sports Sci. J., 5: 1–6.

[2] Graf, D.L., Pratt, L.V., Hester, C.N. and Short, K.R. 2009. Playing active video games increases energy expenditure in children. Pediatrics, 124(2): 534–540.

[3] Siegel, S.R., Haddock, B.L., Dubois, A.M. and Wilkin, L.D. 2009. Active video/arcade games (Exergaming) and energy expenditure in college students. Int. J. Exerc. Sci., 2(3): 165–174.

[4] Barnett, A., Cerin, E. and Baranowski, T. 2011. Active video games for youth: a systematic review. J. Phys. Act. Heal., 8(5): 724–737.

[5] Mills, A. et al. 2013. The effect of exergaming on vascular function in children. J. Pediatr., 163(3): 806–810.

[6] Gao, Z. et al. 2017. Impact of exergaming on young children's school day energy expenditure and moderate-to-vigorous physical activity levels. J. Sport Heal. Sci., 6(1): 11–16, Mar. 2017.

[7] KoenigHarold, G. 2018. Impact of game-based health promotion programs on body mass index in overweight/obese children and adolescents: a systematic review and meta-analysis of randomized controlled trials. Child. Obes.

[8] Hernández-Jiménez, C. et al. 2019. Impact of active video games on body mass index in children and adolescents: systematic review and meta-analysis evaluating the quality of primary studies. Int. J. Environ. Res. Public Health, 16(13): 2424, Jul. 2019.

[9] Moholdt, T., Weie, S., Chorianopoulos, K., Wang, A.I. and Hagen, K. 2017. Exergaming can be an innovative way of enjoyable high-intensity interval training. BMJ open Sport Exerc. Med., 3(1): e000258–e000258, Jul. 2017.

[10] McDonough, D.J., Pope, Z.C., Zeng, N., Lee, J.E. and Gao, Z. 2018. Comparison of college students' energy expenditure, physical activity, and enjoyment during exergaming and traditional exercise. J. Clin. Med., 7(11): 433, Nov. 2018.

[11] Staiano, A.E., Beyl, R.A., Hsia, D.S., Katzmarzyk, P.T. and Newton, R.L., Jr. 2017. Twelve weeks of dance exergaming in overweight and obese adolescent girls: Transfer effects on physical activity, screen time, and self-efficacy. J. Sport Heal. Sci., 6(1): 4–10, Mar. 2017.

[12] Butte, N.F. et al. 2018. A youth compendium of physical activities: activity codes and metabolic intensities. Med. Sci. Sports Exerc., 50(2): 246.

[13] McArdle, W.D., Katch, F.I. and Katch, V.L. 2006. Essentials of exercise physiology. Lippincott Williams & Wilkins.

[14] Piercy, K.L. et al. 2018. The physical activity guidelines for Americans. Jama, 320(19): 2020–2028.

[15] Kozey, S., Lyden, K., Staudenmayer, J. and Freedson, P. 2010. Errors in MET estimates of physical activities using 3.5 $ml \cdot kg^{-1} \cdot min^{-1}$ as the baseline oxygen consumption. J. Phys. Act. Heal., 7(4): 508–516.

[16] Byrne, N.M., Hills, A.P., Hunter, G.R., Weinsier, R.L. and Schutz, Y. 2005. Metabolic equivalent: one size does not fit all. J. Appl. Physiol., 99(3): 1112–1119.

[17] Ainsworth, B.E. et al. 2011. Compendium of Physical Activities: a second update of codes and MET values. Med. Sci. Sport. Exerc., 43(8): 1575–1581.

[18] Wilms, B., Ernst, B., Thurnheer, M., Weisser, B. and Schultes, B. 2014. Correction factors for the calculation of metabolic equivalents (MET) in overweight to extremely obese subjects. Int. J. Obes., 38(11): 1383.

[19] Chris Ferguson, Egon L. van den Broek, Herre van Oostendorp: On the role of interaction mode and story structure in virtual reality serious games. Computers & Education 143 (2020) https://dblp.org/rec/journals/chb/LiuL20.

[20] Sa Liu and Min Liu. 2020. The impact of learner metacognition and goal orientation on problem-solving in a serious game environment. Computers in Human Behavior 102: 151–165. https://dblp.org/rec/journals/csi/GarciaPLC20.

[21] Ivan A. Garcia, Carla L. Pacheco, Andrés León and José Antonio Calvo-Manzano. 2020. A serious game for teaching the fundamentals of ISO/IEC/IEEE 29148 systems and software engineering—Lifecycle processes—Requirements engineering at undergraduate level. Computer Standards & Interfaces 67.

[22] Salma Beddaou. 2019. L'apprentissage à travers le jeu (Serious game): L'élaboration d'un scénario ludo-pédagogique. Cas de l'enseignement-apprentissage du FLE. (Learning through the game (Serious game): The development of a play-pedagogical scenario. Case of FLE teaching-learning). Université Ibn Tofail, Faculté des Lettres et des Sciences Humaines, Marocco 2019.

[23] Zhipeng Liang, Keping Zhou and Kaixin Gao. 2019. Development of virtual reality serious game for underground rock-related hazards safety training. IEEE Access, 7: 118639–118649.

[24] Anna Sochocka, Miroslaw Solarski and Rafal Starypan. 2019. "Subvurban" as an example of a serious game examining human behavior. Bio-Algorithms and Med-Systems, 15(2).

[25] David Mullor, Pablo Sayans-Jiménez, Adolfo J. Cangas and Noelia Navarro. 2019. Effect of a Serious Game (Stigma-Stop) on Reducing Stigma Among Psychology Students: A Controlled Study. Cyberpsy., Behavior, and Soc. Networking 22(3): 205–211.

[26] Jonathan, D. Moizer, Jonathan Lean, Elena Dell'Aquila, Paul Walsh, Alphonsus Keary, Deirdre O'Byrne, Andrea Di Ferdinando, Orazio Miglino, Ralf Friedrich, Roberta Asperges and Luigia Simona Sica. 2019. An approach to evaluating the user experience of serious games. Computers & Education, 136: 141–151.

[27] Shadan Golestan, Athar Mahmoudi-Nejad and Hadi Moradi. 2019. A framework for easier designs: augmented intelligence in serious games for cognitive development. IEEE Consumer Electronics Magazine, 8(1): 19–24.

CHAPTER-3

Intelligent Application for the Selection of the Best Fresh Product According to its Presentation and the Threshold of Colors Associated with its Freshness in a Comparison of Issues of a Counter in a Shop of Healthy Products in a Smart City

Iván Rebollar-Xochicale,[1] *Fernando Maldonado-Azpeitia*[1,*] and *Alberto Ochoa-Zezzatti*[2]

1. Introduction

Around the world, we can find data about food waste and some of its most important causes. To mention some data, every year in the Madrid region, 30% of products destined for human consumption are lost or wasted by improper handling in the food supply chain (CSA) comments Gustavsson et al. (2011). A study in the United States by the Natural Resources Defense Council (NRDC) found that up to 40% of food is lost from the producer's farm to the consumer's table, Gunders (2012).

Losses of perishable products vary among countries around the world, in some countries, such as in China, they even increase. Reports indicate that only 15% of fresh products are transported under optimum temperature conditions, despite the knowledge that these types of products require refrigerated handling (Pang et al., 2011). They also comment that fruits and vegetables are the most affected type of food, where 50% of what is harvested is not consumed and this is mostly due to insufficient temperature control. Approximately one-third of the world's fruits and vegetables are discarded because their quality has fallen and because of this it lacks acceptance and puts food safety at risk.

1.1 Situation of perishable foods

In Mexico, 20.4 million tons of food is wasted annually, these data only correspond, according to the World Bank (2018), to 79 foods representative of Mexico's food basket, which implies large environmental impacts due to the excessive use of water and carbon dioxide generation. This represents the waste of about 34% of the national food production, which if considered the rest of the food could reach about 50% of the total national production produced. Additionally, it was

[1] Universidad Autónoma de Querétaro, Mexico.
[2] Doctorado en Tecnologia, UACJ; Mexico.
* Corresponding author: luis.maldonado@uaq.mx

observed in this study that approximately 72% of losses occur between pre-harvest and distribution. That is, in the early stages of the production chain and by taking them to their target market, in retail, which could be the result of bad consumption habits.

Much of these losses are related to inadequate management of temperature control during CSA processes (production, storage, distribution and transport, and at home) (Jedermann et al., 2014). Similar studies have shown that, in many cases, food security is frequently affected by poor temperature management (Zubeldia et al., 2016). Environmental conditions, mainly temperature, have a great impact on the overall quality and shelf life of perishable foods, according to Do Nascimento Nunes et al. (2014).

These are just some statistics that reveal a scenario where CSAs have deficiencies, in addition to providing sufficient support to strengthen the importance of control and monitoring of the cold chain, not only to solve the problem of food spoilage but also to address general challenges associated with world food security. Good temperature management is the most important and easiest way to delay the deterioration and waste of these foods (Do Nascimento et al., 2014).

Franco et al. (2017) comment that there is no doubt that our way of life depends on our ability to cool and control the temperature of storage spaces and means of food distribution.

1.2 Cooling technology

Today there are several alternatives in refrigeration systems that can be implemented either in commerce or industry. Gauger et al. (1995) classified the different refrigeration systems by an average range according to 6 criteria: state-of-the-art, complexity, size and weight, maintenance, useful life, and efficiency.

Table 1: Definition of numerical ratings for the evaluation criteria.

Evaluation Criteria	Rating of 1	Rating of 5
State-of-the-Art	Only Theory	Completely Mature
Complexity	Very Complex	Very Simple
Size and Weight	High	Low
Maintenance	High	Low
Service Life	Short	Long
Efficiency	Bad	Good

Source Alternative technologies for refrigeration. Gauger et al. (1995)

The 6 criteria to evaluate the refrigeration systems, each one of them was qualified for the different turns where it is necessary, such as commercial air conditioning, domestic and mobile, commercial and domestic refrigeration.

For commercial refrigeration, which is the focus of this work, Gauger et al. (1995) evaluated refrigeration technologies from best to worst according to the criteria mentioned as shown in the following Table.

The refrigeration technologies with the best qualification and, therefore, the most suitable for application in the commercial sector are steam compression and absorption.

Steam compression technology is currently the most widely used refrigeration system for food preservation and air conditioning, both for domestic, commercial and mobile use. This system uses gases, such as chlorofluorocarbon (CFC) and hydrochlorofluorocarbon (HCFC), as cooling agents. These types of gases have excellent thermodynamic properties for cooling cycles, as well as being economical and stable (Gauger et al., 1995).

The favourable environment for the storage of fruits and vegetables is low temperature and high humidity. This is reasonably achievable by steam compression cooling with relatively low investment and lower energy consumption. Dilip (2007) reports that this type of refrigeration

Table 2: Classification of technologies in commercial refrigeration from best to worst.

Classification	Refrigeration Technology	Evaluation
1	Steam Compression	4.70
2	Absorption	3.80
3	Reverse Stirling	3.15
4	Solid Sorption	3.10
5	Reverse Brayton	3.00
6	Pulse Tube/Thermoacustic	2.80
7	Magnetic Cooling	2.05
8	Thermoelectric	2.05

Source Alternative technologies for refrigeration. Gauger et al. (1995)

system achieves a favorable environment for the storage of fruits and vegetables since the shelf life of perishable foods stored under these circumstances increases from 2 to 14 days compared to storage at room temperature, so for CSA this technology is very favorable. However, the gases used by this technology, such as the CFCs and HCFCs used as refrigerants for many years, have depleted the ozone layer, while fluorocarbons (FC) and hydrofluorocarbons (HFCs) have a high global warming potential (GWP) and cause global warming phenomena. For this reason, the use of alternative technologies, such as absorption, Lychnos and Tamainot-Telto (2018), has been targeted.

The absorption cooling system is attractive for commercial refrigeration and air conditioning. If levels of complexity and maintenance can be reduced, it could also be attractive for domestic applications (Gauger et al., 1995). For Wang et al. (2013) this type of cooling is considered as a green technology that can provide cooling for heating, ventilation and air conditioning, especially when silica gel is adopted due to its great suitability in effective contributions to reduce greenhouse gas emissions. Absorption systems use natural refrigerants, such as water, ammonia and/or alcohols, that do not damage the ozone layer and have little or no impact on global warming (Lychnos et al., 2018). However, certain drawbacks have become obstacles to their actual applications and commercialization. For example, the discontinuous operation of the cycle, the large volume and relative weight of traditional refrigeration systems, the low specific cooling capacity, the low coefficient of performance, the long absorption/desorption time, and the low heat transfer efficiency of the adsorbent bed, explain Wang et al. (2018).

On the other hand, Bhattad et al. (2018) reflect that one of the greatest challenges in today's world is energy security. There is a great need for energy in refrigeration and air conditioning applications. Although, due to limited energy resources, research is being conducted in the area of improving the efficiency and performance of thermal systems.

According to Zhang et al. (2017), today the economic growth and technological development of each country depends on energy. Heating, ventilation, air conditioning and domestic and commercial refrigeration consume a large amount of energy. The refrigeration system has great potential for energy savings. Lychnos et al. (2018), present in their work the development of a prototype with hybrid refrigeration systems that combines steam compression and absorption technologies. Preliminary tests showed that it can produce a maximum of 6 kW of cooling power with both systems running in parallel. It is designed as a water cooler with an evaporating temperature of 5°C and a condensing temperature of 40°C.

For countries such as Mexico, promoting energy savings would become a competitive advantage, even more so in the commercial sector for the micro-enterprise which, as mentioned above, often have limited electricity supply at their points of sale and marketing of their perishable products.

2. Methodology

The design methodology to be used to carry out the project will be the "Double Diamond". The Design Council, an organization that advises the English government on the fundamental role of design as a creator of value, argues that designers from all disciplines share perspectives at various points during the creative process, which they illustrate as "The Double Diamond". This design process is based on a visual map that is divided into 4 stages: Discover, Define, Develop and Deliver.

The purpose of choosing this methodology is to discover the best solutions by testing and validating several times since the creative process is iterative and with this, the weakest ideas are discarded.

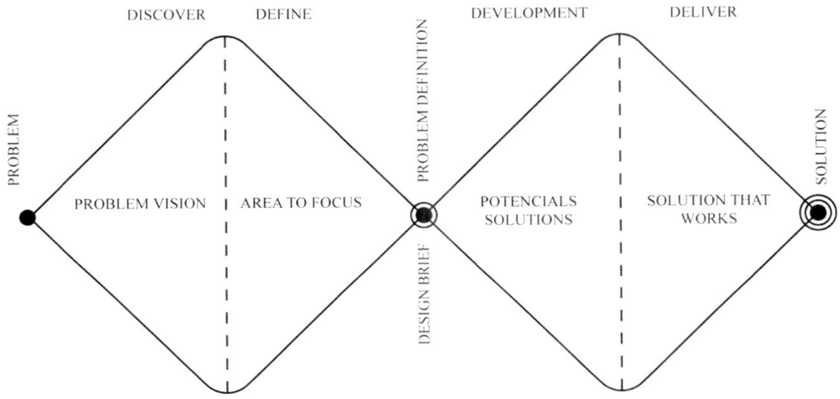

Figure 1: Double Diamond Methodology. Source Desing Council UK.

2.1 Discover

The first stage of this model describes how to empathize with users to deeply understand the problem or problems they are seeking to solve. For this purpose, field visits were made with different micro-enterprises within the food sector in the municipality of San Juan del Río, Querétaro, to which an interview was conducted to obtain data on the management and marketing of their products.

In order to obtain relevant data for the application of this interview and to calculate the size of the sample with an 80% confidence level, a visit was made to the Secretary of Economic Development of San Juan del Río to investigate the number of micro-enterprises operating in the municipality; it was found that there are no data on the number of micro-enterprises at either the municipal or state level because they are such small enterprises that the vast majority are not registered with the Ministry of Finance and Public Credit, making it very difficult to have reliable data on micro-enterprises. Due to this, the interview was applied to 10 micro-enterprises in the municipality of San Juan del Río, Querétaro. The questionnaire is presented below.

1. Company and business.
2. Place of operation of the company.
3. What kind of products do you sell?
4. Do you know at what temperature range they should be kept?
5. Do you produce or market the products you handle?
6. If you sell, do you receive the products at storage temperature?
7. What type of packaging do perishable products have?
8. How long does it take to move perishable products from where they are manufactured to where they are exhibited or delivered?

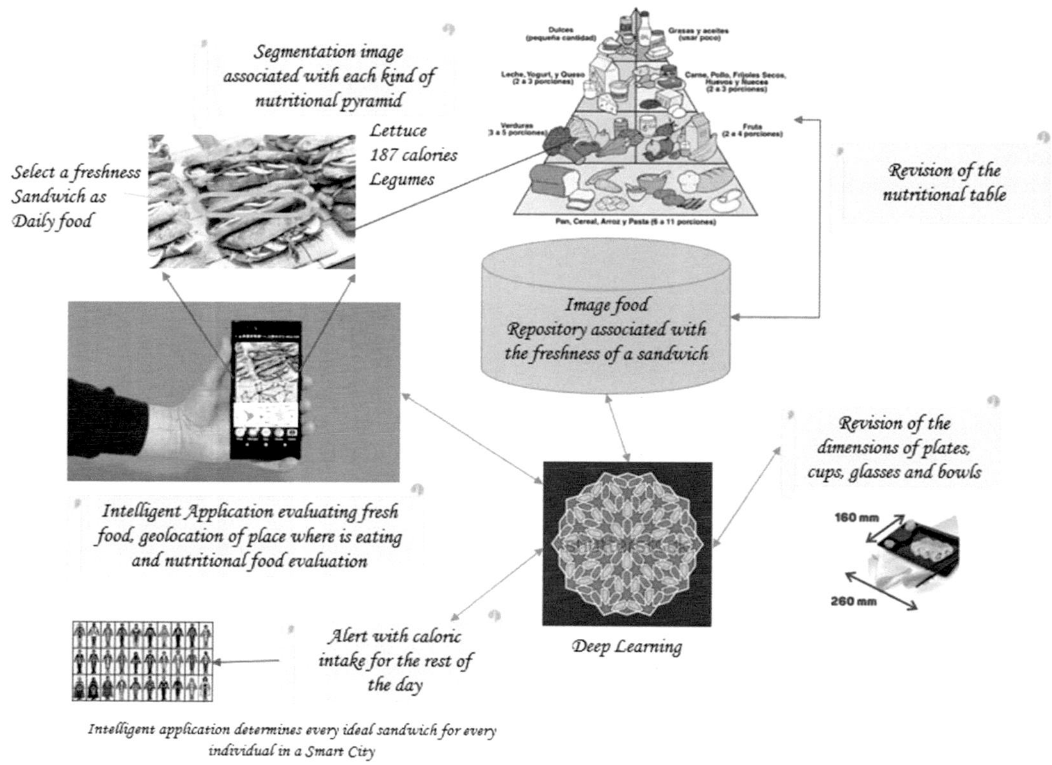

Figure 2: Conceptual diagram of the implementation of an intelligent system that determinates the greatest freshness in the presentation threshold and color analysis of various sándwich issues in a stock of a store selling healthy products. Source own preparation.

9. How much product do you store?
10. How much product do you handle during distribution?
11. What type of vehicle do you use to distribute your products?
12. Do you have the option of home delivery?
13. What tool do you use to preserve the product during shipment?
14. Is there any product that is wasted during distribution? What factors cause this decline?
15. What do you do with the product that is not sold during your working day?
16. What is the amount you plan to invest in a specialized tool to help keep the product better during distribution?

With this interview, we sought to know the situation of micro-enterprises in relation to the distribution, handling and storage of perishable products, as well as the level of loss of them.

2.2 Define

For this stage, the objective was to carry out an analysis of the perception of the quality of perishable products among consumers in the municipality of San Juan del Río, in the state of Querétaro, in order to obtain data that will allow us to know how relevant the freshness, good presentation and first instance perception of the quality of food products are for consumers, and whether this impacts on the purchasing decision. As a measuring instrument, a questionnaire was designed to assess whether the quality, freshness and presentation of food is relevant to people when purchasing raw

Table 3: Survey to Analyze Perception of Quality.

N°	Questions	Answers
1	Please tell us your age range	1) 18 – 28 2) 29 – 38 3) 39 – 48 4) 49 – 60 5) 60 or more
2	Please indicate your gender	1) Female 2) Male
3	Please tell us your highest grade	1) Secondary 2) High School 3) University 4) Posgraduate
4	Please tell us your occupation	1) Goverment employee 2) Private sector employee 3) Freelance 4) Entrepreneur/Independent
5	What is your level of concern about the quality of the food you eat? (1 not at all concerned to 5 rather worried)	1) not at all concerned 2) somewhat worried 3) concerned 4) very concerned 5) rather worried
6	How do you value your confidence in the handling of the food you buy in food businesses? Storage, transportation, refrigeration (1 very suspicious to 5 very confident)	1) very suspicious 2) something suspicious 3) is indistinct 4) something trusting 5) very confident
7	Of the following persons or organizations that handle food, rate from 1 to 5 the degree of information you believe they have about food quality, in terms of optimal temperature management for storage and transportation/distribution (1 nothing informed to 5 well informed) • Products or farmers (Producers of cheese, flowers, fruits and vegetables (food business suppliers)) • Large food chains (Mcdonalds, Dairy Queen, Toks, etc.) • Small food businesses (Food trucks, food carts, local bakeries, etc.) • Food logistics companies (Uber eats, rappi, no apron) • Chain of super markets (Soriana, walmart, etc.)	1) nothing informed 2) poorly informed 3) informed 4) very informed 5) well informed
8	Rate from 1 to 5, 1 being irrelevant and 5 quite relevant, the aspects you consider when buying a food for the first time. • Correct handling of the product during the distribution chain. • Information about the product and how it was produced. • Affordable price. • The quality of the product can be observed (colour, texture, etc.) • The packaging of the product is not in poor condition (knocked, dented, broken, torn, etc.)	1) irrelevant 2) not vey relevant 3) relevant 4) very relevant 5) quite relevant
9	1 being no preference and 4 total preference, value your preferred channel to buy basic basket through the following ways: • Internet/Application (Walmart online, rappi, etc.) • Local market (popular market, central stores, etc.) • Super market (Comer, Soriana, Walmart, etc.) • Small food businesses (corner shops, grocery stores, fruit shops, etc.)	1) no preference 2) low preference 3) preference 4) total preference
10	1 being no preference and 4 total preference, value your channel of preference to buy prepared food through the following ways: • Internet/Application (Rappi, Uber Eats, Sindelantal, etc.) • Local Market (Food Area) • Large food chains (Toks, McDonalds, Starbucks) • Small food businesses (mobile businesses, foodtrucks, local bakeries, local food businesses)	1) no preference 2) low preference 3) preference 4) total preference

Source Own preparation

and prepared foods. In the same way, the aim is to find out what consumers trust so much in the different types of businesses that sell food as raw material for preparing dishes, which we call "basic basket" in the questionnaire and the other type we call prepared foods, which are already prepared dishes that sell the types of business. This questionnaire was applied online through the platform of "google surveys" for consumers in the municipality of San Juan del Rio at random.

In order to determine the sample, INEGI data were taken, which indicates that the number of inhabitants between 18 and 60 years of age in 2015 is 268,408, which we can consider as the total number of potential consumers in our population. The calculation was made with a confidence level of 90% and a margin of error of 10% so the result was 68 to obtain a reliable sample of the total population.

The first part of the questionnaire consists of 4 questions to know the demographic data of the participants which help us to categorize them according to their age range, gender, maximum degree of studies and professional occupation. The second part of the questionnaire is made up of 6 questions that focus on obtaining data that provided us with a panorama to better understand whether there is consumer concern regarding the quality of the perishable products they buy, the level of confidence in the businesses that commercialize these products, more relevant aspects for the purchase decision, level of preference in the different businesses that commercialize both raw and prepared foods. For each of the reagents that make up the previous survey, answers with Likert scales were established to make the way of answering the respondents more dynamic.

2.3 Develop

For this area of the second diamond, a hybrid cooling system with steam compression and absorption systems was tested to validate its operation. These tests were carried out in a laboratory of the company Imbera, located in the municipality of San Juan del Río in the state of Querétaro, with a controlled environment with a maximum temperature of 34°C, and a relative humidity of 59%. A prototype was developed for the tests.

An evaluation protocol was developed for the tests to obtain the following information:

- Pull down time (time it takes the system to generate ice blocks for temperature conservation).
- Temperatures reached by the cooling system during pull down.
- Duration of the ice blocks without power supply.
- Temperatures reached by the cooling system without electrical energy.

The objective of this protocol was to delimit the categories of perishable products that can be optimally conserved by a hybrid refrigeration system.

After this, a preliminary cost analysis of materials and manufacturing processes was carried out in order to know which are the best adapted to the investment capacities of the users. In this exercise, two specific manufacturing processes were analyzed: roto-moulding and thermoforming, since the plastic material is the best option for the manufacture of the conservation tool due to the variety that exists, and therefore the versatility of qualities that it offers. The following table describes prices of materials and tools.

Table 4: Comparison of manufacturing processes.

	Roto Moulding	Thermoformed
Cost Moulds	$415,000	$44,000
Cost Parts	$8,531.27	$2,118.04
Cost of Refrigeration System	$3,404.34	$3,404.34
Cost of Electrical System	$6,981.83	$6,981.38
Total	$18,917.44	$12,504.21

Source: Prepared by the authors.

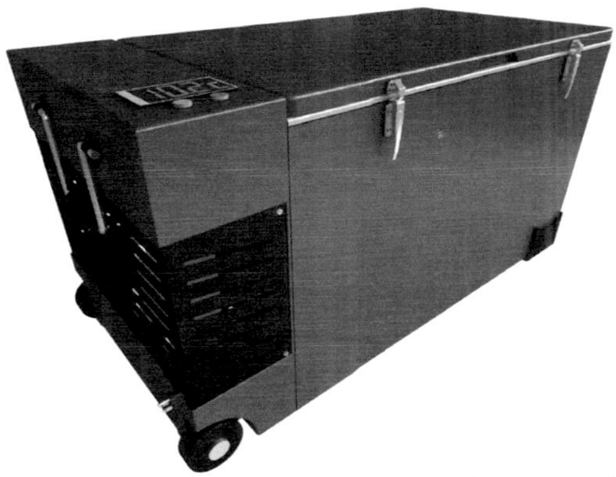

Figure 3: Conservation tool concept. Source Own preparation.

The most appropriate manufacturing process is thermoforming because the estimated price is within the range that users are willing to spend for the conservation tool. With this information, the design of the conceptual proposal of the conservation tool was developed with the considerations described above.

The concept of the conservation tool for small businesses is made of high-density polyethylene. This plastic is commonly used in packaging, safety equipment, and construction, to offer lightness to maneuver and resistance to withstand vibrations and shocks during distribution routes. The dimensions of the equipment are 100 cm wide, 40 cm deep and 40 cm high, with a capacity of 74 liters. These dimensions make it easy to transport, i.e., it can be placed in any compact vehicle or cargo vehicles.

As for the hybrid refrigeration system, it will be composed of a steam compression system and an absorption system, the cooling agent of the absorption system will be water which will be contained in two plastic tanks. Inside the plastic tanks passes a copper tube that is part of the refrigeration system by steam compression to freeze the water. In order for the steam compression

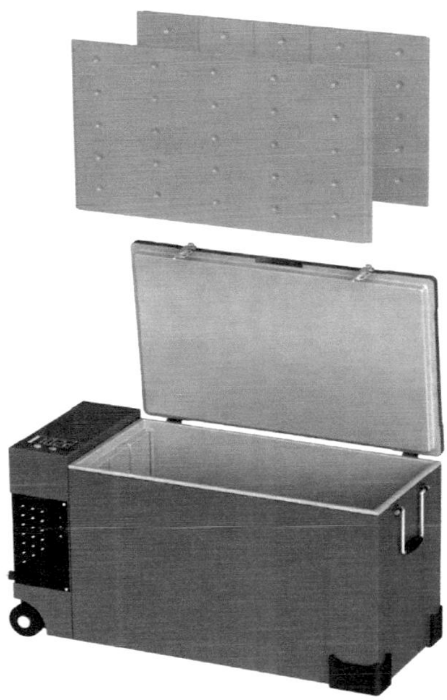

Figure 4: Refrigeration system concept. Source Prepared by the authors.

system to work, it must be connected to the electrical energy during the night and the correct formation of the ice blocks is guaranteed so that the absorption refrigeration system works correctly during the distribution routes.

The internal walls of the equipment, as well as those of the water containers, have slots through which metal separators can slide to generate different spaces for different product presentations. To close this stage, a prototype was made that complies with the functionality of the concept in order to be able to evaluate it in the next stage of the methodology.

Figure 5: Product Organization. Source Prepared by the authors.

2.4 Delivery

For this last stage of the double diamond model, the perishable food logistics strategy was validated with the prototype of the conservation tool with the products developed in the Amazcala campus of the Autonomous University of Querétaro and marketed in the tianguis that is established at the central campus of the same university. The products that were placed to be evaluated within the prototype of the conservation tool are containers with 125 ml of milk and a milk-based dessert (custard), which are produced on that campus.

As can be seen in the Figure above, the samples are identified so that one is placed inside the prototype of the conservation tool and the other outside it. For these milk samples, the following initial data were taken before placing them in the prototype of the conservation tool.

In the case of the milk-based dessert, the measurement will be visual since this product will present syneresis (expulsion of a liquid in a mixture), losing its texture if the cold chain is not maintained correctly during distribution.

The validation protocol for the conservation tool prototype consists of the following steps:

- The conservation tool prototype is mounted on the truck at 8:30 am.
- The products to be marketed are mounted on the truck for a period of 30 minutes.
- Products leave the Amazcala campus for the point of sale at the downtown campus at 9:00 am.
- The van arrives around 10:15 am at the Centro campus, at the engineering faculty, unloads the products at the point of sale. After this, it is withdrawn to make other deliveries to different points of sale.
- At 1:00 pm the van returns to the engineering faculty point of sale to pick up the unsold product.
- At 3:15 pm the van arrives at the Amazcala campus, until then the samples are taken and analyzed.

Before this validation protocol, the equipment was left connected the night before from 1:30 a.m. to 8:30 a.m., a period of 7 hours to ensure that the ice blocks were formed correctly.

Figure 6: Milk samples to be analyzed. Source Esau, Amazcala campus.

Table 5: Initial milk parameters.

Parameter	Value
Acidity	16°D
pH	6.7
Temperature	4.1°C
Storage system temperature	5°C

Source Esau, Amazcala campus

The objective was to make comparisons between the use of the conservation tool prototype and the dry box of the truck with which the products are transported from the Amazcala campus to obtain data from the specialized tool and fine-tune the strategy and be able to launch it to the market.

2.5 Ethical considerations

For the last stage of the methodology that is delivery where it is planned to make tests with the tool specialized in micro-enterprises, the perishable foods that will be used during these activities will not be consumed by people and will be available in an appropriate manner. On the other hand, the information obtained during the discovery and definition stages will be data related only to the working activity of micro-enterprises and not personal data or confidential information about micro-entrepreneurs.

3. Project Development

The micro-enterprises that were considered for research and application of the project are those that are categorized as micro-enterprises, which are formed by no more than 15 workers including the owner according to INEGI, within the trade sector in the branch of food and/or perishable products. These companies will be approached with an interview questionnaire to learn about and analyze the tools and procedures they use during their supply chain. With this information, it is sought that the candidate companies to the project have the following or at least one of the following characteristics:

- They do not have a clear and established logistics for the distribution of their perishable products.
- The tools with which they distribute and maintain the conservation of perishable products are not adequate and mistreat the presentation of their products.
- Distribution logistics are complex, either because they don't know the ideal tools for conservation and optimal temperature ranges for their products, or because they don't have the economic justification to invest in specialized refrigeration transports.

Having explained the process of selecting companies for the development of the project, below is a list of the activities to be carried out in order to design a logistics strategy for perishable products for the microenterprise:

1. A problem is discovered through the observation of the actors.
2. Review of literature on issues associated with the problem.
3. Application of interviews to population of micro-entrepreneurs on the application of CSA in their business. (The sample population are businesses with turn in the commercialization of prepared or perishable foods in the state of Querétaro, specifically in the municipality of San Juan del Río).
4. Case studies and similar projects.
5. Analysis of the information collected.
6. Exploration of users' needs and aspirations.
7. Definition of the design problem.
8. Definition of project specifications.
9. Stage of development of potential solutions.
10. Weighting of solutions to find the most feasible.
11. Conceptualization of the possible solution.
12. Design of support material for the communication of benefits of the strategy.
13. Prototyping of the product.
14. Execution of the strategy based on the prototype.
15. Validation of the strategy according to KPI.
    - Quantity of perishable products that arrive in optimal conditions of presentation and conservation.
    - Time of handling and organization of the perishable products so as not to break the CF in the CSA.
    - Increase in sales due to adequate management of the conservation and presentation of perishable products.
16. Analysis of results.
17. Conclusions.

3.1 Human resources

- Designer: Development, application and evaluation of the strategy based on the established methodology.
- Industrial Designer: Product design based on the requirements and wishes of the user.
- Refrigeration Engineer: Design and development of a refrigeration system that adapts to the requirements of the product and the project.

3.2 Material resources

- Laptop.
- Internet.
- Work space (table, desks).
- Automobile for field visits.
- Automobiles and user vans for tests and validations.

- Smartphone for communication with users, as well as for taking videos and photographs as a record.
- Various materials for prototype.

4. Results and Discussion

The main objective of this stage was to empathize with micro-entrepreneurs and to analyze the problems they encounter when distributing, handling and storing the perishable products they market.

The questionnaire consists of 16 questions with the purpose of shedding light on the deficiencies in the operation of the company and to know how they are solved.

It was found that the products marketed by these companies range from pastries and confectionery (dairy products) to flowers, sausages, fruits and vegetables. The current logistics process in the distribution channels of the micro-enterprise, in general terms.

The strategy for logistics in the distribution of perishable foods based on a specialized conservation tool, in addition to guaranteeing the quality and presentation of the products, helps micro-entrepreneurs to find added value in their economic activities.

Like the applications for the distribution of prepared foods (uber eats, rappi, etc.) that through their services generate an increase in their profits in businesses through home delivery.

This distribution strategy for perishable foods transfers this same service model, through the specialized conservation tool to successfully promote the incursion of new sales channels for micro entrepreneurs and increase their income.

References

Bhattad Atul, Sarkar Jahar and Ghosh Pradyumna. 2018. Improving the performance of refrigeration systems by using nanofluids: A comprehensive review. Renewable and Sustainable Energy Reviews, 82: 3656–3669.

Dilip Jain. 2007. Development and testing of two-stage evaporative cooler. Building and Environment, 42: 2549–2554.

Do Nascimento Nunes M. Cecilia, Nicomento Mike, Emond Jean Pierre, Badia Melis Ricardo and Uysal Ismail. 2014. Improvement in fresh fruit and vegetable logistics quality. Philosophical Transactions of the Royal Society A, 372: 19–38.

Franco, V., Blázquez, J.S., Ipus, J.J., Law, J.Y., Moreno-Ramírez, L.M. and Conde, A. 2017. Magnetocaloric effect: from materials research to refrigeration devices. Progress in Materials Science, 93: 112–232.

Gauger, D.C., Shapiro, H.N. and Pate, M.B. 1995. Alternative Technologies for Refrigeration and Air-Conditioning Applications. National Service Center for Environmental Publications (NSCEP), 95: 60–68.

Gunders Dana. 2012. Wasted: How America is losing up to 40 percent of its food from farm to fork to landfill. NRDC Issue Paper, 12: 1–26.

Gustavsson Jenny, Cederberg Christel, Sonesson Ulf, Van Otterdijk Robert and Meybeck Alexandre. 2011. Global food losses and food waste. Food and Agriculture Organization of the United Nations, Rom., 92: 1–25.

Jedermann Reiner, Nicometo Mike, Uysal Ismail and Lang Walter. 2014. Reducing food losses by intelligent food logistics. Philosophical Transactions of the Royal Society A, 1: 1–37.

Lychnos, G. and Tamainot-Telto, Z. 2018. Prototype of hybrid refrigeration system using refrigerant R723. Applied Thermal Engineering, 134: 95–106.

Pang Zhibo, Chen Qianf and Zheng Lirong. 2011. Scenario-Based Design of Wireless Sensor System for Food Chain Visibility and Safety. Advances in Computer, Communication, Control and Automation, 1: 19–38.

Wang, Dechang, Zhang, Jipeng, Yang, Qirong, Li, Na and Sumathy, K. 2013. Study of adsorption characteristics in silica gel–water adsorption refrigeration. Applied Energy, 113: 734–741.

Wang, Yunfeng, Li, Ming, Du, Wenping, Ji, Xu and Xu, Lin. 2018. Experimental investigation of a solar-powered adsorption refrigeration system with the enhancing desorption. Energy Conversion and Management, 155: 253–261.

Zhang, Wenxiang, Wang, Yanhong, Lang, Xuemei and Fan, Shuanshi. 2017. Performance analysis of hydrate-based refrigeration system. Energy Conversion and Management, 146: 43–51.

Zubeldia, Bernardino, Jiménez, María, Claros M. Teresa, Andrés José Luis and Martin-Olmedo Piedad. 2016. Effectiveness of the cold chain control procedure in the retail sector in Southern Spain. Food Control, 59: 614–618.

CHAPTER-4

Analysis of Mental Workload on Bus Drivers in the Metropolitan Area of Querétaro and its Comparison with three other Societies to Improve the Life in a Smart City

Aarón Zárate,[1] *Alberto Ochoa-Zezzatti,*[2] *Fernando Maldonado*[1,*] and *Juan Hernández*[2]

1. Introduction

Mental workload is investigated in ergonomics and human factors and represents a topic of increasing importance. In working environments, high-cognitive demands are imposed on operators, while physical demands have decreased (Campoya Morales, 2019).

These figures make it possible to measure the serious public health problem that causes road accidents in the world and in our country and the strong negative impact that it generates in the society and the economy. Hence, in 2011, the WHO generated a program that is called the Decade of Action for Security Vial 2011–2020, through which it summoned several countries to generate actions with the purpose of mitigating this problem.

In our country, in 2011, the National Road Safety Strategy 2011–2020 and 2013 was promoted within the National Development Plan, the Road Safety Specific Action Program 2013–2018 (PAESV). The goal of reducing the mortality rate caused by road accidents to 50% was proposed, as well as minimizing injuries and disabilities through 6 strategies and 16 lines of action concentrated in 5 main objectives:

1. To generate data and scientific evidence for the prevention of injuries caused by road accidents
2. To propose a legal framework on road safety that includes the main risk factors present in road accidents

[1] Universidad Autónoma de Querétaro
[2] Universidad Autónoma de Ciudad Juárez
* Corresponding author: luis.maldonado@uaq.mx

3. To contribute to the adoption of safe behaviors of road users to reduce health damage caused by road accidents
4. To promote multisector collaboration at the national level for the prevention of road accident injuries
5. To standardize prehospital medical emergency care of injuries.

2. Implementing Case-based Reasoning to Improve our Order Picking Model

According to the INEGI data projected in the National Road Safety Profile of the Ministry of Health, in the period from 2011 to 2015, it is observed that the mortality rate has not increased, nor significantly decreased, which denotes apparent control as a result of the actions implemented in various programs. However, these results are insufficient for the fulfillment of the goal established in 2011 (to reduce the rate of mortality by 50%).

On the other hand, the Mexican Institute of Transportation conducted a study called "Efficiency and/or effectiveness of road safety measures used in different countries (2013)" (Dominguez, Karaisl, 2013), in which, through a questionnaire applied to 22 countries, it is possible to identify 23 main implemented security measures and the effects obtained by conducting a Cost-Benefit analysis. For the identification of road safety strategies, of the 22 questioned countries, only one (Japan) does not use economic analysis methods liked Cost-Benefit Analysis (CBA) and Cost-Effectiveness Analysis (ACE). The rest of the countries go to accident records and compare the cost of implementing road safety measures against the impact coming from the human capital or the costs derived from the accidents (low productivity, costs per hospitalization, costs per repair and replacement, etc.).

In this study, the lack of data is considered as the biggest barrier for a Cost-Benefit analysis, which ultimately results in the implementation of security measures with insufficient effectiveness.

3. Methodology

The development of this project will be based on a variant of the methodology of the Double Diamond, proposed by the Green Dice company of the United Kingdom, which they called Triple Diamond.

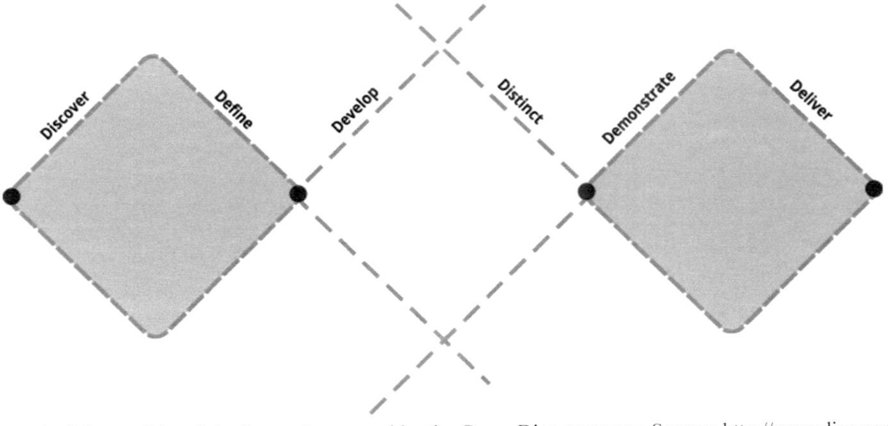

Figure 1: Methodology of the triple diamond proposed by the Green Dice company. Source: http://greendice.com/double-diamond.

Unlike the Double Diamond methodology, the Triple Diamond Methodology incorporates a third intermediate stage, moving the stage of Development to the intermediate diamond and adding two stages, the Distinction and the Demonstration. This methodology is selected because it integrates a series of steps typical of the development and implementation of a project, an essential stage in any innovation project: Distinction. In this stage, the aspects that show that the project has the character of innovation are highlighted. Next, each of the stages of the Triple Diamond methodology process are described in more detail.

Discovery

The discovery stage is the first in the methodology. Start with the idea initial and inspiration. In this phase, the needs of the user are stated and the following activities are executed;

- Market research
- User research
- Information management
- Design of the research groups

Definition

In the definition stage, the interpretation of the user needs is aligned with the business or project goals. The key activities during this phase are the following:

- Project planning
- Project Management
- Project closure

Development

During this phase is when the planned activities for the development of the project are performed, based on the established plan, iterations and internal tests in the company or group of developers.

The key activities during this stage are the following:

- Multi-disciplinary work
- Visual Management of the project development
- Methods Development
- Tests

Distinction

The stage of distinction reveals and establishes the characteristics that distinguish the project proposal of the rest of the proposals, also determines the strategy to continue to ensure that the target customers will actually choose to seek the product or service that is developed in the project. The key tasks during this stage are the following:

- Definition of critical and particular characteristics
- Definition of introduction strategy
- Development of market strategy

Demonstration

In the demonstration stage, prototypes are made to evaluate the level of fulfillment of the project's purpose and to ensure that the design meets the problem for which it was created. Unlike the iterations that are made in the stage of development, the tests that are carried out during the demonstration stage are already made with the final customer. The key tasks in this phase are the following:

- Requirements compliance analysis
- Prototype planning and execution
- Execution of tests with end users
- Evaluation of the test results
- Definition of design adjustments

Delivery

The delivery is the last stage of the Triple Diamond methodology. During this stage, the product or service developed is finalized and launched to the market and/or is delivered to the final customer. The main activities that occur during this stage they are the following:

- Final tests, approval and launch
- Evaluation of objectives and feedback cycles

4. Project Development Process

The following describes the specific activities that are part of the project's development based on the triple diamond methodology.
Discovery Stage:

- Identification of the problem.
- Dimensioning of the problem found and its effects, based on international, national and local statistics.
- Research on the methods currently used to try to attack the problem, the resources used and their effectiveness. The investigation has been conducted through interviews with experts on mobility of the UAQ faculty of engineering, and it is planned to also attend the Mobility Secretariat of the municipality of Querétaro to capture the specific needs for consultation.
- Identification of project actors, including experts in mobility and development strategies of data management platforms. The inclusion of these actors will be agreed voluntarily and their participation will be limited only to the consultation and audit of the proposal.
- For the research of the users of the mobile application, meetings of focus groups will be developed to evaluate the usability of the same. Two rounds of evaluation of the participation of the users in the generation of the reports will be carried out through the implementation of a Beta version. Users will voluntarily download the application, and inside it they will find the information concerning the privacy warnings of their data, as well as an informed consent button where it is confirmed that they know the purpose of the investigation and the conditions of use:

  a. Design of a solution proposal to the problem, raise details and technical specifications of the project. In this stage, the relevant parameters are defined to fulfill the project objective:
     - High-level project design for determination of phases and parts

  b. Definition of scope and project partners

  c. Planning of project development

  d. Definition of resources for project development. The resources will be basically for the development activities of the platform and will be supported by personal contacts and also financed with personal resources.

4.1 Stage of Development

- Structuring of the database. It is taken as an initial part of the development as a strategic step to optimize the development of the mobile application and its performance designed for the user (data usage, optimization of device memory and processing resources)
- Development of the first mobile application prototype to evaluate the user's experience during its use through focus groups
- Definition of Story Boards for the mobile application
- Development of the consultation platform
- Platform performance tests
- Establishment of servers and information management methods.

4.2 Stage of Distinction

- Development of the gamification strategy for the mobile application.
- Definition of the business scheme. The proposal will be based on a flexible scheme, which will allow expanding the scope and personalization of the platform depending on the geographical area where it is to be implemented. The platform will be developed as an independent entity, without any link with governmental institutions or private interests, but it will be open to adaptations in order to respond to the needs and particularities of whoever decides to adopt it.
- Definition of modality for presenting the information in the consultation platform (georeference).

Demonstration stage:

- Carrying out the first integration tests of the platform
- Analysis of the generated information, its processing and presentation to assess the level of value and practicality that they present
- Analysis of compliance with initial requirements
- Project validation

5. Development

5.1 Nature of driver behavior

They will be analyzed in a general demographic factor that influences the behavior of the drivers.

- The problem from the perspective of public health.

The programs that have arisen worldwide and their adoption through programs to attack the problem.

- Information Technology applied to Road Safety.

In this section the available tools of information and how they have been used in terms of road safety will be reviewed.

- Big Data. A brief overview of the term and its application to the draft. It will also explain how this concept is increasingly important in terms of commercial value and business opportunity.
- Mobile applications focused on driver assistance and prevention of accidents.

We will give a tour of the main applications currently available in the market and their contribution to the solution of same problem.

5.2 *Analysis of the mental workload of driver behavior*

Several factors influence the behavior of driver's moment of driving in their vehicles on public roads, they all influence directly or indirectly in the way in which drivers live in the Road environment in which they move:

- Cultural factors. The culture is directly involved with the behavior of drivers. A general culture where respect for others is promoted allows that way of proceeding to be transferred to a road culture of greater harmony. This mentality, in turn, is influenced by a series of social and economic factors, which is why, in countries considered the first world, you can aspire to a road culture of greater respect and harmony.
- Infrastructure. A city whose infrastructure was developed without proper planning generates a conflict environment for drivers, generating emotional stress that results in the modification of their behavior. The road infrastructure of a city exerts an influence on the motorist, modifying his temperament, behaviors, and answers, making him participate or propitiating the road chaos (Casillas Zapata, 2015).
- Type of vehicle. The features and dimensions of the vehicle that is being driven influence the way the driver reacts. Characteristics such as the response to acceleration or maneuverability, as well as large dimensions of the vehicle, generate a state in the driver of courage prone to intrepidity (List, 2000).
- Public politics. The use of public policies that regulate effectively the behaviors that generate road accidents generates an environment of the permissiveness of unsafe behaviors, propitiating conditions of a high probability of accidents. Considering a study conducted directly with the bus drivers of the four samples, we took into account those that were designed with several focus groups, which are concentrated in Figure 2, this information collected, collects various aspects related to mobbing, work stress and social isolation in each group of bus drivers, these samples consisting of 217 individuals (57 women and 160 men): Sample 1—Querétaro: 42 (F: 6; M: 36), Sample 2—Salvador de Bahía: 62 (F: 15; M: 47), Sample 3—Palermo:

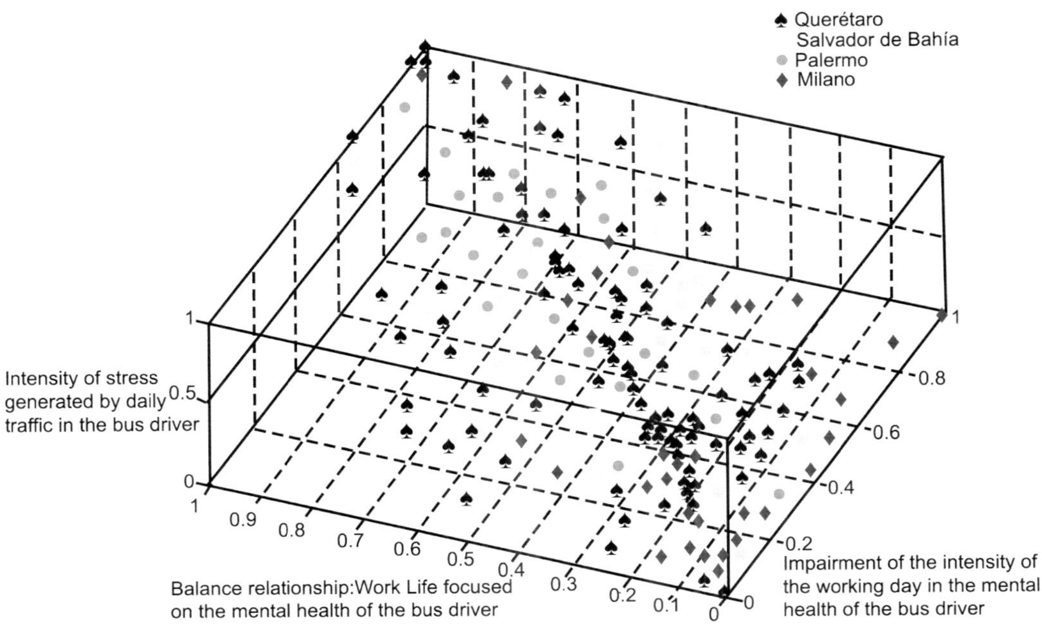

Figure 2: Visual representation of the analyzed sample, characterizing the diverse socio-economic aspects, the mobbing including the social blockade and the reflected labor performance.

57 (F: 14; M: 33) and Sample 4—Metropolitan Area of Milan: 167 (F: 23; M: 44). In the case of Queretaro bus drivers (See Figure 2), which is the group that presents greater differences in the salary relation concerning the working day and this lies in its place of origin, the bus drivers who suffer more mobbing come from Oaxaca, Guerrero, and Veracruz; an intermediate group of bus drivers from Coahuila, Zacatecas and Durango try to group to negotiate with the majority group and finally the children of the most recent wave coming from the Federal District, State of Mexico and Morelos, they even become intimidators in their respective routes of transport, because they have the greatest social capital of the group and tend to be accustomed to longer working days, so this group can be considered completely heterogeneous in its relations with the majority group.

Through public policies of the European Union, it is very easy to identify that the group of Milano bus drivers is first found, but it is very dispersed in the Salvador de Bahia samples where the work stress is greater, in Queretaro where the work-life relationship is not the most appropriate and in the little or no recognition that exists for the bus drivers in Palermo, which implies that there are more strikes than in the rest of the groups.

5.3 The problem from the perspective of Public Health

The effects of car accidents worldwide are devastating from many perspectives, but especially that of Public Health, since it affects society emotionally, functionally and economically. In many cases, deaths and chronic injuries end up generating dysfunctional families, by losing one of their members, or by allocating part of their patrimony to the support of any of them (Ponce Tizoc, 2014).

Similarly, in the economic aspect, automobile accidents have an economic impact of 1% to 3% in the respective GNP of each country, which amounts to a total of more than $ 500 billion (Hijar Medina, 2014), considering the material costs and the decrease in the productivity of the population.

5.4 Information Technology applied to Road Safety

The use of Information Technology to solve many social problems has played an important role in the development and improvement of conditions of life. In the case of Road Safety, they have tried to implement different strategies based on Information Technology solutions, such as the use of urban monitoring and surveillance elements and the processing of data from mobile phones for the determination of congested roads and re-definition of routes, mainly in fleets of cargo vehicles.

These types of benefits have also been raised through network vehicles, a technology that is growing and that will eventually allow an exchange of data between vehicles and urban networks, allowing to improve the traffic dynamics and street safety. The only drawback for them at the moment is a technology in development, which will begin to be implemented in some cities of the world, and eventually in countries like Brazil and Italy.

6. Mobile Applications Focused on Driver Assistance and Accident Prevention

Currently, there are endless applications focused on the assistance of drivers in order to prevent accidents and help you reach your destination without inconveniences. Most applications have been developed to generate a network of information that allows you to be alerted to various circumstances, such as the volume of traffic in the user's route, dangers, obstacles and even checkpoints or presence of patrols. Some others, focus their functionality on the assistance to the user before an emergency. The rest of the applications available in the market are currently focused on education about traffic rules and best practices of driving. These types of applications have characteristics that can make them attractive for children and adolescents, providing a means of education, which may have a positive impact on the road culture of future generations at the steering wheel, as is proposed to a Smart City in Figure 3.

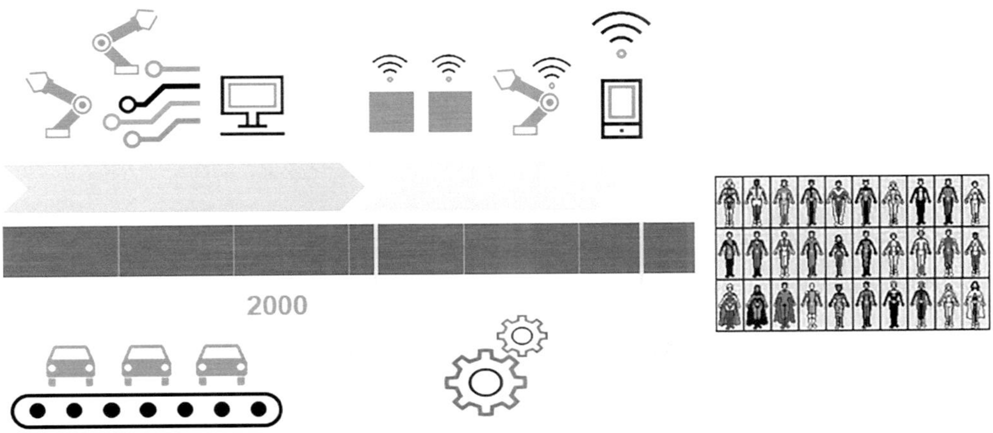

	With a public mental health politics	Without this
Queretaro	7.8	7.1
Salvador de Bahia	6.7	5.4
Palermo	7.3	6.9
Milano	9.4	9.1

Figure 3: Comparative of technology use in a model of Smart City to improve the Mental health in bus drivers of four societies including the human factor.

Of all these applications, the most common worldwide is the so-called "Waze", which refers to the ways that a user can take establishing its destination within the application, taking into account the conditions of said route determined mainly by reports of other users of the same application. Recently, Volvo became the first automaker to launch radar-based security systems in automobiles with the launch of the XC90 Hybrid in India in 2016. Features such as Airbags and ABS have already begun to become a standard safety feature and the future; we expect more of these characteristics to become standard. The most specific implementation is associated when generally more than one vehicle is moving at different speeds, and the collision points can be higher than two, as can be seen in Figure 4.

In the end, the facility that exists today to generate applications with various purposes, and the great boom that the use of mobile phones has had in different societies, allows us to visualize these types of conditions to use them as an excellent source for data mining. This data was provided to map them over the street map view of the city. We found that around 3,000 total accidents in the city

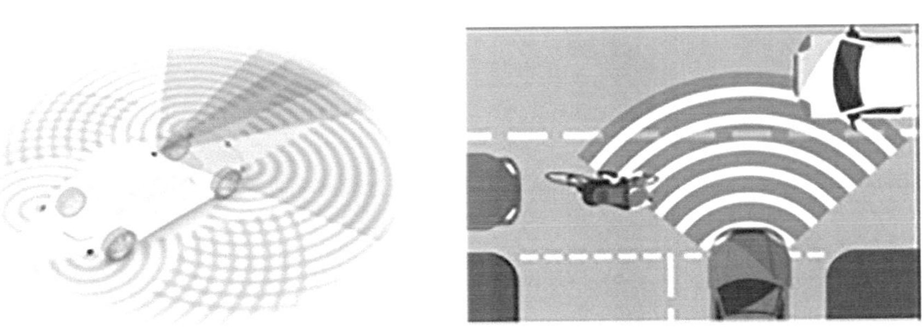

Figure 4: Model of IoT based on sensors for the control of the car and a radar added to identify obstacles in the proximity of up to one kilometer to identify automatic braking and implementation of a multi-objective security model for a smart city.

involve at least one motorcycle vehicle in them. Using a Kriging model, it is possible to determine the correct tendencies map in the future, as is shown in Figure 5.

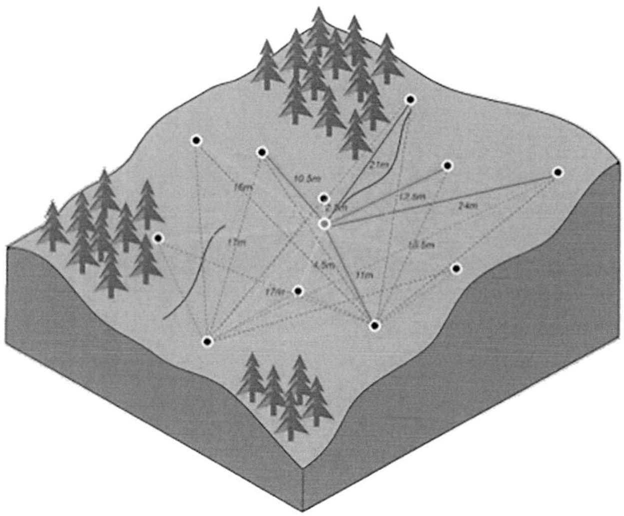

Figure 5: A kriging model used to determine changes in an Ecological model and its Visual representation associated with the modeling of loss of forest in a landscape. Source: http://desktop.arcgis.com/es/arcmap/10.3/tools/3d-analyst-toolbox/how-kriging-works.htm.

We fetched exclusively this data to map it in QGIS. To make all this information fit in the map in an ordered way, we took the spreadsheet file and randomly generated the latitude and longitude of the points of the map. All the other data fields were preserved as they came. Also, the spreadsheet software we used, LibreOffice Calc, to generate the random positions needed to be limited to all the points to fit in the area of Juarez. Calc doesn't generate decimal point random numbers, required for the cartographic precision in QGIS. To solve this problem, we created an algorithm to create positions. We used two different equations which represent each possible point in our model, as is possible to show in Equation 1 and Equation 2.

$$latitude = \frac{rand^{*}\ (\text{norther–souther})}{10000} \tag{1}$$

$$longitude = \frac{rand^{*}\ (wester\text{–}easter)}{10000} \tag{2}$$

The formulas above are the integer random generator for the positions of the points. The actual cardinal limits for the city are in the order of the same integer number (31.5997 south, 31.7845 north, –106.3077 east, –106.5475 west). To solve the problem, we multiplied all numbers by 10,000 to open the range and to let the Calc random algorithm to calculate with a more flexible grid the position of the points. After the number is calculated, the number is divided by 10,000 again. Once all the points were calculated, the spreadsheet file was parsed as a CSV (comma separated values) file, so the QGIS software can read the data of the records and print the layer of the locations. Once this was done, the Kriging method was applied. QGIS has a method to perform this by itself. Then the points were manually positioned to the closest road one by one. Also, some of the points were in zones that compared to OPM were outside the area of Juarez. These were allocated inside the desired area and sometimes deleted. As is shown in Figure 6.

Once this was done, the Kriging method was applied. QGIS has a method to perform this by itself. The kriging method was applied to adjust the grid, the area size, removing the OSM beneath, so the new calculated layer would not be big. As they are still too small, indicated by the absolute

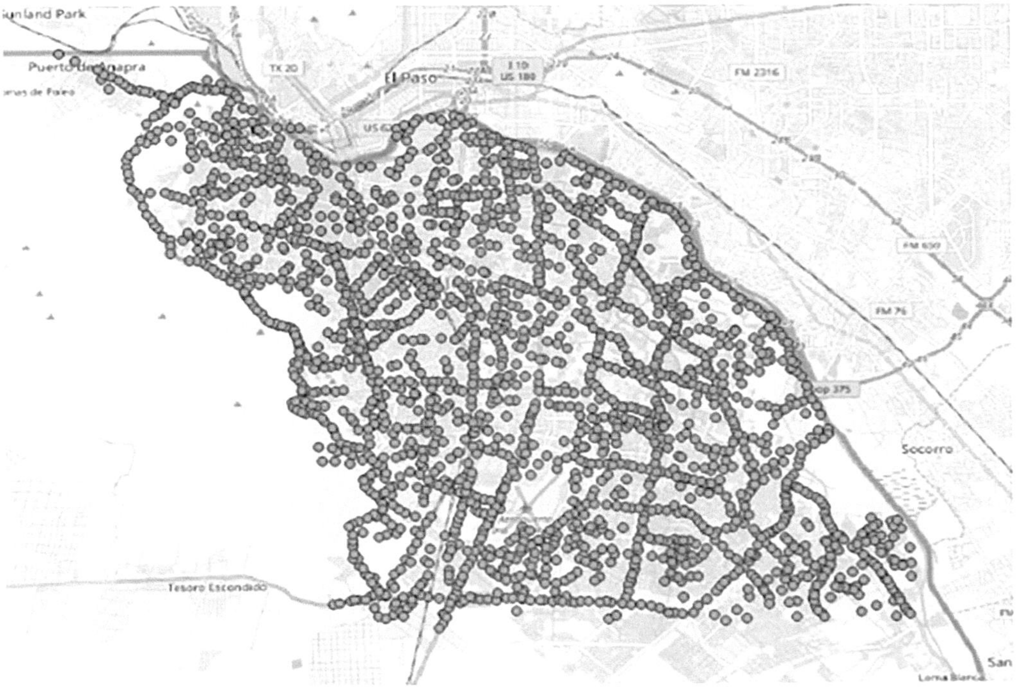

Figure 6: Results of each point in a possible traffic accident in Juarez City.

positions of the points. The new kriging calculated layer was difficult to adjust, as is shown in Figure 7.

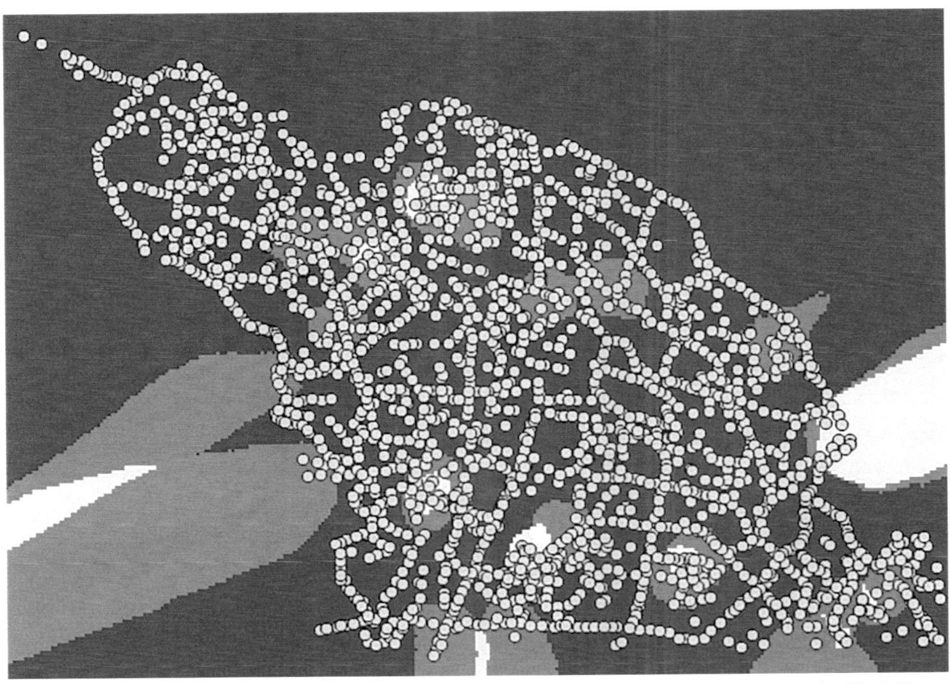

Figure 7: Final analysis of each point where a traffic accident will occur involving deaths associated with Italian scooter. The layer beneath the yellow dots (reddish colors) is the kriging generated layer that shows the most probable places in lighter colors. Some sites outside the city were shaded in the light; these could be errors in the algorithm.

If we compare this map above with the OSM, we can see the areas where the motorcycle accidents are frequent. Besides, we are proposing a tool for decision making using ubiquitous computing to help parents in locating children during the tour in case they had forgotten some object that is required for the class, as can be seen in Figure 8:

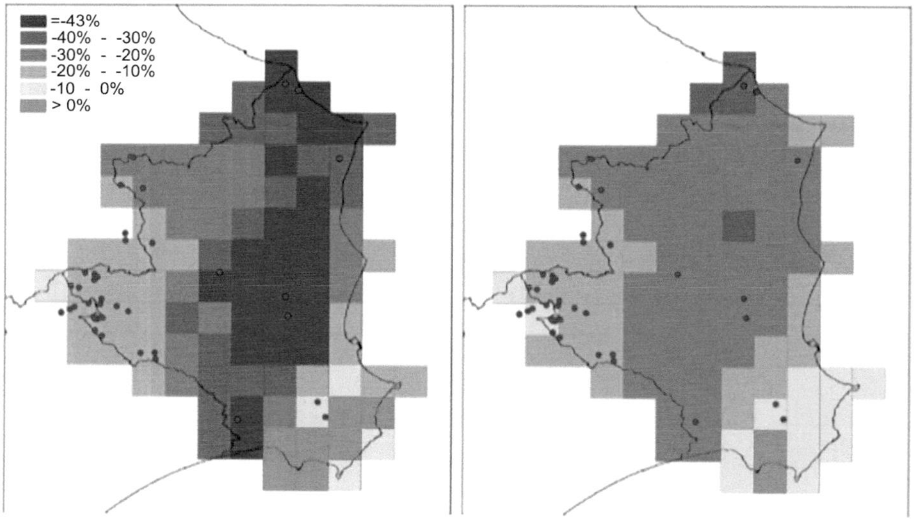

Figure 8: Representation of each node and their visualization in the streets related with our Kriging Model.

In the Google Maps city screenshot of traffic on Friday in the afternoon one can see that the places in dark red, red and orange are the places where traffic is worse than the areas in green, which we can say are fluid and empty. If one compares the areas in reddish with the kriging prediction in Figure 4, for example, the already mentioned Pronaf section and also the crossing bridge to the neighboring city, El Paso, which are very crowded due to security protocols in the United States. In Figure 9, we showed all avenues where the majority of traffic accidents related to Italian Scooters occur.

The kriging applied to the map does not directly predict traffic, but traffic it is directly related to accidents.

7. Conclusions

- Develop a mobile application (first stage) that serves as the interface for capturing reports of risk incidents. Citizens will be able to go through this application installed on their mobile phones, report risk situations that are observed during their daily journeys.
- Design a scheme based on gamification that motivates users to make constant captures of reports of incidences of risk of transit. This scheme must allow a constant feeding of the database.
- Configure a database (second stage) that contains relevant parameters for the definition of specific patterns of traffic risk.
- Process the information contained in the database to present it in a geo-referenced way, classified by type of incident, schedules, type of vehicle and incidence coordinates through a consultation platform (third stage), graphically representing the red dots on a map, as in (Jiang et al., 2015).
- Implementation of a predetermined warning function, which allows to notify other registered users directly about situations that they can put them or other users of public roads at risk, and that have probably gone unnoticed before the notification (conditions of lights, tires, etc.).

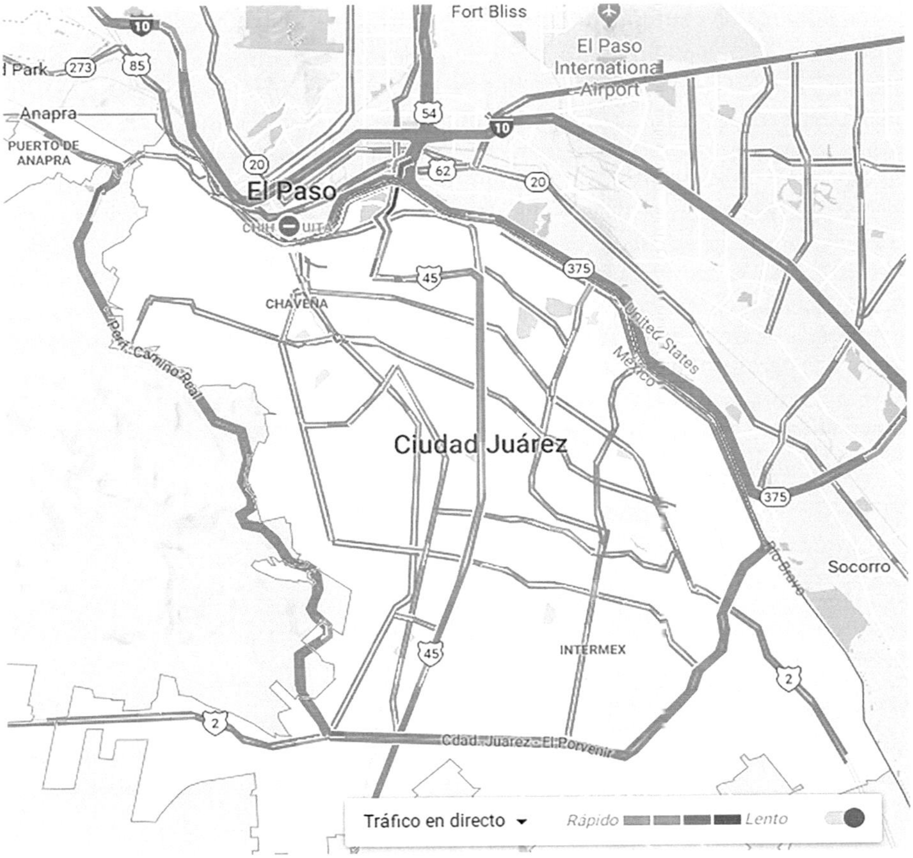

Figure 9: Google Maps traffic map over Juarez area. The green colors are fluid, and orange, red and dark red are jammed sites.

This innovative application can be used in other smart cities, such as Trento, Kansas City, Paris, Milan, Cardiff, Barcelona, and Guadalajara.

8. Future Research

Through the implementation of this project it is intended to generate a platform of consultation that can provide accurate and relevant data in the study of urban mobility research, in such a way as to allow those in charge of generating proposals for improvement in this area to have the information available for its analysis. The configuration of the database will contain the data that are most representative and important to describe the risk factors: date, time, type of vehicle (through the capture of the license plates), coordinates and weather conditions. The information will be presented in the consultation platform in a georeferenced manner. This format will make it possible to identify areas of conflict and a specific analysis can be made through the implementation of filters that help to visualize the specific parameters of study in the map of a specific area. The platform will be limited to providing statistical information, as a way that can be easily tropicalized and adapted to local systems and legislation. Concerning users of the mobile application, to allow a constant feeding of the data, the system based on the gamification will motivate continuously, generating a status scheme within the application and giving them access to various benefits for their collaboration.

Since it is an information platform, the project can be used in different ways:

- Access to the consultation platform by public security authorities may allow better planning and optimization of resources intended for the implementation of accident prevention measures vehicles, since more timely identification of the specific patterns to attack and the statistical data that they locate in temporality and frequency the most representative incidents. For example, it will be able to statistically identify the pathways where they occur with higher recurrence reports for speeding. This pattern will allow them to decide if it is necessary to concentrate patrolling in the detected areas and schedules or days in which most of the events reported are concentrated or if it is necessary to invest in the installation of cameras for speed/red light tickets. In this way, the planning of the resources destined for accident prevention will be greatly optimized. The instances of Government will be able to decide if they should increase the fleet of patrols or make a better distribution according to the incident statistics, generate benefit programs for a good rating in the platform, install surveillance systems at certain points, install traffic lights or speed reducers, and install specific signs in the zones detected, among others.

References

Campoya Morales, A.F. 2019. Different Equations Used During the Mental Workload Evaluation Applying the NASA-TLX Method. SEMAC 2019. Ergonomía Ocupacional. Investigaciones y Aplicaciones. Vol 12 2019. México. Recuperado de: http://www.semac.org.mx/images/stories/libros/Libro%20SEMAC%202019.pdf.

Casillas, Zapata. 2015. La influencia de la Infraestructura Vial del Área Metropolitana de Monterrey sobre el Comportamiento del Automovilista. México.

Dominguez, Karaisl. 2013. Más allá del costo a nivel macro: los accidentes viales en México, sus implicaciones socioeconómicas y algunas recomendaciones de política pública. México.

Hijar, Medina. 2014. Los accidentes como problema de salud pública en México. México.

Jain, S. 2017. What is missing in the Double Diamond Methodology? Recuperado de: http://green-dice.com/double-diamond

Jiang, Abdel-Aty, Hu and Lee. 2015. Investigating macro-level hotzone identification and variable importance using big data: A random forest models approach. Estados Unidos.

List, Schöggl. 2000. Method for Analyzing the Driving Behavior of Motor Vehicles. US6079258. Austria.

Organización Panamericana de la Salud. 2011. Estrategia Mexicana de la Seguridad Vial. México.

Ponce Tizoc. 2014. Diseño de Política Pública: Accidentes de Tránsito Ocasionados por el uso del Teléfono Celular en la Delegación Benito Juárez. México.

CHAPTER-5

Multicriteria analysis of Mobile Clinical Dashboards for the Monitoring of Type II Diabetes in a Smart City

Mariana Vázquez-Avalos,[1] *Alberto Ochoa-Zezzatti*[1] and *Mayra Elizondo-Cortés*[2,*]

In 2015 alone, 23.1 million people were diagnosed with diabetes, according to data gathered by the Centers for Disease Control and Prevention (CDC). This sum of people joined the estimated 415 million people living with diabetes in the world. The fast rise of diabetes prevalence and its life-threatening complications (kidney failure, heart attacks, strokes, etc.) has made healthcare and technology professionals find new ways to diagnose and monitor this chronic disease. Among its different types, type 2 diabetes is the most common found in adults and elderly people. Anyone diagnosed with diabetes requires a strict treatment plan that includes constant monitoring of physiological data and self-management. Ambient intelligence, through the implementation of clinical dashboards and mobile applications, allows patients and their medical team to register and to have access to the patient's data in an organized and digital way. This paper aims to find the most efficient mobile application for the monitoring of type II diabetes through a multicriteria analysis.

1. Introduction

For a disease to be considered chronic, it has to have one or more of the following characteristics: They are permanent, leave residual disability, are caused by irreversible pathological alteration, require special training of the patient for rehabilitation, or may be expected to require a long period of supervision, observation, or care [1]. The last characteristic involves self-management and strict monitoring of the patient's health data, to avoid the development of life-threatening complications [2].

Diabetes checks all these characteristics; therefore, it is considered a chronic metabolic disorder characterized by hyperglycemia caused by problems in insulin secretion or action [3]. Type II diabetes, previously known as non-insulin-dependent diabetes or adult-onset diabetes, accounts for about 90% to 95% of all diagnosed cases of diabetes [4]. Over the years, its prevalence has been increasing all over the world and as a result, it is becoming an epidemic in some countries [5] with the number of people affected expected to double in the next decade.

Furthermore, as it was previously mentioned, diabetes, as a chronic illness, requires constant monitoring of a patient's health parameters. Patient monitoring can be defined as "repeated or

[1] Universidad Autónoma de Ciudad Juárez, Av. Hermanos Escobar, Omega, 32410 Cd Juárez, Chihuahua, México.
[2] Universidad Nacional Autónoma de México, Ciudad Universitaria, 04510, CDMX, México.
* Corresponding author: mayra.elizondo@comunidad.unam.mx

continuous observations or measurements of the patient and their physiological function, to guide management decisions, including when to make therapeutic interventions, and assessment of those interventions" [6]. However, most patients have difficulty adhering to the self-management patterns that diabetes monitoring requires.

This current problematic, as well as the rise of other chronic diseases, has forced the healthcare system to transform how it manages and displays information to meet its needs [7]. New emerging systems for diabetes care have the potential to offer greater access, provide improved care coordination, and deliver services oriented to higher levels of need. The development of tools and implementation of ambient intelligence has brought technical solutions for doctors, as wells as help for the patients in the management of their health [8].

Ambient intelligence in healthcare

Nowadays, new ways to interact with technology in our everyday life continue to develop. Ambient intelligence is an emerging discipline that brings intelligence to our everyday environments and makes those environments sensitive to us [9].

In the ambient intelligence environment, individuals are bounded by embedded intelligent devices' networks to collect information nearby their physical places, in the healthcare environment, these devices are used in medical informatics, decision support, gathered electronic health record and knowledge representation [10].

Health records and the development of clinical dashboards

In the monitoring of diabetes, the continuous measurement of parameters such as blood glucoseare very important. The purpose of an electronic health record system (EHR) is to improve the well-being of patients and to avoid the organizational limitations set by paper-based records [11].

Patients are asked to track their key measures like blood pressure, heart rate, diabetes-relevant data, well-being, or side effects of the medication by taking daily notes on a piece of paper, called a health-data diary. The captured data are expected to show trends in the illness patterns and to help the doctor guide the patient to the best possible health status. Paper base diaries lack proper data representation, feedback, and timely delivery data. Therefore, an easy-to-use and patient-centered data-acquisition system is essential to guide the patient through data capturing and the complex process of self-management [1].

Healthcare is an environment that has been experiencing dramatic progress in computing technology in order to process and distribute all relevant patient information electronically and overall to improve the quality of care [12]. An EHR allows physicians to have easier access to the patient's parameters. Furthermore, the development of clinical dashboards made it faster to read this information.

Dashboards are a tool developed in the business sector, where they were initially introduced to summarize and integrate key performance information across an organization into a visual display as a way of informing operational decision making. A clinical dashboard is designed to "provide clinicians with the relevant and timely information they need to inform daily decisions that improve the quality of patient care. It enables easy access to multiple sources of data being captured locally, in a visual, concise and usable format." [13].

These technologies have been implemented to mHealth (mobile health) apps, which have been defined as software that is incorporated to smartphones to improve health outcome [14], that make it possible for the patient to visualize their information and promotes self-monitoring to record their healthcare diary.

A great variety of apps have been developed to assist patients in the management of diabetes mellitus type2 [15]. Some of the features involved in these self-management apps are medication adherence, self-monitoring of diabetes, insulin dose calculators and promoting physical activity and healthy eating [16].

Some of the barriers that can be found in the use of mobile applications for diabetes management are cost, insufficient scientific evidence, not being useful in certain populations, data protection, data security, and regulatory barriers. In the first one, the cost is a potential barrier for any new medical technology. It can be hard for some patients to use these technologies due to the high cost of smartphones and lack of internet services. Regarding the insufficient scientific evidence, there aren't enough studies that show the effectiveness of these apps. On not being useful in certain populations, it's mentioned that a lot of these apps may not be useful for the elderly, non-English speakers and physically challenged. Also, the protection of the information uploaded to an application, as well as the proper use of this software and digital tools are important [16].

Application domain

The different devices that exist in ambient technology, such as mobile apps that ease the monitoring of diabetes in everyday life, are helpful for patients and their medical team. Each software has different features that make one better than the other, in the following work seven different mobile applications will be analyzed to determine which one has the best performance.

Problem statement

When searching on their phone, a person with type II diabetes must choose among a large number of applications. To choose the best one, four evaluation criteria (functionality, usability, user information, and engagement with user), which were then divided into subsequent sub-criteria, were considered for the selection between seven different mobile applications available in online mobile stores.

2. Methodology

There are different mobile applications (apps) that are used to keep track of the important parameters of a patient with diabetes, as well as encourage self-monitoring.

The seven chosen to be compared by a multicriteria analysis (the analytical hierarchy process) are shown in Table 1.

Diabeto Log is a mobile app developed by Cubesoft SARL, and it allows the user to see their blood glucose tests and medications intake in a single screen (Figure 1). This app was designed specifically to help the user see evolutions and compare data from one day to another. It also registers parameters such as the number of hypoglycemias, number of hyperglycemias, estimated A1c, number of insulin units used.

Table 1: Mobile apps reviewed using multicriteria analysis.

Mobile Applications	
Name	**Developer**
Diabetes Tracker: Diabeto Log	Cubesoft SARL
gluQUO: Control your Diabetes	QUO Health SL
BG Monitor	Gordon Wong
Bluestar Diabetes	WellDoc, Inc.
OnTrack Diabetes	Vertical Health
Glucose Buddy+ for Diabetes	Azumio Inc.
mySugr	mySugr GmbH

Figure 1: Diabetes Tracker: Diabeto Log.

The purpose of GluQuo, an application developed by QUO Health, is to avoid hyperglycemia andhypoglycemia, control sugar levers, integrate all exercise data, and note carbohydrate rations. It also can connect to Bluetooth glucometers (Figure 2). In addition, it generates a report, allows the user to set up insulin reminders, keeps track of exercise and food intake with graphics, and helps in tracking diabetes with glucose dashboards.

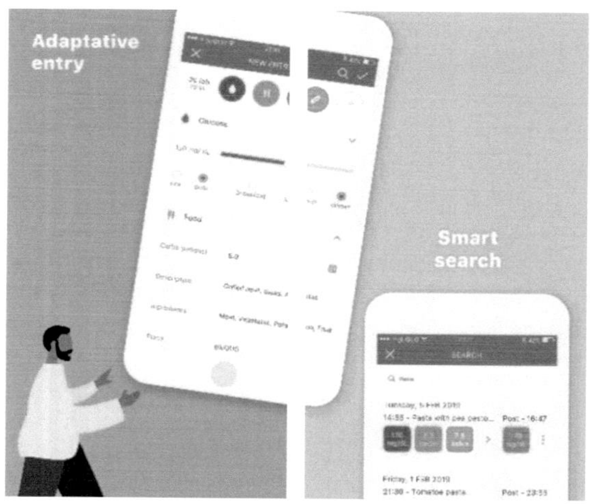

Figure 2: GluQuo: Control Your Diabetes.

BG Monitor is a diabetes management app that has a clean user interface and filtering system that allows the user to find what they are looking for (Figure 3). It provides statistics that show blood glucose levels and can help to identify trends and make adjustments to insulin dosages. This app also has reminders to check blood or give insulin and creates reports so the user can email them from within the app.

Bluestar diabetes is advertised as a digital health solution for type 2 diabetes. It provides tailored guidance driven by artificial intelligence (Figure 4). It collects and analyses data to provide precise, real-time feedback that supports healthier habits in users and more informed conversations

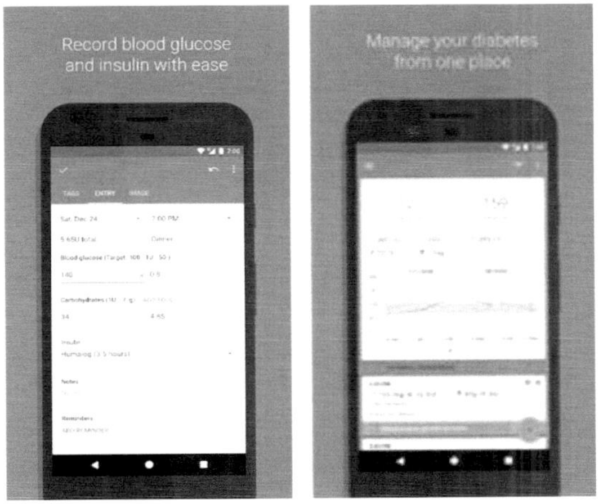

Figure 3: BG Monitor.

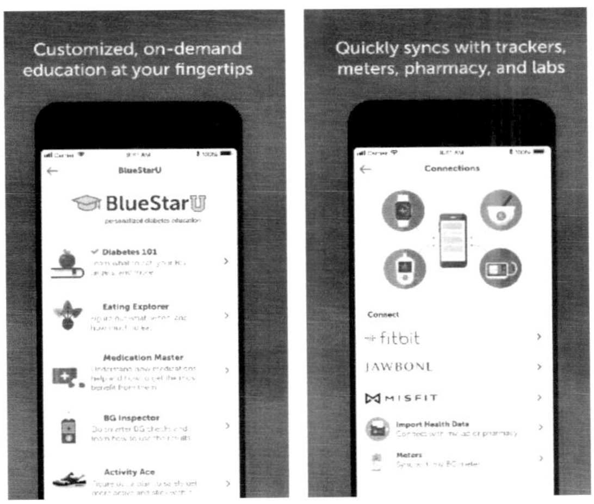

Figure 4: Bluestar Diabetes.

during care team visits. In addition to glucose, it tracks exercise, diet, lab results, symptoms and medication.

Ontrack Diabetes is a mobile app developed by Vertical Health. Some of its features are that it tracks blood glucose, haemoglobin A1c, food, and weight. It generates detailed graphs and reports that the user can share with their physician. It allows to easily keep track of a user's daily, weekly and monthly glucose levels (Figure 5).

Glucose Buddy + helps in the management of diabetes by tracking blood sugar, insulin, medication,and food (Figure 6). The user can get a summary for each day as well as long term trends. It can be accessed as a mobile application, but it is also available for desktop and tablet.

MySugr is an application that allows a digital logbook, shows personalized data analysis such as the estimated A1C (Figure 7). It also has Bluetooth data syncing to glucometers.

A) Analytic Hierarchy Process (AHP)

The analytic hierarchy process was developed by Thomas L. Saaty as a tool to manage qualitative and quantitative multicriteria [17]. There are several steps involved in the AHP. The first step is to define the problem and determine the kind of knowledge sought [18].

Several factors where evaluated to determine the best option among the different diabetes apps. The criteria and sub-criteria are listed in Table 2. These same criteria and sub-criteria are displayed

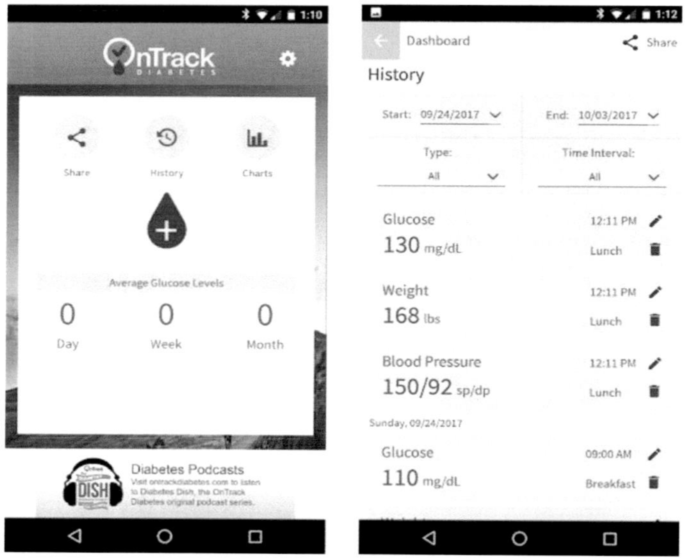

Figure 5: Ontrack Diabetes.

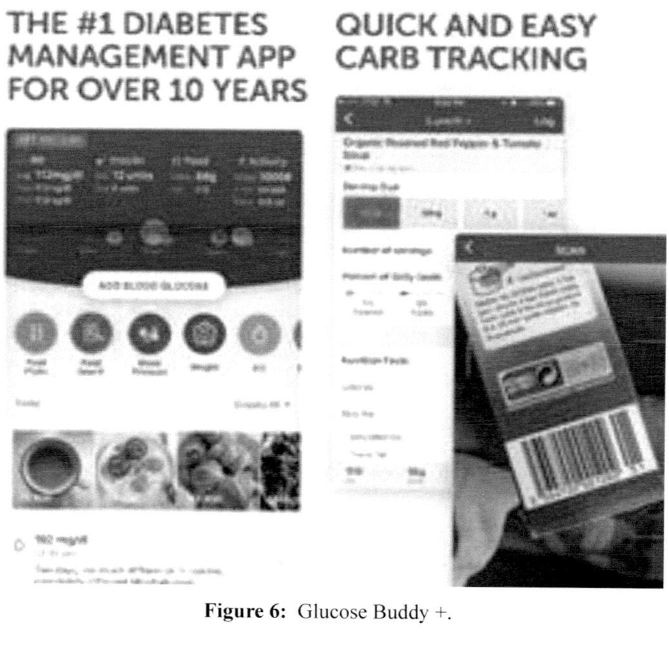

Figure 6: Glucose Buddy +.

in the diagram of Figure 8. The information is structured with the goal of the decision at the top, through the intermediate levels (criteria on which subsequent elements depend) to the lowest level (which is usually an asset of the alternatives).

When searching about these mobile applications, each one has different parameters in the sub-criteria mentioned above. The values for each criterion are shown in Tables 3–6, where each letter represents an application: A is Diabeto Log, B gluQUO, C is OnTrack Diabetes, D is Bluestar Diabetes, E is BG Monitor, F is Glucose Buddy+, and G is mySugr.

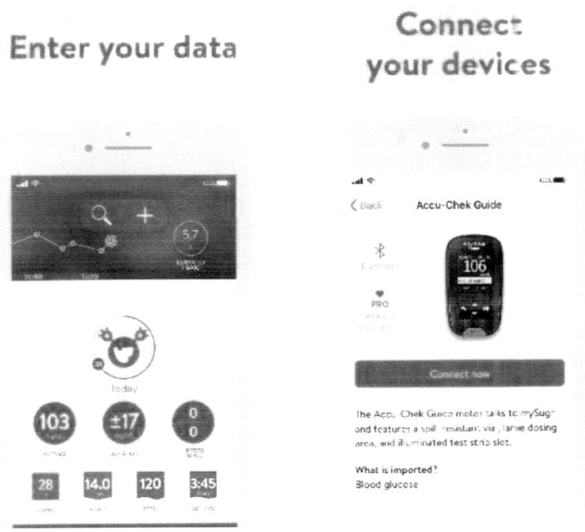

Figure 7: mySugr.

Table 2: Sub-criteria and criteria used for the MCDA

Criteria	**Sub-criteria**
Functionality	• Operating System • Interfaces • Capacity that it occupies in memory • Update • Rank
Usability	• Languages Available • Acquisition Costs • Target user groups
User information	• User rating • Number of consumer ratings • Number of downloads
Engagement with user	• Reminders/alerts • Passcode • Parameters that can be registered

In the next step, with the information available it is possible to construct a set of pairwise comparison matrices. To construct a set of pairwise comparison matrices, each element in an upper level is used to compare the elements in the level immediately below concerning it.

To make comparisons, we need a scale of numbers that indicates how many times more important or dominant one element is over another element concerning the criterion or property for which they are compared [18]. Each number represents the intensity of importance, where 1 represents equal importance and 9 extreme importance (Table 7).

With this scale of numbers, it is possible to construct the pairwise comparison matrices for the criteria and each sub-criteria, this is showed in Table 8.

Once the comparative matrix of the criteria is done, the following step is to obtain the weight trough a standardized matrix. This is done by calculating the sum of each column, then the normalization of the matrix by diving the content of each cell by the sum of its column, and finally the calculation of the average rows, see Table 9.

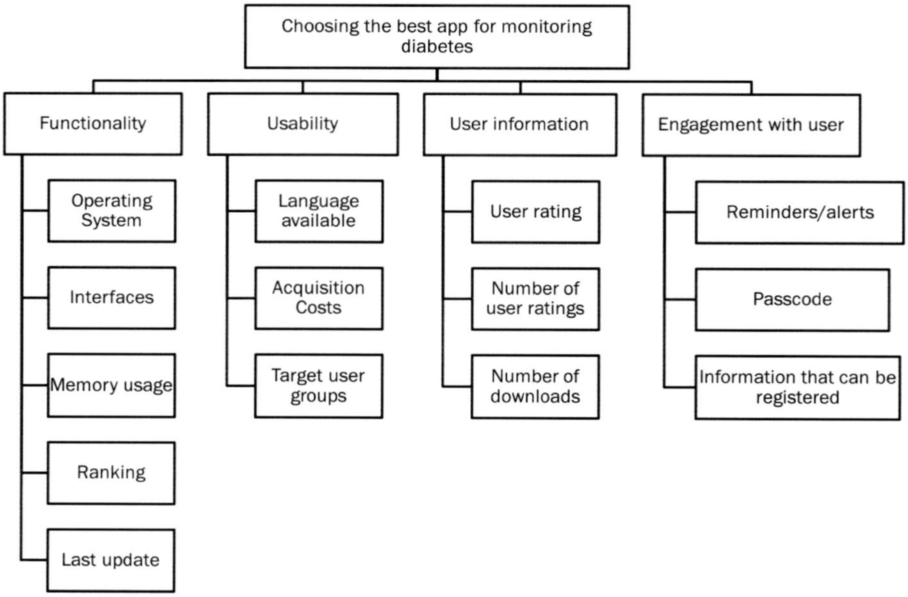

Figure 8: Diagram of The Criteria and Sub-criteria.

Table 3: Sub-criteria of the Functionality criterion.

	Criterion: Functionality				
	OS	**Int.**	**Memory**	**Rank**	**Update**
A	iOS	Yes	55.8 MB	1,290	2019
B	Both	Yes	49.1 MB	1,355	2019
C	Android	Yes	1.8 MB	N/A	2018
D	Both	Yes	45MB	381	2019
E	Android	No	2.8 MB	N/A	2017
F	iOS	No	193.6 MB	103	2019
G	Both	Yes	109 MB	132	2019

Table 4: Sub-criteria of the Usability criterion.

	Criterion: Usability		
	Languages Available	**Acquisition Costs**	**Target User Groups**
A	5	Freeware	Patient
B	2	Freeware	Patient
C	1	Freeware	Both
D	1	Freeware	Both
E	1	$119 MX	Both
F	31	$39 MX	Patient
G	22	Freeware	Patient

Table 5: Sub-criteria of the User information criterion.

	Criterion: User Information		
	User rating	**Number of Consumer Ratings**	**Number of Downloads**
A	4.8	14	420
B	3.3	10	300
C	3.6	6,088	182,640
D	4.2	143	4,290
E	4.6	94	2,820
F	4.8	30	900
G	4.8	284	8,520

Table 6: Sub-criteria of the Engagement with user criterion.

	Criterion: Engagement with user		
	Reminders	**Passcode**	**Parameters**
A	Yes	Yes	7
B	Yes	Yes	7
C	No	Yes	6
D	Yes	Yes	7
E	Yes	No	4
F	Yes	No	7
G	Yes	Yes	10

Table 7: Scale used to determine intensity of importance in pairwise comparison.

Intensity of importance	**Definition**	**Explanation**
1	Equal Importance	Two activities contribute equally to the objective
2	Weak or slight	
3	Moderate importance	Experience and judgement slightly favor one activity over another.
4	Moderate plus	
5	Strong importance	Experience and judgement strongly favor one activity over another.
6	Strong plus	
7	Very strong or demonstrated importance	An activity is favored very strongly over another; its dominance demonstrated in practice.

Table 8: Pairwise comparison matrix of the main criteria with respect to the goal.

Criteria Comparative Matrix				
Criteria	**Functionality**	**Usability**	**User info.**	**Engagement with user**
Functionality	1	3	6	5
Usability	1/3	1	5	4
User info.	1/6	1/5	1	1/3
Engagement	1/5	1/4	3	1
Total	1.7	4.45	15.00	10.33

Table 9: Normalized pairwise comparison matrix and calculation of the weight criteria.

Criteria	**Normalized Matrix**				**Weight**
Functionality	0.58823	0.67415	0.40000	0.48387	0.53656
Usability	0.19607	0.22471	0.33333	0.38709	0.28530
User info.	0.09803	0.04494	0.06666	0.03225	0.06047
Engagement	0.11764	0.05617	0.20000	0.09677	0.11765

As a conclusion, the user information is the least important criterion with a 0.060479 weight, followed by the engagement with user with a 0.11765 weight, then usability with a 0.28530 weight, and finally functionality is the most important criterion with a 0.53656 weight (Figure 9).

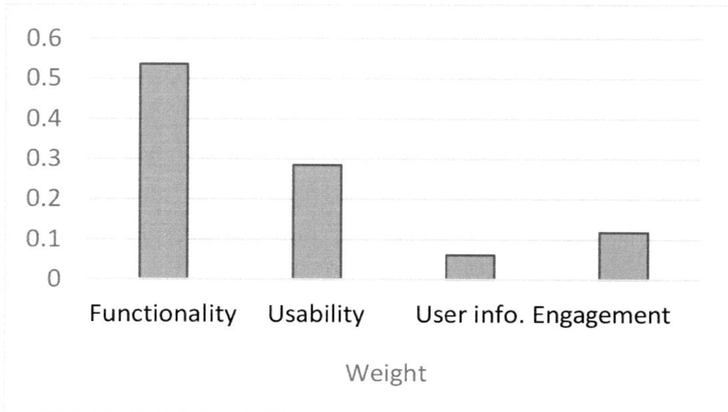

Figure 9: Weights of the criteria with respect to the goal.

To know if the comparison done and the weights obtained are consistent it is important to check the system consistency. The first step of this check is to calculate the weight sums vector:

$$\{Ws\} = \{M\} \cdot \{W\}$$

The weight sums vector is in Table 10.

$$\lambda \text{max} = \frac{4.36 + 4.33 + 4.07 + 4.06}{4} = 4.205 \tag{1}$$

$$\text{Consistency Index} = CI = \frac{\lambda \text{max}}{n - 1} = 0.068 \tag{2}$$

$$\text{Consistency Ratio} = \frac{CI}{RI} = \frac{0.068}{0.99} = 0.06 \tag{3}$$

The value of consistency index is 0.068, while the value of consistency ratio is 0.06. The consistency ratio is CR < 0.1, thus it is acceptable, and the system is consistent.

Table 10: Values obtained when calculating weight sums vector.

$M \cdot W$
2.34353
1.23710
0.24617
0.47769

Now, to choose the best mobile application among the seven alternatives we must obtain the weight for each of the seven alternatives in each criterion. The first criterion analyzed is functionality (Table 11 and Table 12).

Table 11: Pairwise comparison matrix for the sub-criteria with respect to functionality.

Functionality					
Subcriteria	**O/S**	**Int.**	**Mem.**	**Update**	**Rank**
O/S	1	6	3	5	7
Interfaces	1/6	1	1/6	5	1/6
Memory	1/3	6	1	6	7
Update	1/5	1/5	1/6	1	1/4
Ranking	1/7	6	1/7	4	1
Total	1.84	19.20	4.48	21.00	15.42

Table 12: Normalized matrix of the pairwise comparison of the sub-criteria with respect to functionality.

Subcriteria	**Normalized Matrix**					**Weight**
O/S	0.5426	0.3125	0.6702	0.2380	0.4540	0.4434
Interfaces	0.0904	0.0520	0.0372	0.2380	0.0108	0.0857
Memory	0.1808	0.3125	0.2234	0.2857	0.4540	0.2913
Update	0.1085	0.0104	0.0372	0.0476	0.0162	0.0440
Ranking	0.0775	0.3125	0.0319	0.1904	0.0648	0.1354

With the values obtained, the next step is to obtain the prioritization of the functionality sub-criteria (Table 13 and Figure 10).

Table 13: Prioritization of the sub-criteria with respect to functionality.

Subcriteria	**Weight**	**Prioritization**
O/S	0.4434	0.2379
Interfaces	0.0857	0.0459
Memory	0.2913	0.1562
Update	0.0440	0.0236
Ranking	0.1354	0.0726

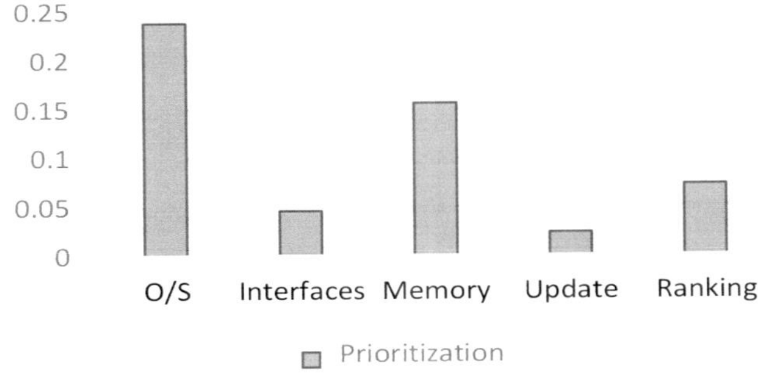

Figure 10: Prioritization obtained of each sub-criterion of functionality.

For the sub-criteria of usability, the same steps as before are used (Tables 14–16 and Figure 11).

Table 14: Pairwise comparison matrix of the usability criterion.

Usability			
Subcriteria	**Language**	**Cost**	**Target Group**
Language	1	6	5
Cost	1/6	1	7
Target Group	1/5	1/7	1
Total	1.36	7.14	13

Table 15: Normalized matrix of the comparison matrix.

Sub-criteria	Normalized Matrix			Weight
Language	0.7352	0.8403	0.3846	0.6533
Cost	0.1225	0.1400	0.5384	0.2669
Target Group	0.1470	0.0200	0.0769	0.2439

Table 16: Prioritization of the usability criterion.

Sub-criteria	Weight	Prioritization
Language	0.6533	0.1590
Cost	0.2669	0.0649
Target Group	0.2439	0.0593

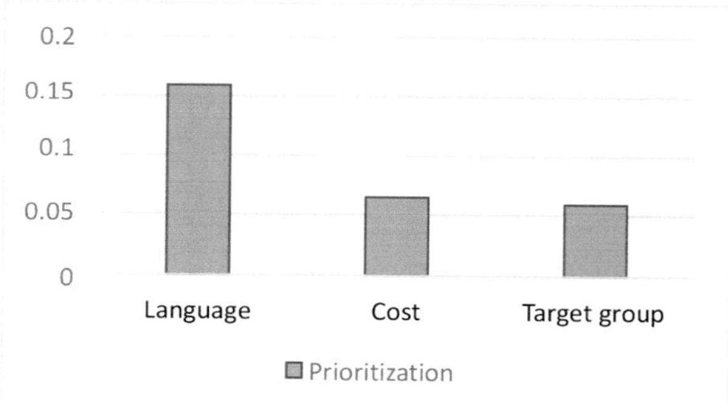

Figure 11: Prioritization obtained of each sub-criterion of the usability criterion.

The same steps are followed for the user information criterion (Tables 17–19 and Figure 12).

Table 17: Pairwise comparison of the user information.

User Information			
Sub-criteria	**Rating**	**No. Rating**	**No. Downloads**
Rating	1	7	5
No. Rating	1/7	1	4
No. Down.	1/5	1/4	1
Total	1.34	8.25	10

Table 18: Normalized matrix of the comparison matrix of the user information criterion.

Sub-criteria	Normalized Matrix			Weight
Rating	0.7462	0.8484	0.5000	0.6982
No. Rating	0.1066	0.1212	0.4000	0.2092
No. Downloads	0.1492	0.0303	0.1000	0.0931

Table 19: Prioritization of the user information criterion.

Sub-criteria	Weight	Prioritization
Rating	0.6982	0.0793
No. Rating	0.2092	0.0237
No. Downloads	0.0931	0.0105

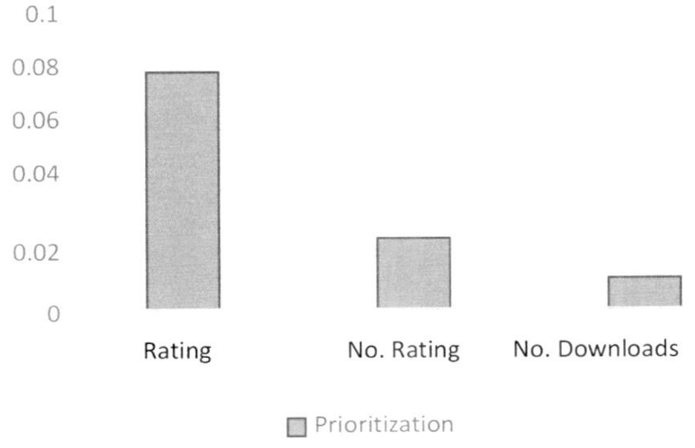

Figure 12: Prioritization obtained of each sub-criterion of the user information criterion.

Finally, for the criterion of user information the weights and prioritization are obtained (Tables 20–22 and Figure 13).

Once the priority values are obtained the next step is to get the weight of each mobile application for each sub-criteria. The weight of each alternative is obtained following the same steps as before.

Table 20: Pairwise comparison matrix for the sub-criteria with respect to the engagement with user.

Engagement with user			
Sub-criteria	**Alerts**	**Passcode**	**Parameters**
Alerts	1	1/4	1/8
Passcode	4	1	1/7
Parameters	8	7	1
Total	13	8.25	1.26

Table 21: Normalized matrix of the pairwise comparison matrix.

Normalized Matrix			Weight
0.0769	0.0303	0.0992	0.0688
0.3076	0.1212	0.1133	0.1807
0.6153	0.8484	0.7936	0.7557

Table 22: Prioritization of the sub-criteria with respect to the engagement with user.

Sub-criteria	Weight	Prioritization
Alerts	0.0688	0.0047
Passcode	0.1807	0.0125
Parameters	0.7557	0.0524

Figure 13: Prioritization of each sub-criterion of the engagement with user criterion.

The criterion of functionality has the sub-criteria of operating systems (Table 23 and Figure 14), interfaces (Table 24 and Figure 15), capacity that it occupies in memory (Table 25 and Figure 16), last time it was updated (Table 26 and Figure 17), and the ranking it has among other medical applications (Table 27 and Figure 18).

Table 23: Comparative matrix of the alternatives with respect to the operating system.

Sub-criteria: Operating System								
Alt.	**A**	**B**	**C**	**D**	**E**	**F**	**G**	**Weight**
A	1	1/7	1/5	1/7	1/5	1	1/7	0.02960
B	7	1	7	1	7	7	1	0.26396
C	5	1/7	1	1/7	1	1/5	1/7	0.05014
D	7	1	7	1	7	7	1	0.26396
E	5	1/7	1	1/7	1	1/5	1/7	0.05014
F	1	1/7	5	1/7	5	1	1/7	0.07823
G	7	1	7	1	7	7	1	0.26396

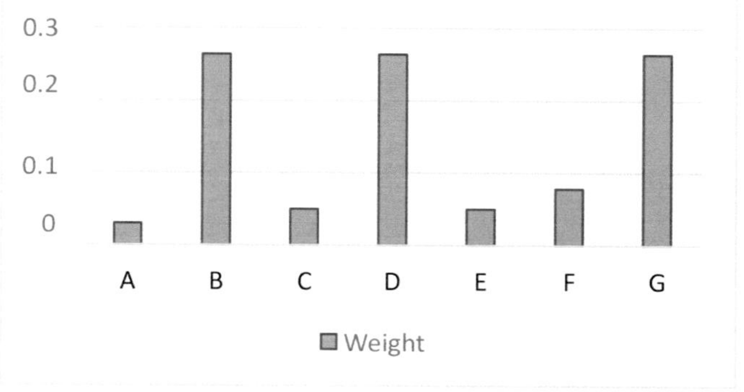

Figure 14: Weight of each alternative with respect of the operating system.

Table 24: Comparative matrix of the alternatives with respect to the interfaces.

Sub-criteria: Interfaces								
Alt.	**A**	**B**	**C**	**D**	**E**	**F**	**G**	**Weight**
A	1	1	1	1	7	7	1	0.1891
B	1	1	1	1	7	7	1	0.1891
C	1	1	1	1	7	7	1	0.1891
D	1	1	1	1	7	7	1	0.1891
E	1/7	1/7	1/7	1/7	1	1	1/7	0.0270
F	1/7	1/7	1/7	1/7	1	1	1/7	0.0270
G	1	1	1	1	7	7	1	0.1891

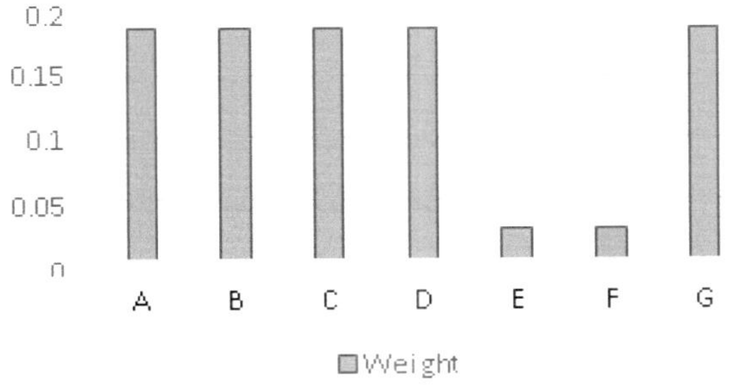

Figure 15: Weight of each alternative with respect of the interfaces.

Table 25: Comparative matrix of the alternatives with respect to the memory it occupies.

Sub-criteria: Capacity that it occupies in memory								
Alt.	**A**	**B**	**C**	**D**	**E**	**F**	**G**	**Weight**
A	1	1/6	1/7	1/6	1/7	5	5	0.06967
B	6	1	1/6	1/3	1/6	5	5	0.10696
C	7	6	1	6	3	5	5	0.35485
D	6	3	1/6	1	1/6	5	6	0.13595
E	7	6	1/3	6	1	7	7	0.26886
F	1/5	1/5	1/7	1/5	1/7	1	1/3	0.02497
G	1/5	1/5	1/6	1/6	1/7	3	1	0.03870

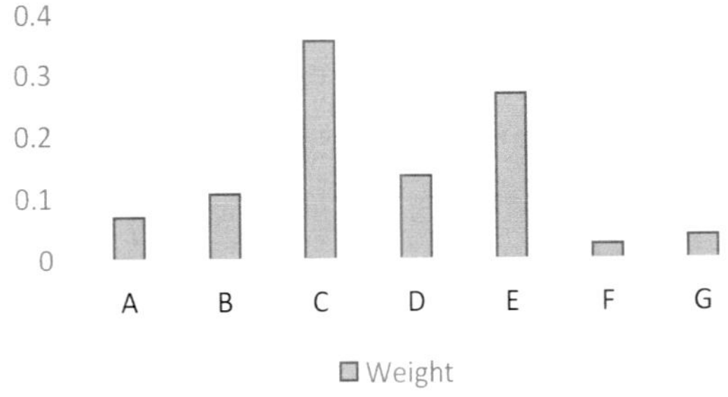

Figure 16: Weight of each alternative with respect of the capacity it occupies in memory.

Table 26: Comparative matrix of the alternatives with respect to the last update.

Sub-criteria: Last Update								
Alt.	**A**	**B**	**C**	**D**	**E**	**F**	**G**	**Weight**
A	1	1	5	1	6	1	1	0.1841
B	1	1	5	1	6	1	1	0.1841
C	1/5	1/5	1	1/5	5	1/5	1/5	0.0519
D	1	1	5	1	6	1	1	0.1841
E	1/6	1/6	1/5	1/6	1	1/6	1/6	0.0272
F	1	1	5	1	6	1	1	0.1841
G	1	1	5	1	6	1	1	0.1841

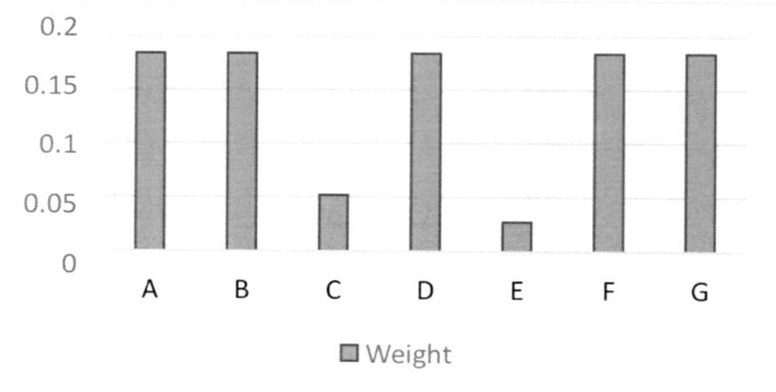

Figure 17: Weight of each alternative with respect of the last update.

Table 27: Comparative matrix of the alternatives with respect to the ranking.

Sub-criteria: Ranking								
Alt.	**A**	**B**	**C**	**D**	**E**	**F**	**G**	**Weight**
A	1	3	8	1/6	8	1/7	1/7	0.0897
B	1/3	1	8	1/6	8	1/7	1/7	0.0735
C	1/8	1/8	1	1/8	1	1/8	1/8	0.0176
D	6	6	8	1	8	1/4	1/4	0.1593
E	1/8	1/8	1	1/8	1	1/8	1/8	0.0176
F	7	7	8	4	8	1	3	0.3210
G	7	7	8	4	8	3	1	0.3210

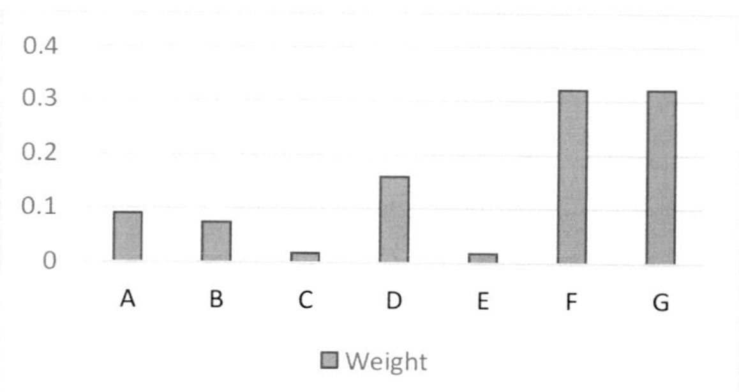

Figure 18: Weight of each alternative with respect of the rank.

The next criterion is usability, it has the sub-criteria of languages available (Table 28 and Figure 19), acquisition costs (Table 29 and Figure 20), and target user groups (Table 30 and Figure 21).

Table 28: Comparative matrix of the alternatives with respect to the languages available.

Sub-criteria: Languages available								
Alt.	**A**	**B**	**C**	**D**	**E**	**F**	**G**	**Weight**
A	1	4	5	5	5	1/8	1/7	0.1315
B	1/4	1	3	3	3	1/8	1/7	0.0718
C	1/5	1/3	1	1	1	1/8	1/7	0.0338
D	1/5	1/3	1	1	1	1/8	1/7	0.0338
E	1/5	1/3	1	1	1	1/8	1/7	0.0338
F	8	8	8	8	8	1	3	0.4179
G	7	7	7	7	7	1/3	1	0.2769

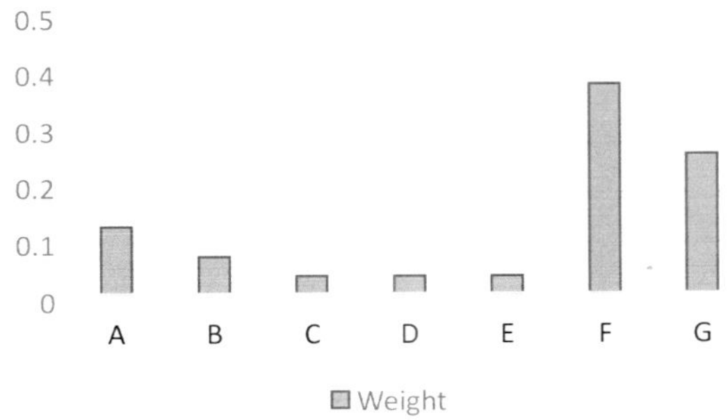

Figure 19: Weight of each alternative with respect of the languages available.

Table 29: Comparative matrix of the alternatives with respect of the acquisition costs.

Sub-criteria: Acquisition Costs								
Alt.	**A**	**B**	**C**	**D**	**E**	F	G	**Weight**
A	1	1	1	1	6	5	1	0.1840
B	1	1	1	1	6	5	1	0.1840
C	1	1	1	1	6	5	1	0.1840
D	1	1	1	1	6	5	1	0.1840
E	1/6	1/6	1/6	1/6	1	1/6	1/5	0.0279
F	1/5	1/5	1/5	1/5	6	1	1/5	0.0558
G	1	1	1	1	5	5	1	0.1800

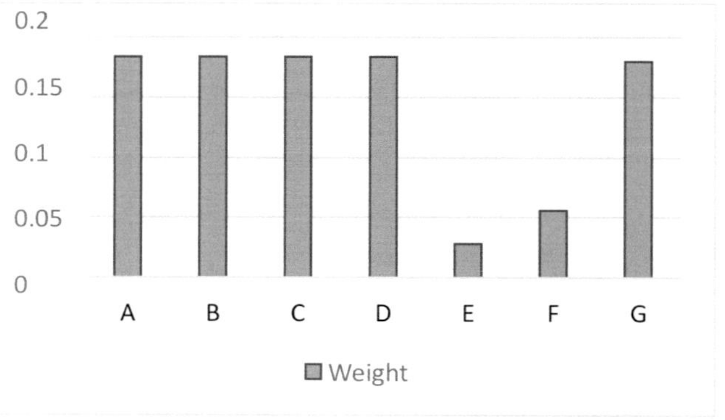

Figure 20: Weight of each alternative with respect of the acquisition costs.

Table 30: Comparative matrix of the alternatives with respect of the target user groups.

Sub-criteria: Target user groups								
Alt.	**A**	**B**	**C**	**D**	**E**	**F**	**G**	**Weight**
A	1	1	1/5	1/5	1/5	1	1	0.0526
B	1	1	1/5	1/5	1/5	1	1	0.0526
C	5	5	1	1	1	5	5	0.2631
D	5	5	1	1	1	5	5	0.2631
E	5	5	1	1	1	5	5	0.2631
F	1	1	1/5	1/5	1/5	1	1	0.0526
G	1	1	1/5	1/5	1/5	1	1	0.0526

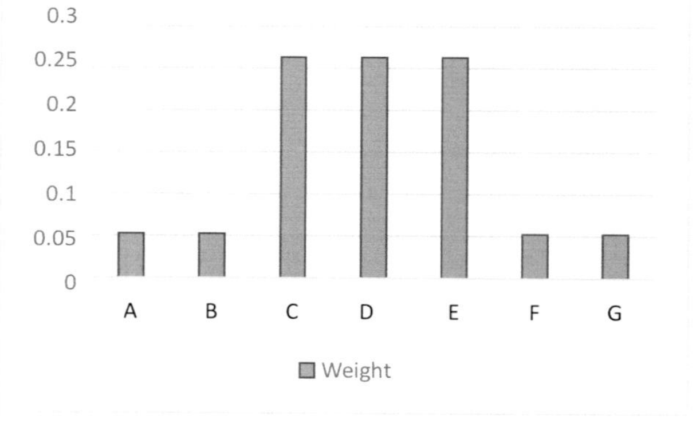

Figure 21: Weight of each alternative with respect of the target user groups.

The user information criterion has the sub-criteria of user rating (Table 31 and Figure 22), number of user ratings (Table 32 and Figure 23), and number of downloads (Table 33 and Figure 24).

Table 31: Comparative matrix of the alternatives with respect of the rating.

Sub-criteria: Rating								
Alt.	**A**	**B**	**C**	**D**	**E**	**F**	**G**	**Weight**
A	1	7	6	4	2	1	1	0.2228
B	1/7	1	1/3	1/6	1/6	1/7	1/7	0.0247
C	1/6	3	1	1/5	1/6	1/7	1/7	0.0365
D	1/4	6	5	1	1/4	1/5	1/5	0.080
E	1/2	6	6	4	1	1/3	1/3	0.1366
F	1	7	7	5	3	1	1	0.2495
G	1	7	7	5	3	1	1	0.2495

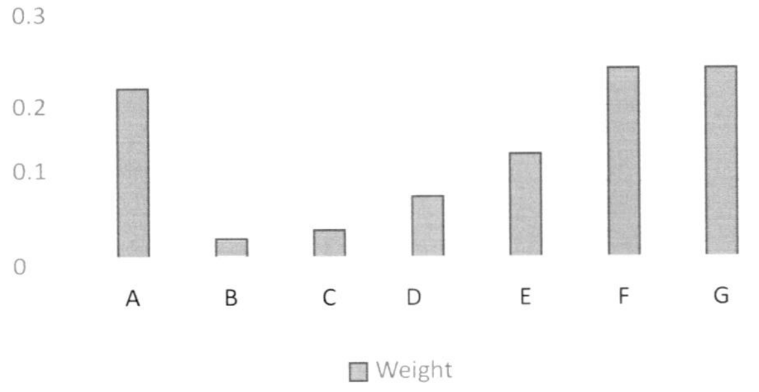

Figure 22: Weight of each alternative with respect of the rating.

Table 32: Comparative matrix of the alternatives with respect of the user ratings.

Sub-criteria: Number of user ratings								
Alt.	**A**	**B**	**C**	**D**	**E**	**F**	**G**	**Weight**
A	1	2	1/8	1/5	1/5	1/4	1/6	0.0315
B	1/2	1	1/8	1/5	1/5	1/4	1/6	0.0244
C	8	8	1	8	8	8	8	0.4559
D	5	5	1/8	1	4	6	1/5	0.1342
E	5	5	1/8	1/4	1	4	1/5	0.0928
F	4	4	1/8	1/6	1/4	1	1/6	0.0595
G	6	6	1/8	5	5	6	1	0.2014

Figure 23: Weight of each alternative with respect to the number of user ratings.

Table 33: Comparative matrix of the alternatives with respect of the number of downloads.

Sub-criteria: Number of downloads								
Alt.	**A**	**B**	**C**	**D**	**E**	**F**	**G**	**Weight**
A	1	2	1/9	1/6	1/5	1/3	1/7	0.0281
B	1/2	1	1/9	1/7	1/6	1/4	1/8	0.0208
C	9	9	1	9	9	9	9	0.4357
D	6	7	1/9	1	4	5	1/7	0.1459
E	5	6	1/9	1/4	1	5	7	0.1425
F	3	4	1/9	1/5	1/5	1	7	0.0899
G	7	8	1/9	7	1/7	1/7	1	0.1366

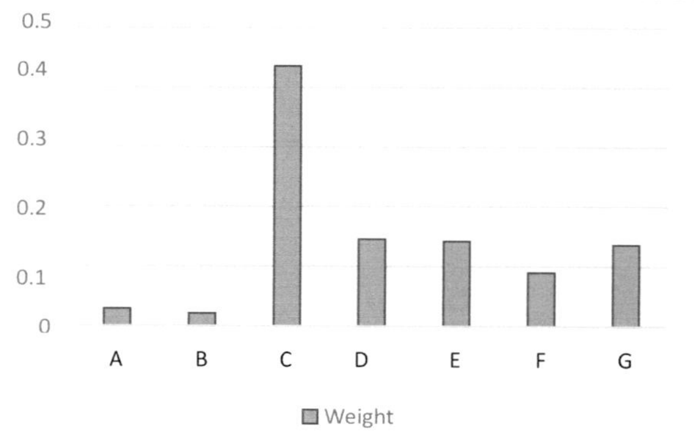

Figure 24: Weight of each alternative with respect of the number of downloads.

For the engagement with user criterion, the sub-criteria to be analysed are the following: if the application send alerts or reminders to the user (Table 34 and Figure 25), if it allows a passcode before seeing the information that the user uploads on the mobile application (Table 35 and Figure 26), and the number of parameters that can be registered to maintain a management of diabetes (Table 36 and Figure 27).

As the last step of the analytic hierarchy process, the values obtained before, shown in Table 37, are analyzed. Each sub-criterion is numbered by the order of appearance in Table 2. Table 38 presents the priority values of each sub-criterion.

Table 34: Comparative matrix of the alternatives with respect of the reminders.

Sub-criteria: Reminders/alerts								
Alt.	**A**	**B**	**C**	**D**	**E**	**F**	**G**	**Weight**
A	1	1	7	1	1	1	1	0.1627
B	1	1	7	1	1	1	1	0.1627
C	1/7	1/7	1	1/7	1/7	1/7	1/7	0.0235
D	1	1	7	1	1	1	1	0.1627
E	1	1	7	1	1	1	1	0.1627
F	1	1	7	1	1	1	1	0.1627
G	1	1	7	1	1	1	1	0.1627

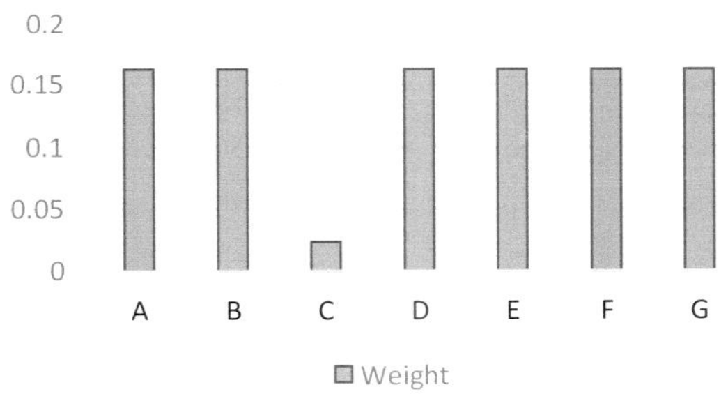

Figure 25: Weight of each alternative with respect of the reminders.

Table 35: Comparative matrix of the alternatives with respect of the passcode.

Sub-criteria: Passcode								
Alt.	**A**	**B**	**C**	**D**	**E**	**F**	**G**	**Weight**
A	1	1	1	1	7	7	1	0.1891
B	1	1	1	1	7	7	1	0.1891
C	1	1	1	1	7	7	1	0.1891
D	1	1	1	1	7	7	1	0.1891
E	1/7	1/7	1/7	1/7	1	1	1/7	0.0270
F	1/7	1/7	1/7	1/7	1	1	1/7	0.0270
G	1	1	1	1	7	7	1	0.1891

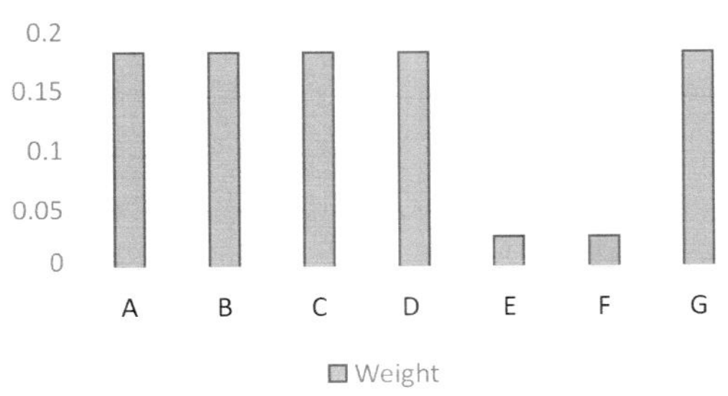

Figure 26: Weight of each alternative with respect of the passcode.

Table 36: Comparative matrix of the alternatives with respect of the number of parameters.

Sub-criteria: Number of parameters								
Alt.	**A**	**B**	**C**	**D**	**E**	**F**	**G**	**Weight**
A	1	1	2	1	4	1	1/4	0.1208
B	1	1	2	1	4	1	1/4	0.1208
C	1/2	1/2	1	1/2	4	1/2	1/5	0.0759
D	1	1	2	1	4	1	1/4	0.1208
E	1/4	1/4	1/4	1/4	1	1/4	1/5	0.0360
F	1	1	2	1	4	1	1/5	0.1178
G	4	4	5	4	5	5	1	0.4075

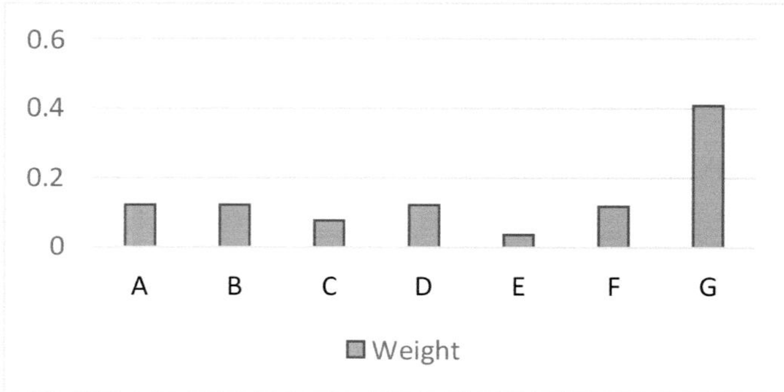

Figure 27: Weight of each alternative with respect of the number of parameters.

Table 37: Weights obtained for each alternative in each sub-criterion.

mySugr	Buddy+	BG	Bluestar	OnTrack	gluQUO	Diabeto	
0.2639	0.0782	0.0501	0.2639	0.0501	0.2639	0.0296	1
0.1891	0.0270	0.0270	0.1891	0.1891	0.1891	0.1890	2
0.03870	0.0249	0.2688	0.1359	0.3548	0.1069	0.0696	3
0.1841	0.1841	0.0272	0.1841	0.0519	0.1841	0.1841	4
0.3210	0.3210	0.0176	0.1593	0.0176	0.0735	0.0897	5
0.2769	0.4179	0.0338	0.0338	0.0338	0.0718	0.1315	6
0.1800	0.0558	0.0279	0.1840	0.1840	0.1840	0.1840	7
0.0526	0.0526	0.2631	0.2631	0.2631	0.0526	0.0526	8
0.2495	0.2495	0.1366	0.0800	0.0365	0.0247	0.2228	9
0.2014	0.0595	0.0928	0.1342	0.4559	0.0244	0.0315	10
0.1366	0.0899	0.1425	0.1459	0.4357	0.0208	0.0281	11
0.1627	0.1627	0.1627	0.1627	0.0235	0.1627	0.1627	12
0.1891	0.0270	0.0270	0.1891	0.1891	0.1891	0.1891	13
0.4075	0.1178	0.0360	0.1208	0.0759	0.1208	0.1208	14

Table 38: Prioritization for each sub-criterion.

Sub-criteria	**Prioritization**
Operating System	0.2379
Interfaces	0.0459
Memory	0.1562
Update	0.0236
Ranking	0.0726
Languages	0.1590
Costs	0.0649
User Groups	0.0593
User rating	0.0793
No. user ratings	0.0237
No. downloads	0.0105
Reminders	0.0047
Passcode	0.0125
Parameters	0.0524

In Table 39 are the values obtained on the ultimate prioritization, each value was calculated with the weight the alternative for each sub-criterion and with the priority values of that same sub-criterion.

Table 39: Ultimate prioritization for each alternative.

Alternatives	Last Prioritization
Diabeto Log	0.1015827
GluQUO	0.1365309
OnTrack	0.1361640
Bluestar	0.1620315
BG Monitor	0.0973677
Glucose Buddy+	0.1539829
mySugr	0.2144584

Therefore, using the values obtained with the analytical hierarchy process as reference (see Figure 28), the best alternative for a mobile application used to manage diabetes is mySugr, followed by Bluestar, then Glucose Buddy+. Meanwhile, the least preferred option is BG Monitor.

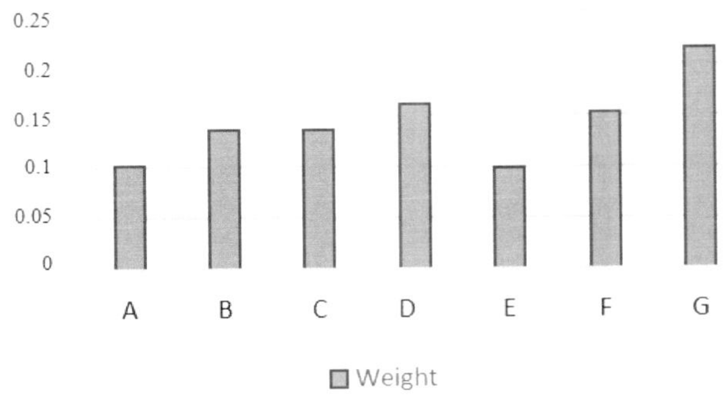

Figure 28: Ultimate prioritization values for each alternative.

B) Grand Prix

The Grand Prix model is a tool that allows choosing amongst an array of alternatives considering the human and economic factor. It can be applied to the selection of the best mobile application for the monitoring of diabetes.

As mentioned before, according to the AHP method, the best mobile app is mySugr with a score of 0.2144. The second-best option is Bluestar with a 0.1620 score, followed by Glucose Buddy+ with a 0.1539 score. It can be observed that, for these three options, the values obtained aren't far apart from each other. These three options are shown in Figure 29.

Considering the economic factor, mySugr and Bluestar are freeware, whilst Glucose Buddy+ is a paid option. Analysing this information, it can be observed that mySugr is still the best option from an economic standpoint.

On the other hand, considering the human factor, we sought the application that is easy to use for a wide demographic, especially older adults who are the main population affected by this disease. Also, it is available in a variety of languages and in both operating systems, so that it can reach a bigger amount of people. From these three options, the one that fits these characteristics is mySugr.

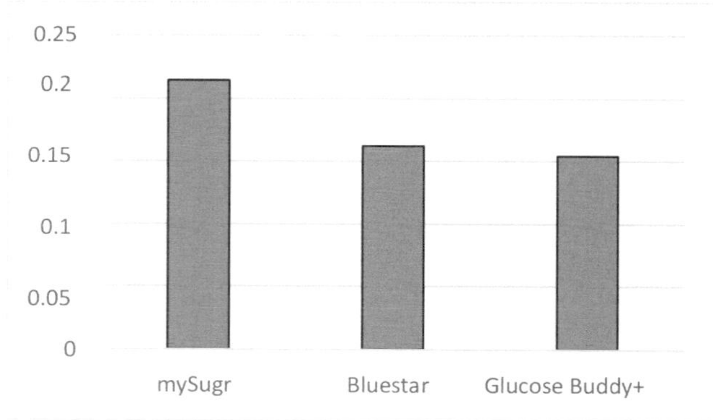

Figure 29: Three best options with AHP.

However, if considering the languages, Glucose Buddy+ is available in 31 different languages, while Bluestar only in one. This means that Bluestar is not as accessible as Glucose Buddy+.

With the Grand Prix model, it can be observed that amongst the top three options mySugr is still the best, but from the other two are still good choices.

C) MOORA Method

The MOORA (Multi-Objective Optimization on the basis of Ratio Analysis) is a method introduced by Brauers and Zavadskas. It consists of two components: The Ratio System and Reference Point approach. The basic idea of the ratio system part of the MOORA method is to calculate the overall performance of each alternative as the difference between the sums of its normalized performances which belongs to benefit and cost criteria.

The first step of the MOORA method is to construct a decision matrix of each alternative and the different criteria:

$$X = \left[x_{ij}\right]_{mxn} \begin{bmatrix} x_{11} & \cdots & x_{1n} \\ \vdots & \ddots & \vdots \\ x_{m1} & \cdots & x_{mn} \end{bmatrix} \tag{4}$$

This decision matrix shows the performance of different alternatives concerning the various criteria (Table 40).

Next, from the decision matrix, the normalized decision matrix is obtained (Table 41). The following equation is used to obtain the values.

$$x_{ij} = \frac{x_{ij}}{\sqrt{\sum_{i=1}^{m^2} x_{ij}}}\ ; i=1,2, \ldots, m \text{ and } j = 1,2, \ldots, n \tag{5}$$

The normalized decision matrix is weighted and shown in Table 42.

The overall performance of the alternatives is measured by

$$y_i = \sum_{j=1}^{g} w_j x_{ij} - \sum_{j=g+1}^{n} w_j x_{ij} \tag{6}$$

Table 40: Decision matrix.

A7	A6	A5	A4	A3	A2	A1	
2	1	1	2	1	2	1	C1
2	1	1	2	2	2	2	C2
109	193.6	2.8	45	1.8	49.1	55.8	C3
132	103	5000	381	5000	1355	1290	C4
2019	2019	2017	2019	2018	2019	2019	C5
22	31	1	1	1	2	5	C6
0	39	119	0	0	0	0	C7
1	1	2	2	2	1	1	C8
4.8	4.8	4.6	4.2	3.6	3.3	4.8	C9
284	30	94	143	6088	10	14	C10
8520	900	2820	4290	182640	300	420	C11
2	2	2	2	1	2	2	C12
2	1	1	2	2	2	2	C13
10	7	4	7	6	7	7	C14

Table 41: Normalized decision matrix.

A7	A6	A5	A4	A3	A2	A1	
0.500000	0.250000	0.250000	0.50000	0.250000	0.500000	0.250000	C1
0.426401	0.213201	0.213201	0.426401	0.426401	0.426401	0.426401	C2
0.456861	0.811453	0.011736	0.188613	0.007545	0.205797	0.23388	C3
0.018018	0.014059	0.682481	0.052005	0.682481	0.184952	0.17608	C4
0.378045	0.378045	0.37767	0.378045	0.377857	0.378045	0.378045	C5
0.572443	0.806625	0.02602	0.02602	0.02602	0.052040	0.130101	C6
0.000000	0.311432	0.950268	0.0000000	0.000000	0.000000	0.000000	C7
0.2500000	0.250000	0.5000000	0.5000000	0.500000	0.250000	0.250000	C8
0.418151	0.4118151	0.400728	0.365882	0.313613	0.287479	0.418151	C9
0.0465793	0.00492030	0.0154171	0.0234536	0.998504	0.001640	0.002296	C10
0.0465793	0.0049203	0.0154171	0.0234536	0.998504	0.001640	0.002296	C11
0.400000	0.4000000	0.4000000	0.4000000	0.200000	0.400000	0.400000	C12
0.426401	0.213201	0.213201	0.426101	0.426401	0.426401	0.426101	C13
0.536056	0.375239	0.214423	0.375239	0.321634	0.375239	0.375239	C14

Table 42: Weighted normalized decision matrix.

A7	A6	A5	A4	A3	A2	A1	
0.1188	0.0594	0.0594	0.1188	0.0594	0.1188	0.0594	C1
0.019572	0.009786	0.009786	0.019572	0.019572	0.019572	0.019572	C2
0.071362	0.126749	0.009786	0.029461	0.001178	0.032146	0.036532	C3
0.000425	0.000332	0.001833	0.001227	0.016107	0.004365	0.004155	C4
0.027446	0.027446	0.0274189	0.027446	0.0274325	0.027446	0.027446	C5
0.091018	0.128253	0.004137	0.004137	0.004137	0.008274	0.020686	C6
0.000000	0.0202120	0.0616724	0.0000000	0.000000	0.0000000	0.000000	C7
0.014825	0.014825	0.02965	0.02965	0.02965	0.014825	0.014825	C8
0.033159	0.033159	0.031778	0.029014	0.02487	0.022797	0.033159	C9
0.0011039	0.0001166	0.0001166	0.0005559	0.0236645	0.0000389	0.0000544	C10
0.0004891	0.0000517	0.0000517	0.0002463	0.0104843	0.0000172	0.0000241	C11
0.00188	0.00188	0.00188	0.00188	0.00094	0.00188	0.00188	C12
0.00533	0.002665	0.002665	0.00533	0.00533	0.00533	0.00533	C13
0.028089	0.019663	0.011236	0.019663	0.016854	0.019663	0.019663	C14

Where, g and $(n\text{-}g)$ are the number of criteria to be maximized and minimized, respectively. And $w_j x_{ij}$ is the weight of the criterion. The results are in Table 43 and Figure 30.

Table 43: Overall performance of the alternatives.

	$\sum_{j=1}^{g} w_j x_{ij}$	$\sum_{j=g+1}^{n} w_j x_{ij}$	yi	Ranking
A1	0.1642678	0.0784590	0.085808	4
A2	0.2248360	0.0503174	0.174518	3
A3	0.1341652	0.1054533	0.027811	5
A4	0.2828456	0.0041372	0.278708	1
A5	0.0656681	0.1921217	-0.12675	7
A6	0.2311133	0.2134249	0.017688	6
A7	0.3268881	0.086612	0.210276	2

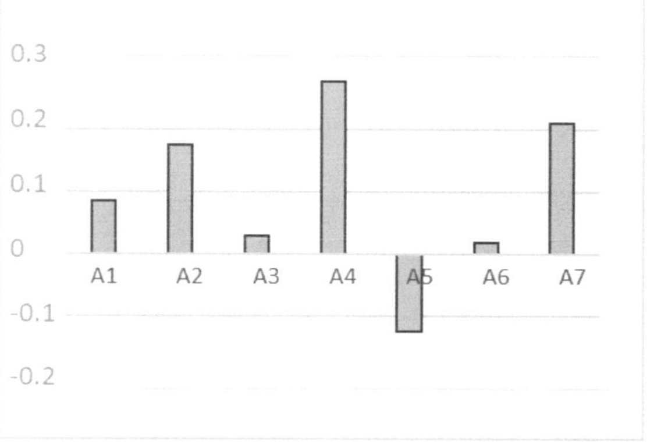

Figure 30: Value of y for each alternative.

3. Multivariate Analysis

Cluster analysis groups individuals or objects into clusters so that objects in the same cluster are more like one another than they are to objects in other clusters. The attempt is to maximize the homogeneity of objects within the clusters while also maximizing the heterogeneity between clusters.

Cluster analysis classifies objects on a set of user-selected characteristics. The resulting clusters should exhibit high internal homogeneity and high external heterogeneity. Thus, if the classification is successful, the objects within clusters will be close together when plotted geometrically, and different clusters will be far apart.

The process followed is to start with all observations as their cluster, use the similarity measure and combine the two most similar clusters into a new cluster, repeat the clustering process and continue combining the two most similar clusters into a new cluster.

The results in the hierarchical clustering can be represented as a dendrogram as shown in the Figure 31, which uses a single link (re-escalated cluster combination).

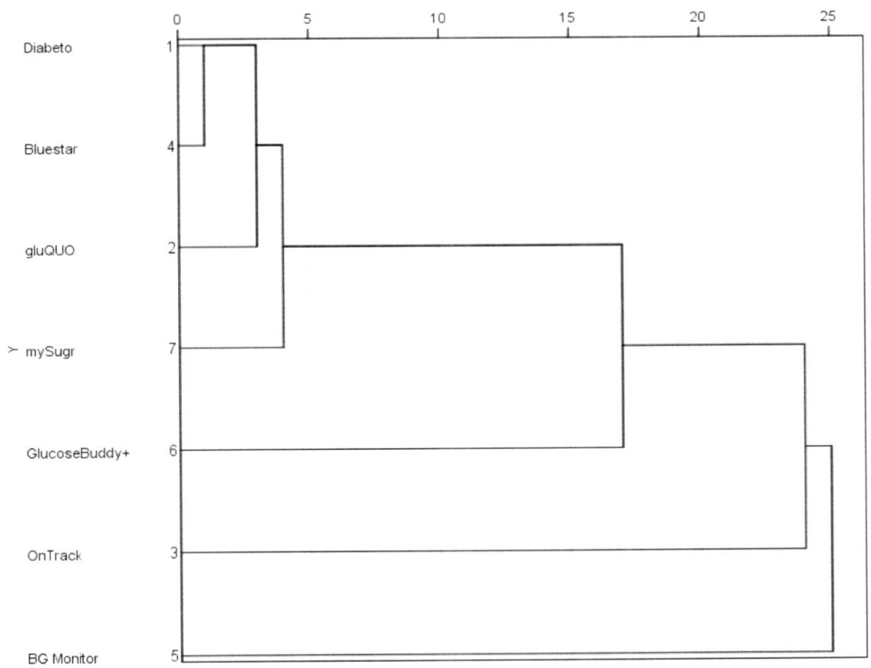

Figure 31: Dendogram.

4. Discussion

There is a wide array of health mobile applications on the market. Since diabetes continues to rise in numbers, mobile apps for management of this disease are popular. Among all these apps seven alternatives were analysed using the AHP and Grand Prix method.

From the first set of criteria, using the AHP method, it was concluded that functionality is the most important factor when looking for a mobile application, while the user information wasn't as important. These values were also used to find the priority values of each sub-criterion. From the functionality sub-criteria, the operating system had the higher values, thus it was deemed as the most important. From the usability sub-criteria, languages had the higher value. On the user information criteria, the user rating sub-criterion was the most important. Furthermore, from the engagement with user criterion, the parameters that could be registered were the most important.

Of each alternative, mySugr was considered the best alternative for the management of diabetes, Bluestar the second and Glucose Buddy+ the third option. The alternative with the lowest value was BG Monitor.

The MOORA method was also used to analyse which of the seven alternatives is the best option. The results from this method are like the ones obtained through the AHP method. The number one option from AHP is the second option using MOORA, while option number two from AHP is the first and best option from MOORA.

5. Conclusions and Future Research

The multiple-criteria decision analysis (MCDA) is an important and useful tool for a variety of fields. The AHP method allows us to compare many alternatives with respect to a set of criteria. The MOORA method is also another type of multiple-criterion decision analysis, and like the AHP method, it compares alternatives to certain criteria. These two methods showed similar results and, as a conclusion, it can be said that the mobile application mySugr is the best amongst the diabetes monitoring apps.

References

[1] Kollman, A., Kastner, P. and Schreier, G. 2007. Chapter, X, Utilizing mobile phones as patient terminal in managing chronic diseases, in web mobile- bases applications for healthcare management. IGI Global, pp. 227–257.
[2] Papatheodorou, K., Papanas, N., Banach, M., Papazoglou, D. and Edmonds, M. 2015. Complications of Diabetes, Journal of Diabetes Research.
[3] American Diabetes Association, Diagnosis and Classification of Diabetes Mellitus, Diabetes Care, 31, 2009, pp. 62–67.
[4] Centers for Disease Control and Prevention, Diabetes [Online]. URL: https://www.cdc.gov/media/presskits/aahd/diabete s.pdf. [Accessed 10 September 2019].
[5] Tabish, S. 2007. Is diabetes becoming the biggest epidemic of the twenty-first century? International Journal of Health Sciences, 1(2): V.
[6] Gardner, R. and Shabot, M. 2006. In Biomedical Informatics, Springer, p. 585.
[7] World Health Organization, Integrated chronic disease prevention and control, Available: https://www.who.int/chp/about/integrated_cd/en/. [Accessed 12 September 2019].
[8] Iakovidis, I., Wilson, P. and Healy, J. (eds.). E-health: current situation and examples of implemented and beneficial e-health applications. Vol. 100. Ios Press, 2004.
[9] Cook, D.J., Augusto, J.C. and Jakkula, V.R. 2009. Ambient intelligence: Technologies, applications, and opportunities. Pervasive and Mobile Computing, 5(4): 277–298.
[10] Dey, N. and Ashour, A. 2017. Ambient intelligence in healthcare: a state-of-the-art. Global Journal of Computer Science and Technology, 17(3).
[11] Tang, P. and McDonald, C. 2006. Electronic Health Record Systems, in Chapter 12: Biomedical Informatics, Springer, New York, NY, pp. 447–475.
[12] Panteli, N., Pitsillides, B., Pitsillides, A. and Samaras, G. 2006. Chapter IV: An e-Healthcare Mobile Applicationin Web Mobile-Based Applications for Healthcare Management (Editor Dr L. Al-Hakim), Book chapter, Idea Group, accepted for publication.
[13] Dowding, D., Randell, R., Gardner, P., Fitzpatrick, P., Dykes, P., Favela, J. and Hamer, S. 2015. Dashboards for improving patient care: review of the literature. International Journal of Medical Informatics, 84(2): 87–100.
[14] Nouri, R., Niakan, S., Ghazisaeedi, M., Marchand, G. and Yasini, M. 2018. Criteria for assessing the quality of mHealth apps: a systematic review. Journal of the American Medical Informatics Association, 25(8): 1089–1098.
[15] Arnhold, M., Quade, M. and Kirch, W. 2014. Mobile applications for diabetics: a systematic review and expert-based usability evaluation considering the special requirements of diabetes patients age 50 years or older. Journal of Medical Internet Research, 16(4): e104.
[16] Shah, V., Garg, S., Viral, N. and Satish, K. 2015. Managing diabetes in the digital age. Clinical Diabetes and Endocrinology, 1(1): 16.
[17] Taherdoost, H. 2017. Decision making using the Analytic Hierarchy Process (AHP): A step by step approach, International Journal of Economics and Management Systems, 2: 244–246.
[18] Saaty, T. 2008. Decision Making with the analytics hierarchy process. International Journal of Services Sciences, 1(1): 83–98.

CHAPTER-6

Electronic Color Blindness Diagnosis for the Detection and Awareness of Color Blindness in Children Using Images with Modified Figures from the Ishihara Test

Martín Montes,[1,*] *Alejandro Padilla,*[2] *Julio Ponce,*[2] *Juana Canul,*[3] *Alberto Ochoa-Zezzatti*[4] and *Miguel Meza*[2]

Color blindness is a condition that affects the cones in the eyes; it can be congenital or acquired and is considered an average disability that affects about 10% of the world's population. Childrens with color blindness have particular difficulties when entering an educational environment with materials developed for people with normal vision. This work focuses on modifying the Ishihara test to apply it to preschool children. The proposed test helps to identify children who suffer from color blindness so that the teacher who guides them in the school can attend them.

1. Introduction

The sense of sight in human beings as in other organisms depends on their eyes; these use two types of cells for the perception of images, i.e., rods and cones (Richmond Products, 2012).

The rods are used to identify the luminosity, i.e., the amount of light received from the environment and the cones are used to identify the color or frequency in the spectrum of light received (Colorblindor, 2018).

In most people there are three types of cones, each one to perceive a basic color, these can be red, green or blue, and the other colors that are generated are the result of various combinations that are received from amounts of light in tune with the frequencies of these basic colors (Deeb, 2004).

The world around us is designed to work with the colors that are perceived with three cones, since most people can perceive the environment with three basic colors, i.e., they are trichromats, however, there are data of people with a fourth type of cone, which allows them to perceive more colors than the average person visualizes, however, these people often have problems describing the environment and tones they perceive, since the world is not made with their sensory perceptions in mind (Robson, 2016).

[1] Universidad Politécnica de Aguascalientes, Calle Paseo San Gerardo, Fracc. San Gerardo, 20342 Aguascalientes, Aguascalientes, México.
[2] Universidad Autónoma de Aguascalientes. México.
[3] Universidad Juárez Autónoma de Tabasco. México.
[4] Universidad Autónoma de Juárez. México.
* Corresponding author: jorge.rodas@uacj.mx

On the other hand, there are also cases of people with a lower color perception, this condition is called color blindness and is considered a medium disability, since the colors with trichromat perception are used in various activities, such as to identify objects in a conversation, to identify dangerous situations, know when to advance in a traffic light, to decide what clothes to buy and to enjoy art forms such as painting or photography (Kato, 2013).

Color blindness can be catalogued in four variants according to the cones available to perceive the environment, these variants can be anomalous trichromacy, dichromacy, and monochromacy or achromatopsia (Colorblindor, 2018).

The most common variant of color blindness is anomalous trichromacy, in which there are all the cones for color perception, but there is a deficiency in some of them. Anomalous trichromacy can be separated depending on its severity into mild anomalous trichromacy, medium or strong and depending on the color in which the deficiency is presented, trichromacy can be divided into deuteranomaly or deficiency in the perception of green, protanomaly in the perception of red and tritanomaly in the perception of the blue (Huang et al., 2011).

Another variant of color blindness that occurs less frequently, but more severely, is dichromacy. In this, one type of cone is absent, i.e., the person cannot perceive one of the basic colors, which causes problems with all colors that have this tone in their constitution, for example, a person who has problems with the green cone will have problems with all forms of green, but also with yellow and brown because they are constituted with the color green as a base. Dichromacy can also be classified depending on the absent cone, it is deuteranopia when the green cone is absent, protanopia when the red cone is absent and tritanopia when the blue cone is absent (Colorblindor, 2018).

Monochromacy, or achromatopsia, is the rarest condition of color blindness, in this, all the cones are absent, therefore, the environment is only perceived in gray scales or luminosity and, although it is very rare, it represents a major difficulty for people who suffer from it, since they cannot live a normal life without the assistance of healthy people. These people cannot drink a liquid from a bottle without first looking for a label confirming its contents, they cannot identify if a food is in good condition before eating it, they cannot choose their clothes or identify a parking place, among other difficulties (Colorblindor, 2018).

About 10% of people suffer from some deficiency or blindness to color, that is, about 700 million people suffer from color blindness, considering that the world population exceeds 7,000 million inhabitants. Table 1 shows the percentages of incidence in the world for men and women with each of the variants of color blindness (Colorblindor, 2018).

As often as in adults worldwide, 1 in 10 children is born colorblind, facing a world that is not designed for him, generating various difficulties even in their learning and school performance (Pardo Fernández et al., 2003). Mexico has a similar situation in color blindness, as detailed in the study of the prevalence of color blindness in public school children in Mexico (Jimenéz Pérez et al., 2013).

Table 1: Prevalece of color blindness in the world (Colorblindor, 2018).

Type	**Variant**	**Prevalence H/M**	
Monochromacy	Acromatopsia	0.00003%	
Dichromacy	Deuteranopia	1.27 %	0.01%
	Protanopia	1.01%	0.02%
	Tritanopia	0.0001%	
Anomalous Trichromacy	Deuteranomaly	4.63%	0.36%
	Protanomaly	1.08%	0.03%
	Tritanomaly	0.0002%	

Children with visual difficulties associated with color blindness may have school problems when performing activities that involve color identification, such as using educational material with colored content, for example, relating content in their textbooks and participating in games, among other difficulties related to visual tasks (Pardo Fernández et al., 2003).

Despite all the difficulties that children with color blindness are exposed to, this condition is one of the vision anomalies that take longer to be detected by parents and teachers, because by the intelligence or support they receive from other people they manage to get ahead with their education (Pardo Fernández et al., 2003).

Teaching materials are often designed for people with normal vision, so it is important to detect vision impairments before school age so that children receive appropriate assistance in their educational processes (Jimenéz Pérez et al., 2013).

The detection of color blindness can be done through the application of various tests that are designed according to the variant that is presented and the colors that are confused in this condition, for this, it is important to reproduce the perception of a colorblind people so that the tests be correctly designed. Several algorithms adjust the model of a digital image represented with red, green and blue parameters (RGB), to simulate color blindness and then allow people with a normal vision to see as people with this medium disability do (Tomoyuki et al., 2010).

Figure 1 shows the color spectrum seen by an average trichromate, Figure 2 by a person with protanopia, Figure 3 by deuteranopia, and Figure 4 by tritanopia, all of them obtained applying color blindness simulation models.

Figure 1: Full color spectrum.

Figure 2: Full color spectrum perceived by a protanope.

Figure 3: Full color spectrum perceived by a deuteranope.

Figure 4: Full color spectrum perceived by a tritanope.

The most common test used by ophthalmologists, based on the principle of detecting confusing colors, is the Farnsworth-Munsell test, in which the patient is presented with discs with colors of the entire color spectrum and is asked to make their arrangement in the correct order, this test is highly accurate in identifying any variant of color blindness and the severity with which it is presented, however, its application is highly complicated and in order to carry it out correctly, specific ambient conditions are required, such as a specific brightness of 25 candles at 6740 K, which describe the lighting conditions at midday (Cranwell et al., 2015). Figure 5 shows a simplified Farnsworth-Munsell test manufactured by Lea Color Vision Testing.

The most commonly used test for the detection of color blindness are Ishihara plates, designed for the detection of protanopia, protanomaly, deuteranopia and deuteranomaly, however, any other variant of color blindness cannot be detected by Ishihara plates as they are designed to detect the colors that confuse people with these variants of color blindness (Ishihara, 1973). Figure 6 shows plate 42 that is seen by people with protanopia or strong protanomaly as a 2 and by deuteranopia and strong deuteranomaly as a 4.

The tests to perform the Ishihara test are widely known and simple to evaluate, since the difficulties to perceive the color are observable when the individual who is submitted to the test, cannot see the number inside or it is difficult to visualize it, in addition, depending on the plate with which you have problems, it is identified which variant of color blindness is presented. Table 2 is used as a comparison list to identify the type of color blindness variant presented with the Ishihara test.

The initial problem with this type of test is the purchase of the same, however, currently, there are several websites of organizations which allow a rapid assessment when there are suspicions of color blindness.

An age-appropriate color blindness test (Jouannic, 2007) includes the possibility of detecting color blindness in toddlers when dealing with figures; however, given the options presented for each plate, the test can still be complicated for a preschooler. One of the images presented in this test is

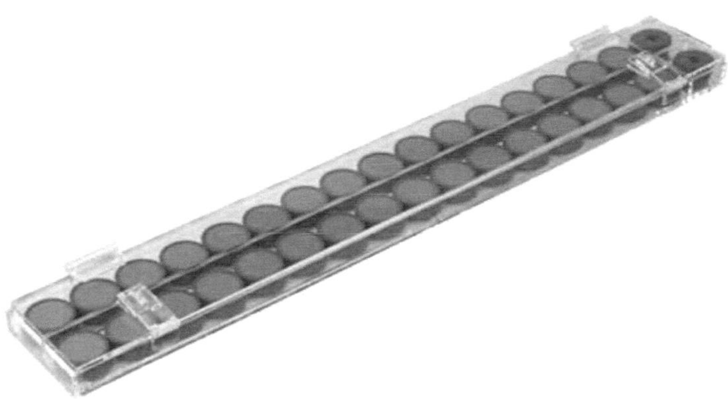

Figure 5: Farnsworth-Munsell Test manufactured by Lea Color Vision Testing (Vision, 2018).

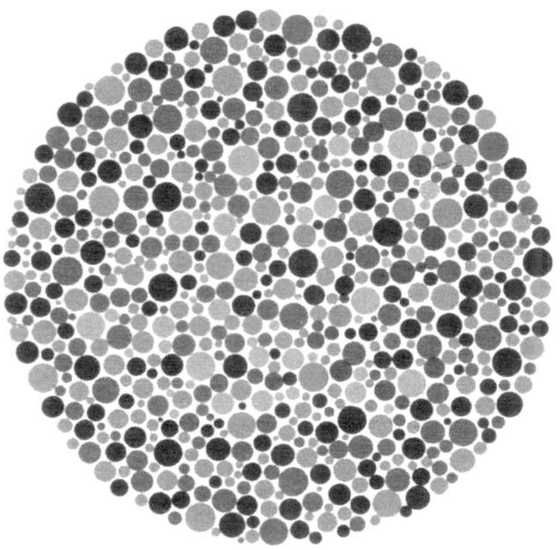

Figure 6: Ishihara Test Plate 42 (Ishihara, 1973).

Table 2: Checklist for evaluation with the Ishihara test of 17 plates, where the X mark that cannot be read (Ishihara, 1973).

<table>
<tr><th>Plate Number</th><th>Plate perceived trychromats</th><th colspan="2">Plate perceived by protanopes or deuteranopes</th><th colspan="2">Plate perceived by monochramats</th></tr>
<tr><td>1</td><td>12</td><td colspan="2">12</td><td colspan="2">12</td></tr>
<tr><td>2</td><td>8</td><td colspan="2">3</td><td colspan="2">X</td></tr>
<tr><td>3</td><td>29</td><td colspan="2">70</td><td colspan="2">X</td></tr>
<tr><td>4</td><td>5</td><td colspan="2">2</td><td colspan="2">X</td></tr>
<tr><td>5</td><td>3</td><td colspan="2">5</td><td colspan="2">X</td></tr>
<tr><td>6</td><td>15</td><td colspan="2">17</td><td colspan="2">X</td></tr>
<tr><td>7</td><td>74</td><td colspan="2">21</td><td colspan="2">X</td></tr>
<tr><td>8</td><td>6</td><td colspan="2">X</td><td colspan="2">X</td></tr>
<tr><td>9</td><td>45</td><td colspan="2">X</td><td colspan="2">X</td></tr>
<tr><td>10</td><td>5</td><td colspan="2">X</td><td colspan="2">X</td></tr>
<tr><td>11</td><td>7</td><td colspan="2">X</td><td colspan="2">X</td></tr>
<tr><td>12</td><td>16</td><td colspan="2">X</td><td colspan="2">X</td></tr>
<tr><td>13</td><td>73</td><td colspan="2">X</td><td colspan="2">X</td></tr>
<tr><td>14</td><td>X</td><td colspan="2">5</td><td colspan="2">X</td></tr>
<tr><td>15</td><td>X</td><td colspan="2">45</td><td colspan="2">X</td></tr>
<tr><td></td><td></td><td colspan="2">Protanope</td><td colspan="2">Deuteranope</td></tr>
<tr><td></td><td></td><td>Strong</td><td>Mild</td><td>Strong</td><td>Mild</td></tr>
<tr><td>16</td><td>26</td><td>6</td><td>(2)6</td><td>2</td><td>2(6)</td></tr>
<tr><td>17</td><td>42</td><td>2</td><td>(4)2</td><td>4</td><td>2(4)</td></tr>
</table>

shown in Figure 7. In this slide, the child is asked whether he sees a letter B behind the mesh, a star behind the mesh, or the mesh itself.

Developing a test that can be done at the preschool level with a group of children would make it possible to raise awareness of the difficulties presented by some of the peers in that group and allow the preschool teacher to identify students who have problems with color perception, in order to adjust the activities to children who might face this type of condition in a preschool group.

The aim of this chapter is to review the issue of color blindness, the difficulties, and guidelines when present in preschool children, while proposing images that can be identified through sight by groups of healthy children and disagree with those perceived by people with some form of

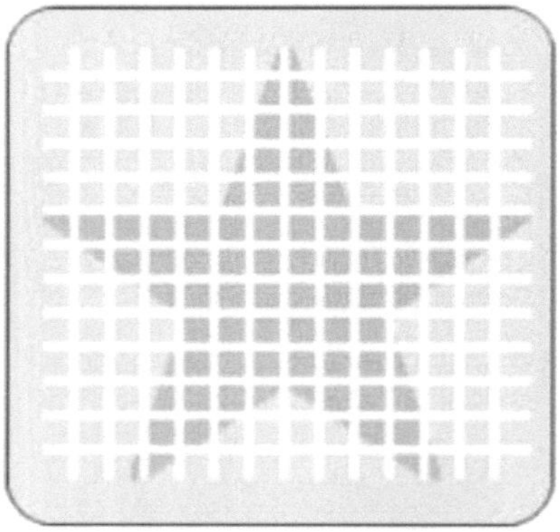

Figure 7: Star test plate in (Jouannic, 2007).

color blindness, in this chapter focusing primarily on the identification of protanopia, protanomaly, deuteranopia, and deuteranomaly, to recommend that the child's parent go to a specialist to confirm the evaluation and present a specialist test such as Farnsworth's test. The tests are also mounted on an application so that they can be taken from home by the preschool-age child's parent or tutor.

2. Backgrounds

The concern for color blindness in children is not an issue that has begun to be studied recently and there are several papers that are presented around this average disability.

Several of these papers focus on identifying the incidence of color blindness in children in certain parts of the world, such as at (Jimenéz Pérez et al., 2013), where color blindness is detected in school-age children in Mexico. The same incidence is being studied in eastern Nepal (Niroula and Saha, 2010). In (Moudgil et al., 2016) a similar study is conducted on children between the ages of 6 and 15 in Jalandhar, in all cases using Ishihara plates. Another group of works seeks to identify problems that children with color blindness have and their detection. One of these studies shows that children with color blindness problems often have difficulties in identifying colors, but these are less than those expected in the color blindness model, as indicated in (Lillo et al., 2001).

The work proposed in (Nguyen et al., 2014) shows the development of a computer interface for school-age children, it uses images suitable for children in a game that can be solved correctly by children with normal vision, however, they need instructions and supervision, making it difficult when working with a group of children.

3. Development

Considering that the most used diagnostic tests and simple to evaluate are linked to the identification of confusing colors, images are designed with these colors using similar plates to those used in the Ishihara test, but instead of using numbers and figures, it is proposed to use drawings similar to those recognized at an early age.

Another difficulty that should be kept in mind is to keep the indications simple, asking the child to confirm if what he or she sees is correct or not. For this purpose, the Microsoft Net Assembly speech interface is used to make the computer tell the child what it would have to see in the developed images. When opening the application, the first thing shown is a description addressed to the applicator or the teacher, which indicates what color blindness is, the purpose of the test and the instructions to follow in the application.

Once the applicator clicks the button that starts the test, the images are presented to the children for three seconds, this following the indications of the original Ishihara test, since it is difficult for children with color blindness to use the contrast and brightness conditions to try to identify the images; at the same time, with the speaker, the child is told what it should see and these indicate to the applicator whether or not they could see the figure.

It is hoped that, with this, the applicator or teacher can identify which children have problems with the colors, inform the parents and take the child to a specialist in order to assess the child's condition, furthermore, the teacher should consider the special condition of the affected children when preparing the material for their classes.

4. Results

The plates developed based on the colors used in the Ishihara plates, with designs that are known to preschool children are shown in Figure 8 to Figure 13.

Table 3 shows the images obtained using a dichromacy color blindness simulation model (protanopia, deuteranopia, and tritanopia), they show how each plate looks like in each of the most critical color blindness variants.

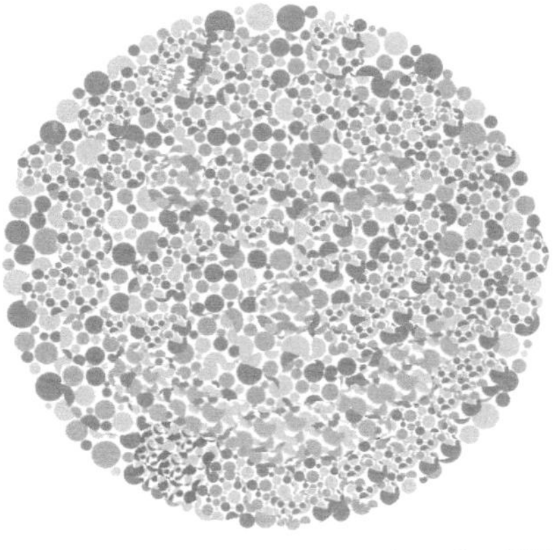

Figure 8: Face pseudochromatic plates for children.

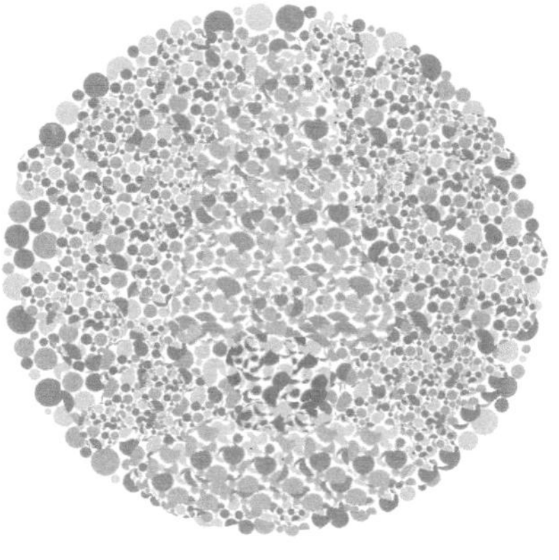

Figure 9: Tree plates pseudochromatic for children.

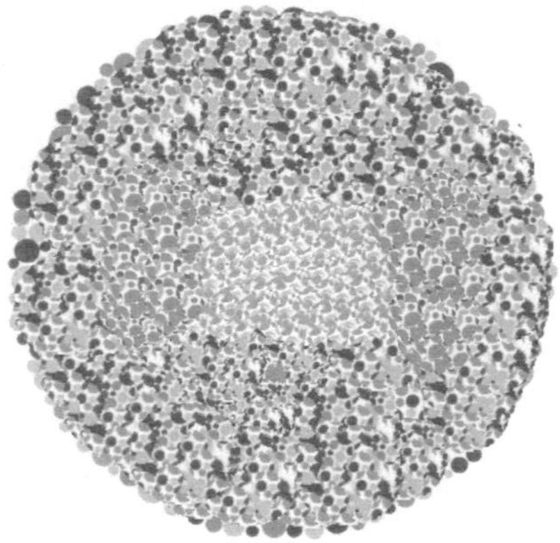

Figure 10: Sweet pseudochromatic plates for children.

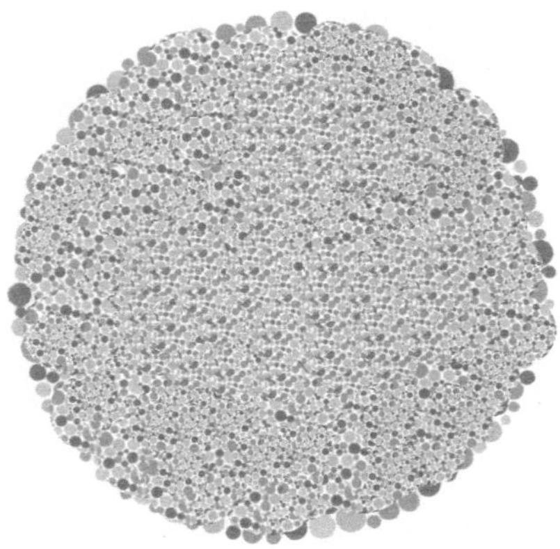

Figure 11: Boat pseudochromatic plates for children.

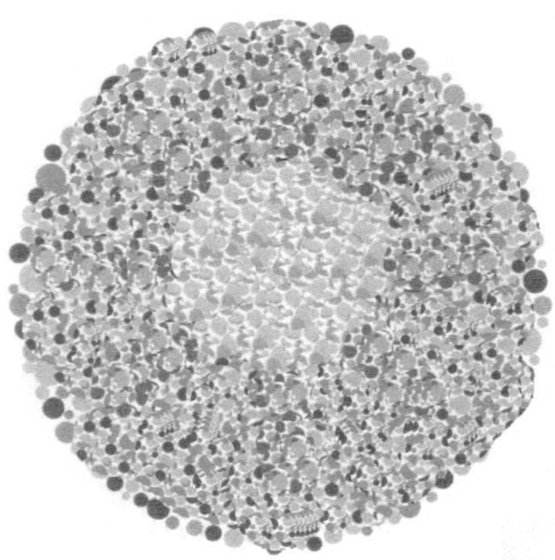

Figure 12: Sun pseudochromatic plates for children.

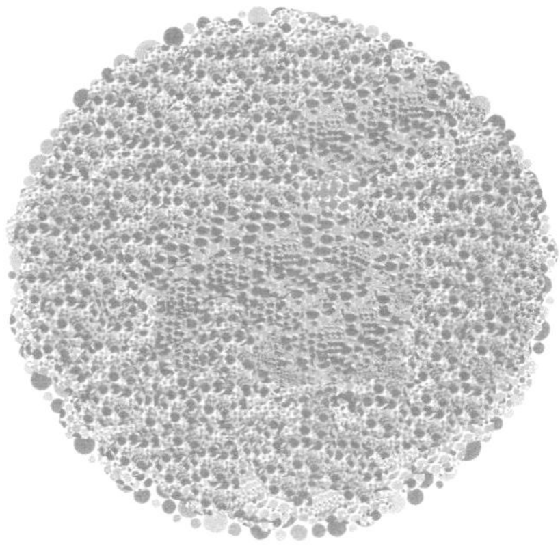

Figure 13: House pseudochromatic plates for children.

Table 3: Perception of proposed pseudochromatic plates for children with different variants of dichromacy

Normal perception	Deuteranopia	Protanopia	Tritanopia
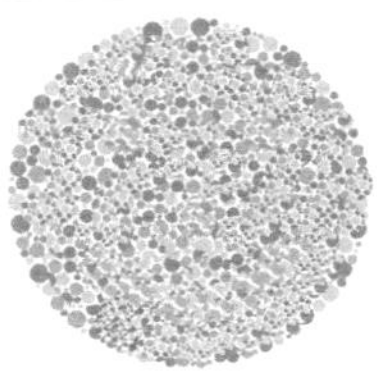	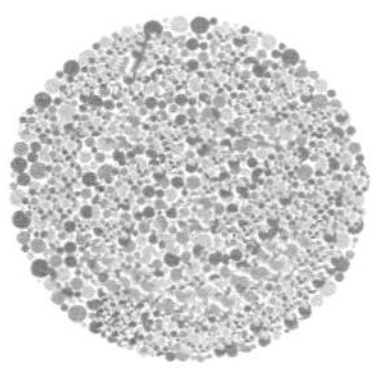	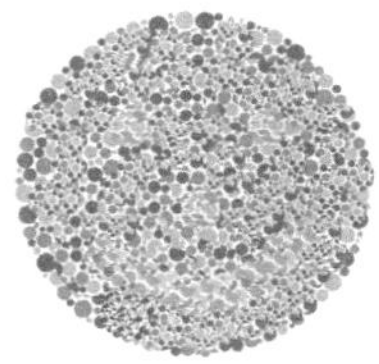	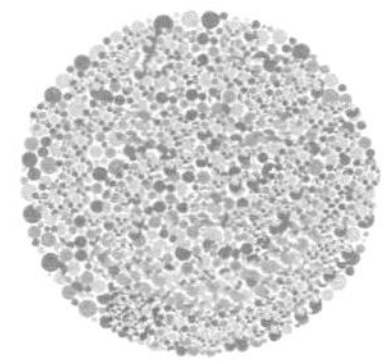
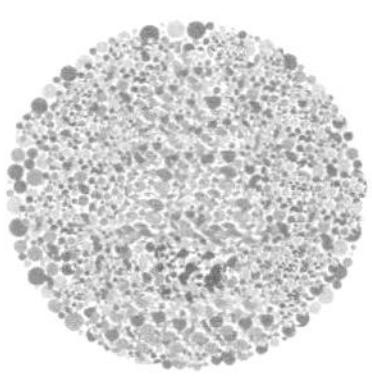		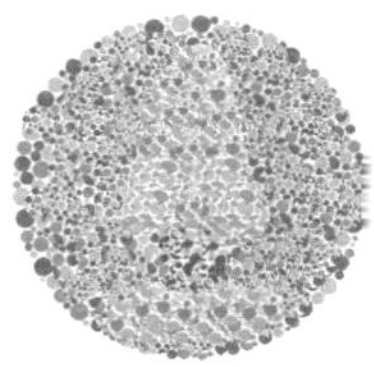	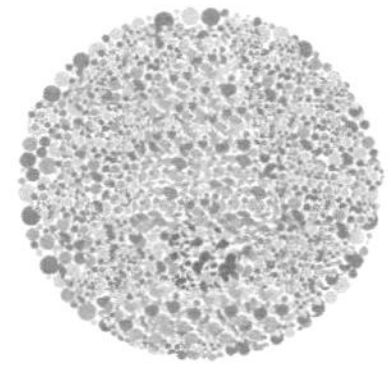

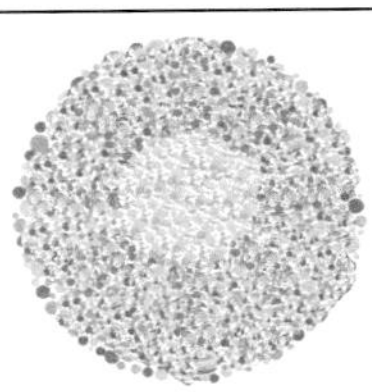	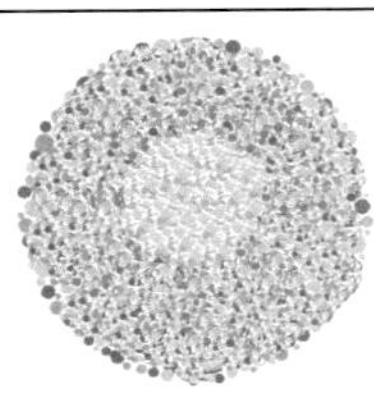	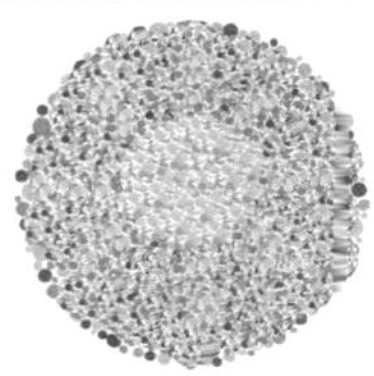	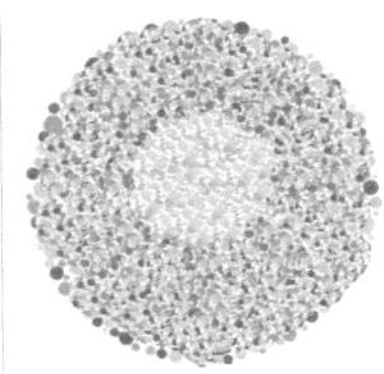
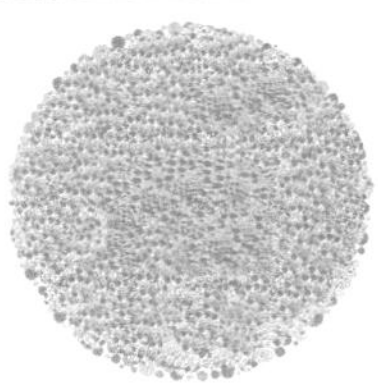	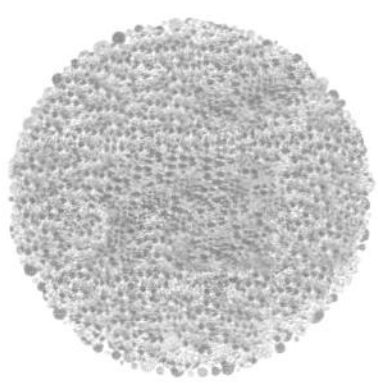	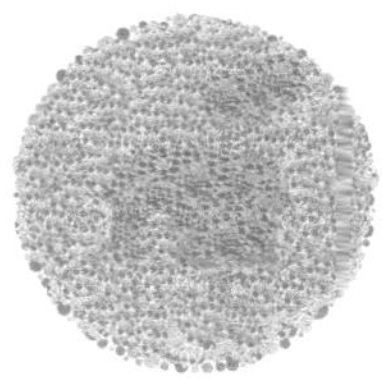	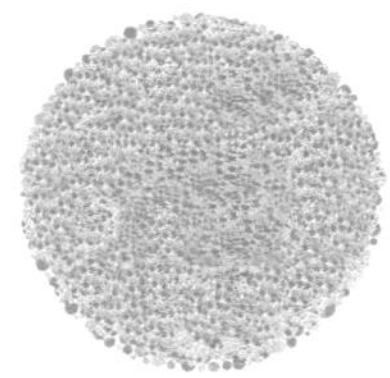

The diagrams of the application generated at each of the moments of the test are shown from Figure 14 to Figure 20. Initially, instructions are shown when opening the application to start the test.

When the applicator or teacher clicks the start test button, images begin to be shown (Figure 15 to Figure 20) as the sound produced with Microsoft Net Assembly tells the children which image they should see and then the applicator selects from the drop-down list whether the plate was viewed correctly or incorrectly.

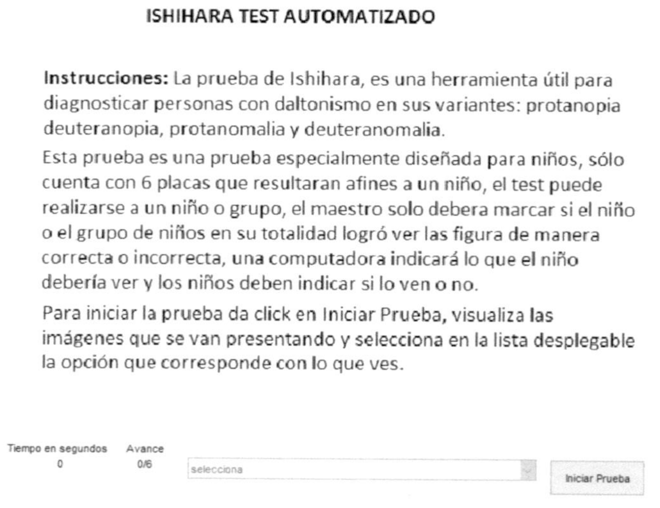

Figure 14: Instructions shown in the application when opening the test.

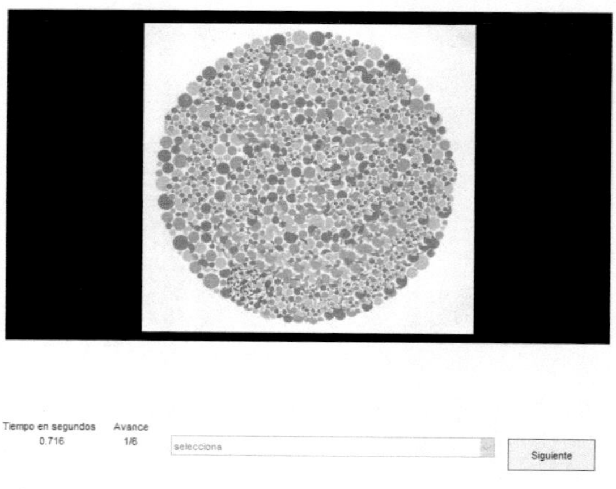

Figure 15: Face shown in the application when the Microsoft Voice Assistant says "You should see a Face"

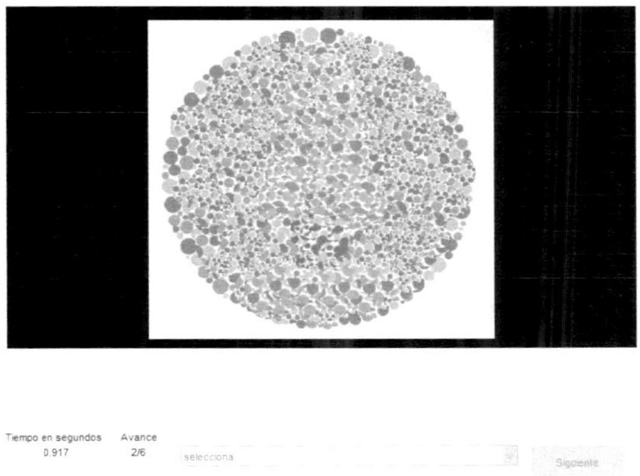

Figure 16: Tree shown in the application when the Microsoft Voice Assistant says "You should see a Tree"

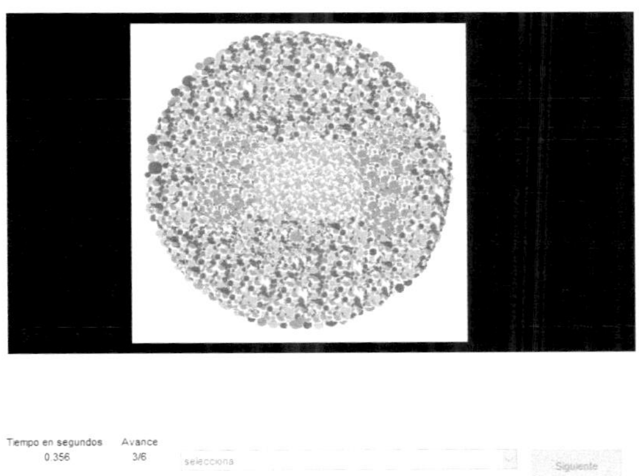

Figure 17: Candy shown in the application when the Microsoft Voice Assistant says "You should see a Candy".

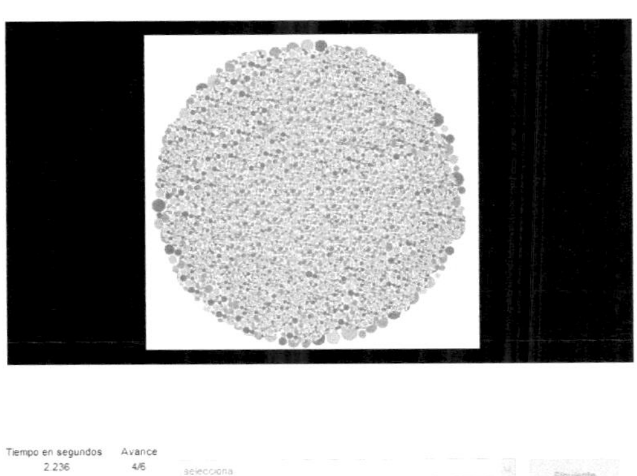

Figure 18: Ship shown in the application when the Microsoft Voice Assistant says "You should see a Ship".

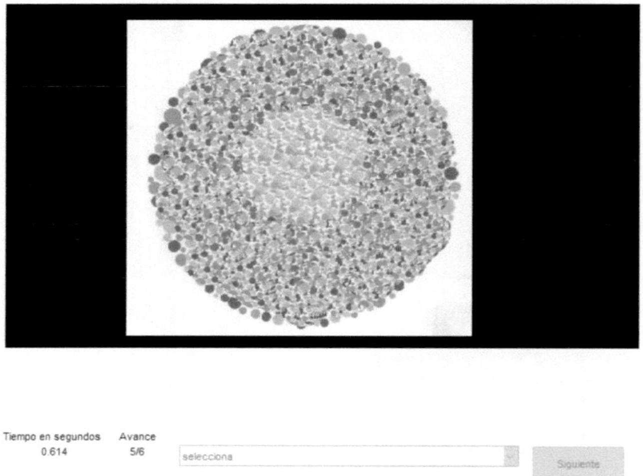

Figure 19: Sun shown in the application when the Microsoft Voice Assistant says "You should see the Sun".

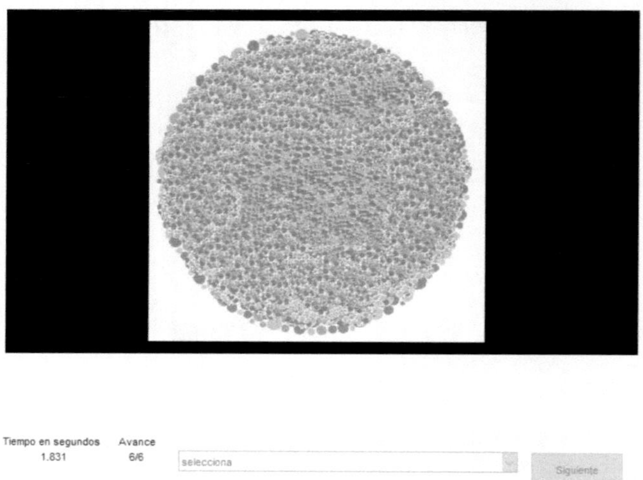

Figure 20: House shown in the application when the Microsoft Voice Assistant says "You should see a House".

When the test is completed, a chart is received indicating that a doctor should be visited visually and by audio, in case of failure in the identification of the plates. The user has the possibility to do the test again.

6. Conclusions

In this work, we manage to design plates with figures that preschool children can identify, as well as using them in an application that presents an audio aid that tells children what they should see on each plate. So, with the help of an applicator that could well be the teacher, children can take

Figure 21: Conclusion of the test indicating that there is deficiency in color perception and a doctor should be visited.

the test. The design of each of the plates is intended to make it difficult for people with problems in the different variants of color blindness that are detected by the Ishihara test, i.e., protanopia, protanomaly, deuteranopia, and deuteranomaly. The graphs show different figures depending on the variants of color blindness that are presented as shown in the results section.

6.1 Future Research

Apply the pilot test with preschool children, to verify that they are familiar with the figures, as well as seek to detect real cases of color blindness using the proposed test.

References

Colorblindor. 2018. Color Blind Essentials. Recuperado de https://www.color-blindness.com/color-blind-essentials.

Cranwell, M.B., Pearce, B., Loveridge, C. and Hurlbert, A.C. 2015. Performance on the farnsworth-munsell 100-hue test is significantly related to nonverbal IQ. Investigative Opthalmology & Visual Science, 56(5): 3171. https://doi.org/10.1167/iovs.14-16094.

Deeb, S.S. 2004. Molecular genetics of color-vision deficiencies. Visual Neuroscience, 21(3): 191–196. Recuperado de http://www.ncbi.nlm.nih.gov/pubmed/15518188.

Huang, C.-R., Chiu, K.-C. and Chen, C.-S. 2011. Temporal color consistency-based video reproduction for dichromats. IEEE Transactions on Multimedia, 13(5), 950–960. https://doi.org/10.1109/TMM.2011.2135844.

Ishihara, S. 1973. Test for Colour-Blindness, 24 Plates Edition, Kanehara Shuppan Co. Ltd., Tokyo.

Jimenéz Pérez, A., Hinojosa García, L., Peralta Cerda, E.G., García García, P., Flores-Peña, Y., M-Cardenas, V. and Cerda Flores, R.M. 2013. Prevalencia de daltonismo en niños de escuelas públicasde México: detección por el personal de enfermería. CIENCIAUANL, 16(64): 140–144.

Jouannic, J. 2007. color blindness test (free and complete). Recuperado el 17 de diciembre de 2019, de http://www.opticien-lentilles.com/daltonien_beta/new_test_daltonien.php.

Kato, C. 2013. Comprehending Color Images for Color Barrier-Free Via Factor Analysis Technique. En 2013 14th ACIS International Conference on Software Engineering, Artificial Intelligence, Networking and Parallel/Distributed Computing (pp. 478–483). IEEE. https://doi.org/10.1109/SNPD.2013.39.

Lillo, J., Davies, I., Ponte, E. and Vitini, I. 2001. Colour naming by colour blind children. Anuario de Psicologia, 32(3): 5–24.

Moudgil, T., Arora, R. and Kaur, K. 2016. Prevalance of Colour Blindness in Children. International Journal of Medical and Dental Sciences, 5(2): 1252. https://doi.org/10.19056/ijmdsjssmes/2016/v5i2/100616.

Nguyen, L., Lu, W., Do, E.Y., Chia, A. and Wang, Y. 2014. Using digital game as clinical screening test to detect color deficiency in young children. En Proceedings of the 2014 conference on Interaction design and children - IDC '14 (pp. 337–340). New York, New York, USA: ACM Press. https://doi.org/10.1145/2593968.2610486.

Niroula, D.R. and Saha, C.G. 2010. The incidence of color blindness among some school children of Pokhara, Western Nepal. Nepal Medical College journal : NMCJ, 12(1): 48–50. Recuperado de http://www.ncbi.nlm.nih.gov/pubmed/20677611.

Pardo Fernández, P.J., Gil Llinás, J., Palomino, M.I., Pérez Rodríguez, A.L., Suero López, M.I., Montanero Fernández, M. and Díaz González, M.F. 2003. Daltonismo y rendimiento escolar en la Educación Infantil. Revista de educación, ISSN 0034-8082, N° 330, 2003, págs. 449–462, (330), 449–462. Recuperado de https://dialnet.unirioja.es/servlet/articulo?codigo=624844.

Richmond Products. 2012. Color Vision Deficiency: A Concise Tutorial for Optometry and Ophthalmology (1a ed.). Richmond Products. Recuperado de https://pdfs.semanticscholar.org/06bf/712526f7e621e7bc7a09e7f9604c5bae6899.pdf.

Robson, D. 2016. Las mujeres con una visión superhumana. BBC News Mundo. Recuperado de https://www.bbc.com/mundo/noticias/2014/09/140911_vert_fut_mujeres_vision_superhumana_finde_dv.

Tomoyuki, O., Kazuyuki, K., Kajiro, W. and Yosuke, K. 2010. Proceedings of SICE Annual Conference 2010 (pp. 18–21). Taipei, Taiwan: Society of Instrument and Control Engineers. Recuperado de https://ieeexplore.ieee.org/abstract/document/5602422.

Vision, L.C. 2018. Color Vision Test Color Vision Test Clinical Evaluation of Color Vision. Recuperado el 23 de diciembre de 2018, de www.good-lite.com1.800.362.38601.888.362.2576Faxwww.good-lite.com.

CHAPTER-7

An Archetype of Cognitive Innovation as Support for the Development of Cognitive Solutions in Smart Cities

Jorge Rodas-Osollo,[1,4,*] *Karla Olmos-Sánchez,*[1,4]
Enrique Portillo-Pizaña,[2] *Andrea Martínez-Pérez*[3]
and *Boanerges Alemán-Meza*[5]

This chapter presents a Cognitive Innovation Model that formalizes the basic components and the interactions between them for the establishment of a Cognitive Architecture (CA). The convenience of walking in the sense of achieving an archetype as support for the implementation of innovative intelligent solutions in Smart Cities and having the client as a means of convenient validation of the representation and processing of the knowledge expressed in the CA in comparison with those executed by humans in their daily activities.

1. Introduction

Smart cities are a vital issue in this constantly changing world, dynamically shaped by science, technology, nature, and society, which implies people still face many challenges, both individual and social, where innovation plays a vital role. These constant challenges are now addressed more frequently through *Cognitive & Innovative Solutions (CgI-S)* which establish new schemes—*innovation*—of how to address them. Our current technological world uses a lot of *pieces of knowledge*, this means *high valuable information*, useful to solve a problem or satisfy a particular need and, of course, to drive innovation. Thus, the satisfaction of who has the problem or need is achieved when the *knowledge* is capitalized by *CgI-S*. Hence, the importance of finding out how to use and take advantage of as much of the *creative expertise* as possible, including imagination, even though its use in a systematic way is a complex challenge, even if only to share it through traditional ways, requires it to be made explicit. Even though dominating the challenge is a fundamental key for the *cognitive era* to progress and its *artificial intelligence technologies*, *machine learning*, *cognitive computing*, etc., coexist daily with humans.

In the *cognitive era*, *Cognitive Architects (Cg.Ar)* together with specialists, from the domain to be treated, make up the *Cognitive & Innovative Solution's Architects & Providers team (CgI-*

[1] Universidad Autónoma de Ciudad Juárez, Av. Hermanos Escobar, Omega, 32410 Cd Juárez, Chihuahua, México.
[2] Digital transformation at ITESM. México.
[3] Stragile Co., Interactive Technology & Solutions Group.
[4] Applied Artificial Intelligence Research Group.
[5] Department of BioSciences, Rice University, Houston, TX 77005, USA.
* Corresponding author: jorge.rodas@uacj.mx

SAP team) to provide *CgI-S* using highly specialized information, experience, creativity, coming from an *ad hoc Collaborative Network (ahCN)*; which allows the team to do an adequate job even with innovation. Also, the *CgI-SAP team* applies science and technology to take advantage of this knowledge in order to achieve the *Capitalization of Experience or Knowledge* in solutions or innovation. It is undeniable that the above represents a complex situation [1, 2] since it requires a complete orchestration of the process, on the part of the *Cg.Ar*, which results in a *CgI-S* which requires technological developments and changes in the processes of the organization where the *Cg.Ar* must work side-by-side with *ahCN*. This arduous labour must be supported by a Cognitive Architecture, particularly apt, when cognitive approaches are required to meet the challenges of the *cognitive era*.

This document is an effort to match situations or needs that should be faced with intelligent technologies and innovation processes, at times when the environment is extremely dynamic being this characteristic very typical within what is now called a *cognitive era*. The above motivates to provide a *Conceptual Model of Cognitive-Innovation (CgI-M)* as an Archetype that has formal support that consists of the *Systematic Process for Knowledge Management (KMoS-REload)*; which formalizes the interaction between an *ahCN*, a *Cognitive Architecture*, and the *CgI-S* implementation process or particular treatment. The remainder of this chapter is structured as follows: In Section §2, sensitive concepts and related work to the subject are described. A general proposal *Conceptual Model of Cognitive Innovation* is presented in Section §3 where the *ad hoc Collaborative Network*, the *Cognitive Architecture*, and the dynamics of the *Systematic Process for Knowledge Management (KMoS-REload)* and its main characteristics are also presented. As an application, Section §4 introduces the start-up of the *KMoS-REload* process through a client study to describe the benefits of using the *Conceptual Model of Cognitive Innovation* and then presents the results of this study. A brief discussion is given in Section §5. Finally, the conclusion and future challenges are presented in Section §6.

2. Sensitive Concepts and Related Work

In the present section, concepts and works are pointed out, both related to the model that is made known in this article, which is essential and summarizes the support towards the establishment of the archetype.

2.1 Informally structured domain

A situation wherein an individual or company must face a challenge of adaptation to the cognitive era is treated in this chapter as a problem or need, and can consist of how to do it (processes), the incorporation of technologies, or both. Generally, who suffers from a situation, problem or need, belonging to the cognitive era, is aware of this situation but does not have the time, ability, or knowledge to determine the nature of the problem and less to give the appropriate treatment or implement actions that resolve it because the activities related to the dynamics and environment of the problem are constantly changing, which implies that the problem cannot be stopped. The organization and processes of such activities could be carried out in acceptable conditions, but to survive in the current environment, innovation is required. This innovation must start from the fact that there is no a knowledge-base where knowledge is formal and explicit, which generates gaps between the dynamics of processes and, even the communication between them. The knowledge of the environment is uncertain, ambiguous, and only some decision-makers and specialists in the domain have it but incomplete and with different degrees of specificity. The *Conceptual Model of Cognitive Innovation (CgI-M)* general proposal establishes how to deal with the situations or needs, mentioned above, through particular treatments in dynamic environments of the *Informal Structure*

Domain (ISD). An *ISD* is a complex domain that can be described by characteristics of how its data, information, and knowledge are, and how are the representation and communication between them in the following way:

- heterogeneous data and information; specialized knowledge with a high degree of informality, partial and non-homogeneous; and
- knowledge that is mostly tacit and without structure.

Besides, the *ISD* interacts with an *ahCN* that must understand the problem, need or business, identify application opportunities and obtain the knowledge requirements of this intricate knowledge

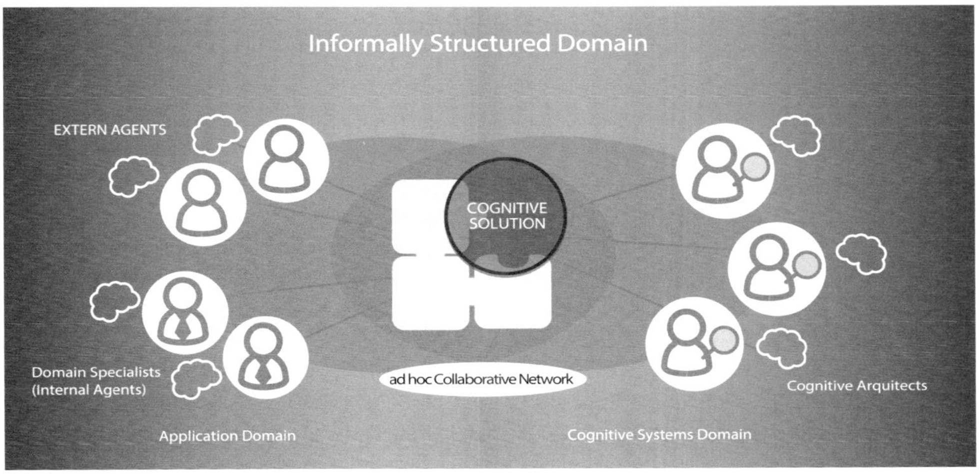

Figure 1: An overview of Informal Structured Domain's Eco System.

ecosystem to propose a convenient, viable and valuable *CgI-S*. In Figure 1, an *ISD* is characterized by exemplifying the context or environment of whoever requires a *CgI-S*.

Finally, in the context of the *ISD*, in particular, the *External Knowledge* under the business concept must include the market and consumers; it is very important to understand from the beginning of the user experience under the integration approach of a value chain.

The pace of business development, the amount of data and knowledge they handle from their clients and the need to insist on the concept of strategic adaptation—as opposed to the traditional strategic planning approach—have forced companies to think about a new approach that is different from the traditional "B2B".

We know the importance of the consumer-oriented approach (Business to Client, B2C), we also know how the focus changes when we talk about collaboration between companies (Business to Business, B2B); however, under the current optics of handling artificial intelligence, machine learning, and cognitive technologies, it is necessary to evolve the last concept to a new one: Business to Business to Client (B2B2C).

Under this concept, the need to understand the biometric profiles of final consumers adds an additional element to the appropriate handling of data or knowledge and its impact on the development of more efficient knowledge or predictive models. Companies supplying goods and services to other companies must now insist and collaborate in understanding the factors that motivate the choice of the offer of one company or another. That is, to be able to add value in a value chain, it is now not only necessary to understand the dynamics of the companies that are served, but the factors that motivate their respective market niches.

2.2 Natural cognition process

The *natural cognition process* can be managed as the capacity of some living beings to obtain information about their environment and, from its processing by the brain, interpret it, give it a meaning and further acted upon. In this sense, cognitive processes depend on both sensory capacities and the central nervous system. Therefore, *Cognition* is understood as the processing of any type of information through mental functions. It is important to mention that this conceptualization is derived from the traditional separation between the rational and the emotional; however, today emotion is seen as a cognitive process too. Thus, the faculties that makeup cognition are multiple from attention, language, metacognition (or knowledge about one's cognition) to emotion:

- Perception. This is a cognitive process by which a mental representation of this information and its interpretation is generated. The process includes the capture of stimuli from the environment by the sensory organs and their transmission to higher levels of the nervous system.
- Attention. It is a general ability to focus cognitive resources, such as selection, concentration, activation, monitoring or expectations, in stimuli or specific mental contents; therefore, it has a regulatory role in the functioning of other cognitive processes.
- Learning and memory. Learning is defined as the acquisition of new information or the modification of existing mental contents (together with their corresponding neurophysiological correlates). Different types of learning have been described, such as classical and operant conditioning models, which are associated with synaptic potentiation mechanisms. Memory is a concept closely related to learning, since it covers the coding, storage and retrieval of information. In these processes, key structures of the limbic system are involved, such as the hippocampus, the amygdala, the fornix, the nucleus accumbens or the mammillary bodies of the thalamus.
- Language is the faculty that allows certain living beings to use methods from simple to complex communication, both oral and written.
- Although emotion, traditionally treated separately from cognition, is understood as equivalent to thought, it has been established that these two processes work in a similar way both at the level of the sympathetic nervous system and the motivation to approach or move away from a stimulus.
- Reasoning and problem solving. Reasoning is a high-level cognitive process that is based on the use of more basic ones to solve problems or achieve objectives around complex aspects of reality. There are different types of reasoning according to how we classify them; if it is done from logical criteria, there is a deductive, inductive and abductive reasoning.
- Social cognition. It includes several models that integrate the theories of attribution and the theory of schemes on the representation of knowledge.
- Metacognition. The faculty that allows one's own cognitive processes to be known, by themselves, and reflected upon. Special attention is paid to metamemory, since the use of strategies to improve learning and memory is very useful to improve cognitive performance.

2.3 Innovation process

Innovation is the change or series of changes, however minimal, through inventions that will always improve and add value to a product, a process, or service. The innovation can occur incrementally or disruptively and has to be tangible in the same process, or product or service resulting from it. Therefore, an *innovation process* is important due to:

- Opportunities for problem-solving: When innovation is fostered, brainstorming arises from attempts to solve existing problems or needs.

- Adapting to change: In the technological world, where the environment changes drastically, change is inevitable and innovation is the means, not only to keep a company afloat, but also to ensure that it remains relevant and profitable.
- Maximization of globalization: Innovation is a necessity to solve the needs and challenges and take advantage of the opportunities that open markets around the world.
- Be in competition: Innovation can help establish or maintain the vanguard of a company, compete strategically within a dynamic world and make strategic moves to overcome the competition.
- Evolution of the dynamics of the workplace: Innovation is essential for the use of demographic data in the workplace, which constantly change, and ensure the proper functioning of the product, service or process.
- Knowing the changing desires and preferences of clients: Currently, clients have a wide variety of products, services or processes at their disposal and are well informed to make their choices. Therefore, it is imperative to keep up with changing tastes and also forge new ways to satisfy the clients.

3. Conceptual Model of Cognitive Innovation Archetype: A General Proposal

In this section, the *CgI-M model* is presented as a starting point to achieve an archetype: an original mould, or pattern, that links the process where elements or ideas intervene in order to establish an architecture to support a *Cognitive Solution*, a *Cognitive Innovation*. The archetype proposes models of knowledge representation that include experiences, behaviours and ways of thinking, collectively shared, which are constructed from theoretical tools—by imitation or 185 similarity to human ones—assimilating and codifying knowledge, defining the existing relationships between concepts of the *Informally Structured Domain* given. Consequently, the archetype produces a physical or symbolic *solution*, tangible or intangible things or processes, or parts of them, that could generate something more of themselves. Figure 2 shows a general outline of the *CgI-M model*.

It is important to indicate that the *CgI-M model* is open, constantly revised, enriched, updated, and that it is currently implemented as a *modus operandi* of a *Cognitive Architects team* to build *Cognitive Solutions*. Subsequent subsections give a review of the parts of the model.

3.1 ad hoc collaborative network

As previously pointed out, *cognitive innovations* use *creative experience* or *specialized knowledge* from various sources that can be entities, agents, systems, etc., who possess knowledge or information, and together set up an *ad hoc Collaborative Network (ahCN)*. An *ahCN* is compound by a triplet where:

$$ahCN = (IK, IS, EK) \tag{1}$$

Equation (1): The *ahCN* is equivalent to a triplet compounded of knowledge or information sources, internal or external, from a given domain; where:

- Internal and External Knowledge (IK & EK) are pieces of Knowledge or Experiences that are present in the Informally Structured Domains (ISD) and that compound a set of abstract representations with the useful purpose of solving or addressing something that happens in their environment that they are stored through experience or they are acquired through the senses. Such pieces are obtained, or come, from Internal or External Agents who belong to different fields of the same domain, and they know about the problem, their environment and the actions that must be carried out in it and they usually can be: specialists, decision makers,

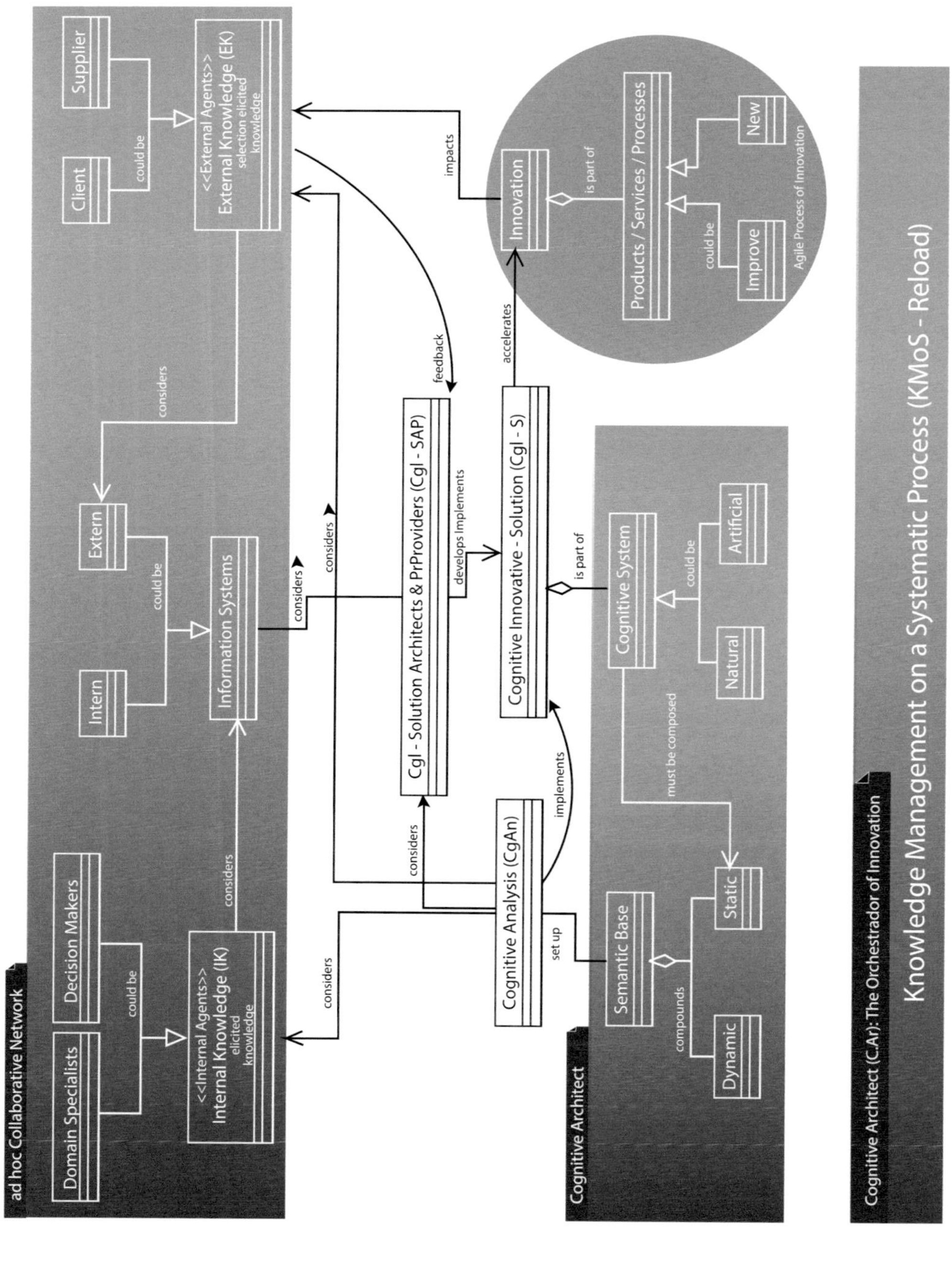

Figure 2: A general outline of the Cognitive and Innovation Model.

stakeholders, competition, workforce, clients, knowledge requirements engineers, cognitive engineers, cognitive architects, etc.

- Information Systems (IS) is the information or data from a system what can be understood as a set of components that interact to collect, organize, filter, process, generate, store, distribute and communicate data. The interaction occurs with a certain number of users, processors, storage media, inputs, outputs and communication networks. A system with access to selected data clouds, databases or research sites about a given domain; and it is important to emphasize that:
  1. the data, information or knowledge sources of the *ahCN* can be largely autonomous, geographically distributed and heterogeneous in terms of their operating environment, culture, social capital or goals, but they work in close collaboration to achieve the best common goals or, at least, compatible ones, and whose interactions could be internal, external or both to ensure proper functioning of the *ahCN* [3];
  2. the Knowledge or Experience belonging to agents, from different fields, is capitalized in the *Cognitive/Innovative Solution* (*CgI-S*);
  3. the Pieces of *Internal Knowledge* (*IK*) are considered as the foundation of the solution and the Pieces of *External Knowledge* (*EK*) are considered as the feedback the solution and influence, motivating the *CgI-SAP team*, to provide the best solution. It is important to note that some *EK* pieces come from neuroscience, biometric profiles, often trivialized by Artificial Intelligence, but generate updated perceptions from the user's evolutionary experience, traditionally presented as insights.

3.2 Cognitive architecture

In this *cognitive era* of surprising changes that have taken place in an extremely fast period, the idea of *Cognitive Architecture* must be properly delimited. Therefore, the authors consider it convenient to achieve the homogenization of concepts or paradigms, related to the cognitive field and hypotheses about the nature of mind, among those who work in this area. The task is hard because every day something new arises, but it is worth trying to go in the same direction. Consequently, we agree with [4] when they point out that *Cognitive Architectures* are hypotheses about fixed structures, and their interactions, intelligent behaviour and underlying natural or artificial systems. In essence, a *Cognitive Architecture* must have a *Semantic Base*, derived from a *Cognitive Analysis*; which, in turn, is the essential component of the *Cognitive System* that must support a *CgI-S*.

Semantic Base. The semantic base formalizes, through a consensus, the relationships between concepts or terms and their attributes belonging to the domain related to *CgI-S*. The terms are registered constituting knowledge through an extended lexicon (KDEL) that classifies them into objects, subjects and verbs and is based on LEL [5]. The externalization of this knowledge allows the achievement of a consensus among the interested parties and, consequently, minimizes the symmetry of ignorance. The concepts and relationships identified generate a matrix called *Piece of Knowledge* (*PoK*). It also facilitates the construction of a conceptual graphic model that provides a visual medium for the semantic base of the domain and facilitates its validation, where an entity-relationship model can be used. Generally, after forming a semantic base, it is common to find that a good amount of terms used in the domain are ambiguous, are not unified and are particular to those who use them. It is important to bear in mind that, although the domain specialists validate the description of the concepts of the lexicon, the graphic conceptual model provides a very complete description of the knowledge of the domain that allows domain specialists to identify possible errors and what lacks in the semantic base; particularly, between the relations of the concepts. This is very important since this model is essential for the design of a *Cognitive Architecture*.

Cognitive System. Set of entities, definitions, rules or principles that when interrelated in an orderly manner contribute to formalizing a cognitive process, at least the irreducible set of components that are used to explain or carry it out.

3.3 Knowledge management on a systematic process (KMoS-REload)

The Systematic Process for Knowledge Management *KMoS-REload* (Figure 3, all details in [5]) is specially designed to interact with *Informal Structure Domains* (*ISD*), supporting the *Cognitive Analysis*, and provides a formal procedure for obtaining, structuring and establishing formal knowledge relationships that serve as a guide for the cognitive architect to: (a) integrate the *Cognitive Architecture* that supports a *Cognitive and Innovative Solution* and avoid ambiguity, incompleteness and inappropriate links between pieces of knowledge in the context of a given *Informally Structured Domain*; and (b) coordinate and operate the *CgI-M model*.

In particular the process performs three sequential phases:

1. Conceptual Modelling Phase which models the *CgI-S* 's domain using a linguistic model and a graphic conceptual model;
2. Strategic Model to visualize the general functionality of the *CgI-S* 's domain; and
3. Tactical knowledge phase, which is in charge of obtaining, discovering, giving structure and enrichment to the knowledge of the *CgI-S*.

In addition, cross-cutting activities are included to identify *tacit knowledge*, and, once this knowledge is explicit, the wrong beliefs are recorded and the relationships between the concepts and their behaviors are traced. Three activities complement the models used in the process:

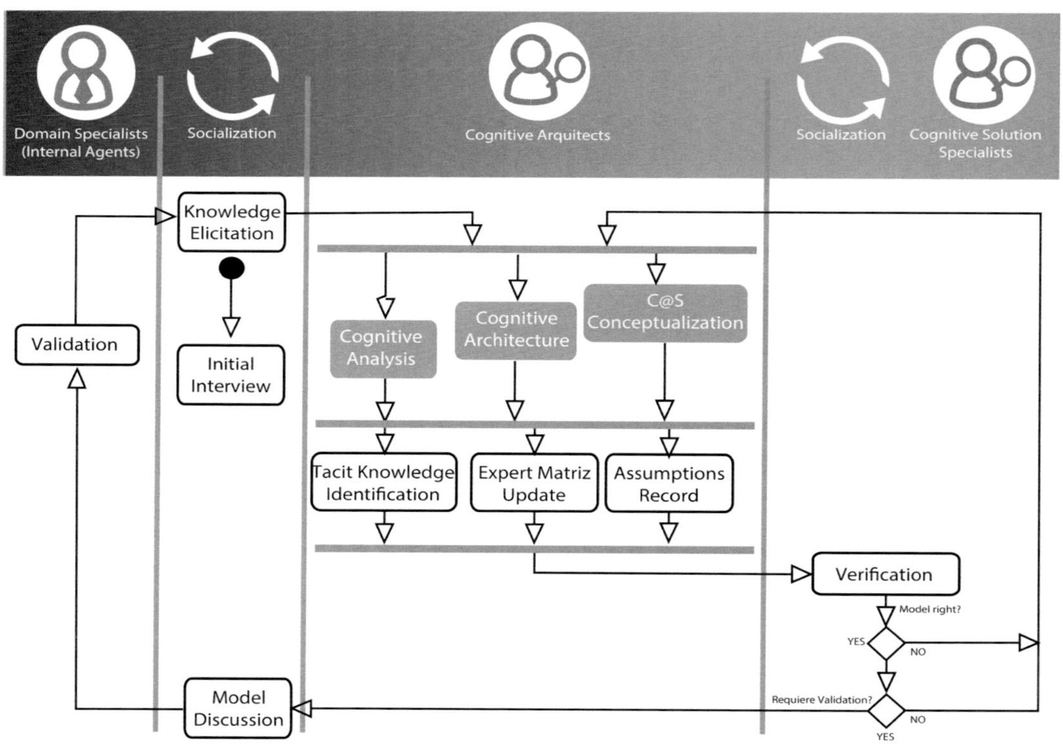

Figure 3: General overview of the *KMoS-REload process* is represented by activities flow diagram.

1. The tacit identification of knowledge, where a discourse analysis is carried out with the objective of identifying the knowledge that is hidden behind the linguistic indicators as presuppositions;
2. The capture and updating of specialized knowledge of the matrix, throughout the process, knowledge is associated with those involved in the domain, which forms a matrix that captures experience in the domain; and
3. The assumption record, a phenomenon that occurs when learning a new domain is to associate our mental scheme with new concepts and relationships, therefore, false assumptions will be made clear as the process progresses, these assumptions must be recorded to facilitate the learning of new members in the project.

The process begins with an initial interview between the Solution's Architects and Providers (*CgI-SAP team*) and the *Internal or External Knowledge* (Domain Specialists) in a session where socialization predominates. Then, the *Tacit Knowledge Identification*, the *Expert Matrix Update* and the *Assumptions Record* are developed in parallel by *C.Ar* (or the *CgI-SAP team*)—a *Cognitive Analysis* is done by them in a socialized way—in order to verify the artifacts and decide if they should continue with the following phases or require a validation of them. In fact, under a lean or agile innovation approach, living iterative processes exist that allow adding value from the validation of their elements and proposals. The validation requires that the *CgI-SAP team* explain the models in order to validate the knowledge. The process-in-turn generates more knowledge, then the cycle starts again, and the process may end when all those involved in the *CgI-M Model* reach an agreement. Finally, the process makes the team aware that, in order to develop a *CgI-S*, it is necessary to understand and formally define the knowledge requirements and the domain that circumscribes them [6]. The details of the *KMoS-REload* process application can be found in [15].

3.4 The cornerstone of the cognitive/innovative solution

What is a solution? In the *CgI-M model* context, a solution means solving a situation, problem or need of an individual or company (the client), through the experience and talents of a highly specialized team of people. Despite the fact that the concept of a solution is simple, the cornerstone of a *Cognitive/Innovative Solution* (*CgI-S*) is the result of processes and actions, obtained collaboratively.

Cognitive/Innovative Solution. The *CgI-S* is the result, given by *Solution Cognitive Architects and Providers* (*CgI-SAP team*), of solving a problem or cognitive need taking into account the connections and relationships of the models obtained from *Cognitive Analysis* (*CgAn*), making use of the *Internal Knowledge* (*IK*) and *External Knowledge* (*EK*), and any other feedback from the *Informal Structure Domain*. Thus, *CgI-S* can be represented as a function of three parameters that can be represented by the Equation (2).

$$CgI - S = F\ (CgAn,\ IK,\ EK) \tag{2}$$

Equation (2): The *CgI-S* is the result of the development and implementation function carried out by the *CgI-SAP team*.

At this moment, it would be convenient that two concepts are in mind: *Open Innovation* and *Corporate Venturing*. Today, companies are learning that their innovation models and proposals can find more value—and much faster—if they find a way to integrate the approach and proposals of their potential clients into their innovation models, by the way, users of their technologies can usually express their needs more easily with respect to technology itself. Today, reorienting its research and development efforts, originally armoured towards an open innovation approach that includes the multiplied vision of its clients, adds much greater innovation potential and more variation. Companies such as *Telefónica* are leading worldwide collaboration initiatives such as these; the term that has been coined to name this type of effort is that of "Corporate Venturing", where companies allocate resources to encourage start-ups or small businesses to develop new concepts, indeed much more economically accessible.

Cognitive Innovative-Solution Architects & Providers (*CgI-SAP*) *team*. This is a team of human talent that performs consultation and analysis of information technology systems, intelligent and cognitive. The *CgI-SAP team* supports all its activities, within the *CgI-M model*, in a *Systematic Process for Knowledge Management* (*KMoS-REload*) to develop cognitive and, therefore, innovative solutions that bring great value to clients. It is well known how engineers or scientists become obsessed with past solutions and how the process of scientific discovery and the engineering design process can lead them to new solutions.

However, there is still much to understand about the cognitive and innovative processes, particularly with respect to the underlying natural cognitive processes. Behind the *KMoS-REload process,* there are theories and methods of several disciplines related to cognition and knowledge, such as cognitive psychology, social psychology, knowledge representation, machine learning to analyze, structure and formalize the complex cognitive processes that occur in the real world, the world of the *Informal Structure Domains*. It implies that the *CgI-SAP team* is highly trained to be empathetic and solve problems of a given *Informally Structured Domain*. Consequently, there are two essential roles carried out by this team: as an architect of solutions, the team must have a balanced combination of technical, social and business skills; as a supplier, the team must offer solutions based on any combination of technologies, processes, analysis, commercialization, internal organizational environment or consulting. Such solutions can be customized for your clients; or, it can provide solutions based on existing products or services.

Regardless of the roles played by the *CgI-SAP team*, the core of its activity is the interaction with the elements of the triplet of Equation (2) and applying science and technology advances to take advantage of all the knowledge that exists around to achieve the *Capitalization of Experience or Knowledge* and to provide a *CgI-S*. It is undeniable that the above represents a complex situation [1, 2], but also an excellent opportunity for the *CgI-SAP team*.

Cognitive Analysis (*CgAn*). The *CgAn* is a process of examining in detail a given *ISD* in order to understand it or explain it. Commonly, one or several strategies or processes are used that enable the knowing and formalizing of the existing relationship between certain types of functions, actions and concepts related to this domain. The main objectives of performing the *CgAn* in a given *ISD* are:

(a) to obtain the best view of your own internal processes, e.g., in a business domain this could be how the market receives its products and services, customer preferences, how customer loyalty is generated or other key questions where precise answers are used to provide a company with a competitive advantage; and
(b) to set up the cognitive architecture established by the semantic base and the components of the appropriate cognitive system.

It is worth mentioning that the *CgAn* often focuses on the realization of a predictive analysis, where the extraction of data and other cognitive uses of the data can generate business and commercial predictions. Therefore, the practical problems surrounding such analyses involve the precise methods used to collect and store data in a special location, as well as the tools used to interpret this data in various ways. *Solution Cognitive Architects & Providers* can provide analysis services and other useful help, but in the end, the practical use of the analysis depends on the people who are part of the domain, where they not only need to know how to collect data but also how to use it correctly.

3.5 Agile process of innovation

The high dynamism and constant change of the world and its markets require that innovation is contained in an agile, continuous, cyclical and constant process of changes and adjustments where the *CgI-S* frees time from the process actors so that they focus on supervisory activities and that can agilely search for new products, services, internal processes or improvements, adaptations or updates to existing ones. Currently, a "complete study of x-ray + computed tomography + magnetic resonance imaging" of the client's environment or its *ISD* is required to identify areas of opportunity and map the process, know the products and services to clarify and be assertive in the client's vision and goals. From the beginning of the *KMoS-REload* that will implement the *CgI-S*, through the *CgAn*, this "complete study" starts and the client will become aware of the intangible good that will be obtained. The *Cognitive Architecture*, since it is being formed, is offering the client content and tentative activities to be carried out. It is necessary to highlight that at the beginning it is impossible to detail the components of the architecture to the minimum since a given *ISD* is unknown and, in the same way, the end of the process is relative since it depends on the client's satisfaction concerning its environment.

As the environment is changing, the cognitive architecture and, therefore, the solution could/must change, that is, innovate. The concept of agile innovation implies an organizational culture that is prepared with the necessary technological architecture to be at the forefront, but also with the appropriate mentality to assimilate the exhausting challenge of permanent change. That is to say, it is useless to have an environment full of cutting-edge technology, when the mentality of the organization remains anchored in past paradigms of work in silos, focused on particular objectives and leading profiles.

Finally, the expertise of real solutions implementation indicates that innovation is implicitly presented—however marginally—and even more, it accelerates the cyclical process of innovation whose impact can occur as an improvement of Products, Services or Processes; or, the generation of new ones.

4. FLUTEC: A Client Study

FLUTEC worldwide company—located on the US-Mexican Border (Juarez City)—designs, builds and sells Heating Ventilation and Air Conditioning (HVAC) modules tailored to meet the particular needs of its clients; that is, each module could be similar but not identical. In fact, a build-to-suit approach for every project makes for a high-cost project.

To find greater benefits from a project requires the improvement of the process to carry out it. The HVAC project process starts when a client issues the basic specifications for its design and ends with the delivery of it. Therefore, it includes a *mare magnum* of aspects to take into account when carrying out a project and, consequently, an erroneous decision directly impacts the time of the general process and even its viability. In addition, the dynamism of the HVAC's singular market motivates the company to find greater benefits, and at the same time obliges it to continuously improve its processes, especially the delivery time of the project budget, the time and the quality of the design process.

Is the implementation of the CgI-M model convenient? The company, and all its processes, are subjected to the HVAC's market to innovate continuously; otherwise, take a risk that compromises the survival of the company. Besides, all characteristics of the *ISD*, indicated in subsection §2.1, are present in FLUTEC environment. Therefore, the HVAC's domain specification can be listed as:

- Seven main processes at least required to achieve it: *Heating, Cooling, Humidifying, Dehumidifying, Cleaning, Ventilating, and Air Movement*;

- Five complex tasks that include each one's activities: *Basic specifications establishment, building characteristics analysing, Air circulation patterns analysing, Appropriate components selection, and Control system analysing*;
- Non-organized and incomplete data is present in it;
- Determination of the criteria and decision making about the achievement of the project is carried out under the umbrella of an *ahCN*; and
- The project has a *unique design* and solves or addresses a *particular situation*.

To deal with the challenges of obtaining the *knowledge requirements* of an HVAC project, typified as belonging to an ISD, the company FLUTEC uses an empirical guide—*DNA document*—composed of general attributes that gather the necessary basic information for each project. This document should be a guide to obtain the knowledge requirements that would allow a good design of an HVAC module. However, being an empirical and, therefore, informal document, it was a very flimsy communication bridge between the *ahCN* within FLUTEC. In addition, there were often delays, reworkings and high-cost problems arising from this *DNA* document and the additional processes of the FLUTEC's processes related to the realization of a project.

Characterization of the CgI-M model through the determination of the peculiar attributes and additional activities related to the HVAC project. Once the Flutec's environment relative to the HVAC's design process has been identified as an *ISD* domain, the *Cognitive Architect* starts the *KMoS-REload* process to characterize and, consequently, establish the *CgI-M model*:

- *Distributed Tacit Knowledge*: Tacit distributed technical knowledge, heterogeneous, diverse degrees of specificity;
- *Incomplete data*: Unorganized and Incomplete Data of all the processes related to the HVAC and that should be used in the development of the *Cognitive Architecture Specification*;
- *ad hoc Collaborative Network*: composed of multiple specialists in the Flutec's domain, the *CgI-SAP team*, decision-makers; and
- any other problems, in particular, must always be addressed when developing an HVAC; therefore, it is a unique project that requires a *CgI-S*.

Results of the use of the CgI-M Model. In order to provide FLUTEC with an adequate *cognitive solution (CgI-S)*, the *CgI-SAP team* identified the elements of the HVAC's process that needed to be improved and established a consistent model that would give it the corresponding support from the following:

- *Analysis of the DNA guide document*. As mentioned above, the DNA guidance document is empirical and lacks the overall vision of the project. Therefore, the analysis should describe the significant assumptions and conceptual relationships of all FLUTEC knowledge. Consequently, the DMP confirmed that the DNA had the following deficiencies:
    - Disorganization;
    - Incomplete: Missing essential information for the proper development of the HVAC project;
    - Incorrect: Existence of fake attributes;
    - Irrelevant information: Informal descriptions had been recorded;
    - Ambiguous information. The initial and basic knowledge requirements were not well described; and
    - Time lost due to searching, as often as necessary, of missing or poorly recorded information.

The analysis allowed the obtaining of knowledge, the formalization of *empirical DNA* domain, and its transformation into a *new and formal solution*.

- *Specialized Explicit Training*. Before applying the *KMoS-REload process*, it was already difficult for FLUTEC engineers to understand the importance of a formal process of obtaining knowledge requirements and, consequently, there was great ignorance about certain elements or concepts belonging to the domain of the project. Once the process was used in the project, the FLUTEC's specialists were trained by the domain modelling phase and were able to assimilate (make tacit) new explicit knowledge, reduce their own ignorance and ambiguity and, as a result, improve the quality of work in *ahCN* learning that:
  - *Knowledge-Requirements Elicitation* could be carried out systematically;
  - The *CgI-M Model* transfers knowledge; and
  - FLUTEC has *preconceived and tacit ideas or expectations* of the project and when they turn *explicit* the redesigns are usually avoided on the project's post-delivery time.
- *Improvement of HVAC-DNA Process*. The *CgI-M Model* was carried out through *KMoS-REload process*, as a result, the models will be established and, therefore, the HVAC-DNA process will be renewed:
  - HVAC project concepts, attributes, relationships between concepts and basic integrity restrictions were formalized, e.g., HVAC design and budget project properties. The externalization, transfer, and consensus are activities carried out within the ahCN with its knowledge in order to integrate a set of *pieces of explicit knowledge* that minimizes the symmetry of ignorance. Thus, the learning curve about the HVAC domain was reduced from a couple of months to a couple of weeks. In addition, the *CgI-SAP team* noticed that the DNA document was not useful during the project process, especially, since it requires a lot of time to be filled and does not meet the goal it is supposed to achieve.
  - The process view as a stream of decisions from the *ahCN* allowed the *CgI-SAP team* to obtain a *Cognitive Architecture* with support of the *KMoS-REload process*.
  - The set of *knowledge-requirements* are derived and integrated into the *CgI-S's specification document*.

A CBR to support a fast delivery of proposals. The *cognitive architecture* from the knowledge, acquired and managed, also allowed to constitute as a part of the *CgI-S*: a robust *Case-Base*, textual files and all necessary to implement a *CBR prototype* in the jCOLIBRI tool [7, 8]. This tool provides a standard platform for developing CBR applications through specialized methods using several Information Retrieval or Information Extraction libraries as Apache Lucene, GATE, etc. Thus, an important goal achieved by the CBR was to demonstrate if FLUTEC could do a time reduction in obtaining the matches between the expectations of the clients with the HVAC project design blueprints and, consequently, a fast delivery of budget proposals.

In summary, the establishment of an adequate cognitive architecture, using the *KMoS-REload process*, manages to capitalize the *knowledge* of the *ahCN* and its expertise, explicitly and formally, to allow: a clear understanding of the *project's ISD*; its assimilation by the *CgI-SAP team*; give a *CgI-S*; and characterize, as a whole, the *CgI-M model*—to a total customer satisfaction—whose remarkable products were a new DNA guide and the CBR prototype.

5. Discussion

We live in a world that changes minute by minute, for better or for worse, due to advances in science and technology strongly framed in artificial intelligence, machine learning, and cognitive computing. The assimilation of the advances is not a trivial issue and, consequently, the companies, the individuals, and the society must find the way to survive at the great speed with which "the future and the present are amalgamated". This issue is not trivial because changes that happen too quickly can often produce a disconnection between a scientific or technological advance and the understanding of its potential by the providers of technological solutions. Scientifically and technologically speaking, there are many examples throughout history of wrong judgments about what the future holds. For example, when business owners introduced electricity into their factories, they stayed with older models of how to organize the use of their machines and industrial processes and lost some of the productivity improvements that electricity enabled. In 1977, the president of Digital Equipment Corporation, the largest computer company of that time, saw no market for home computers. Thirty years later, in 2007, Steve Ballmer, CEO of Microsoft, predicted that the iPhone would not take off [9]. From these examples, it is possible to infer that there are essential reasons for justifying the investment of time, money and effort required to develop a successful bridge to knowledge and technology.

In addition, the problems or needs that belong to an *Informally Structured Domain* must be solved by a solution that will come to innovate, either because it modifies the procedure to address the problem or by itself is a new solution. So, this innovative solution will be cognitive because, in order to obtain it, it must have extracted persistent knowledge; or, of the existing cognitive process or that the solution by itself is of the scope of the Artificial Intelligence, the Machine Learning or Cognitive Computation.

There will be occasions when the problem or need can be addressed without further ado by a product or tool of Artificial Intelligence, Machine Learning or Cognitive Computing; or, in most cases, the cognitive solutions will be tailored to the situation of problem or need. Trivializing tasks of modeling and analyzing the problems derived from "time pressure" or the well-known phrase "we no longer have time" can translate into a real loss of time, making bad decisions and opportunities that are going away or that will never come.

Who can identify the right type of solution for each real situation that arises?

In *Cognitive Architecture*: *Designing for How We Respond to the Built Environment*, Ann Sussman and Justin B. Hollander review the novel trends (2014) in psychology and neuroscience to support architects and planners, of the current world of construction, to better understand the clients, like the sophisticated mammals they are, and in response to a constantly evolving environment. In particular, they describe four main principles that observe a relationship with the cognitive processes of the human being: people are a thigmotactic species, that is, they respond to the touch or surface of external contact; visual orientation; preference for bilateral symmetrical forms; and finally, narrative inclinations, unique to the human being. The authors emphasize that the more we understand human behaviour, the better we can design for it and suggest the obligation to carry out activities of analysis, "preparation of the cognitive scaffolding", before carrying out construction and anticipating the future experience of the client [10].

Similarly, Portillo-Pizaña et al. suggest the importance of considering four stages for the implementation of a process of conscious innovation in an organization: *consciousness, choice, action,* and *evolution*. This conscious process of corporate innovation initially implies that every human being who integrates an innovation effort within an organization understands that any process of change or transformation begins with a state of consciousness where the existing gaps between the current situation are identified and the desired situation under the perspective of a user; subsequently, a decision-making process must be faced that allows an agile iteration, in order to move on to a real commitment to innovation and entrepreneurship—with all the necessary characteristics required of a true entrepreneur—and conclude then with a disposition to agile evolution, without sticking to ideas that are not well received by the market [11]. Thus, to identify the right kind of *cognitive solution* in a real situation it is highly convenient to have a *cognitive architect*. A *Cg.Ar* is a role with multidisciplinary knowledge in areas that should be treated as they are: Artificial Intelligence, Machine Learning, Cognitive Computing, logic, cognitive processes, psychology, sociology, philosophy to mention a few.

5.1 Why the need for a cognitive architect?

In the *Cognitive Industry*, development projects usually require the knowledge and understanding of psychology, artificial intelligence, computer science, neuroscience, philosophy, linguistics, anthropology. These are important disciplines that deal with the process of creating the cognitive basements and structures on which the new schemes of how to go about thinking of innovative products, solutions and things are based.

These disciplines have critical functions that are essential in a *cognitive architecture* job and they rely on one-another to accomplish a given cognitive task.

Currently, many people believe that there is no difference between Computer's Science or Systems and *Cg.Ar* and, even though there are similarities in some subjects, there are clear differences between them. As a matter of fact, they have well-defined roles that make them distinguishable from each other. Both Computer's Science or Systems and *Cg.Ar* are involved in programming and designing apps. However, *Cg.Ar* focuses more on planning and designing cognitive structures, elements and collation of the development work and is more concerned with the knowledge elicitation, management, modeling and functionality of the design, making certain that those structures can support normal and extreme cognitive demands.

Even though computer science or systems engineers are involved in the design process of software solutions, architects take the lead role in terms of the design of the structure. The *Cg.Ar* will initiate and create the design, including the Knowledge Requirements, Cognitive Modelling and processes of the development work, then computers science or systems engineers professionals, when *Cognitive Solutions* will be software, will analyse it to find ways to make the software design possible. The computer engineers could be responsible in finding suitable intelligent algorithms, suggesting modifications and adjustments and evaluating the structural integrity to transform the cognitive architect's vision into realization. To summarize, cognitive architecture's primary concern is making very good models from cognitive blueprints and designing the development work while the computer engineering's responsibility is ensuring that everything that is foreseen in the cognitive blueprints can be implemented functionally and reliably. Computer Engineers and Cognitive Architects may sometimes overlap each other's work but a good relationship between the two professions will make the cognitive software solution job more effective and successful. Today, computer engineers make a point to work harmoniously with *Cg.Ar* to ensure superior quality results and proper design implementations for all stakeholders because they understand that teamwork and cooperation are vital to the success of any cognitive project.

Finally, to highlight the work of the *Cg.Ar* in three words it can be said he is an *orchestrator of innovation* (see subsection §3.5).

5.2 Why the need to establish a model to support a cognitive architecture?

When the *CgI-SAP team* faces a problem and must design and implement a *CgI-S*, it must have to appropriate a reality, which, in initial conditions, overcomes their capacity and comprehension. Therefore, it is very convenient to have a model that simplifies this reality: *CgI-M*. This model will be as detailed as it is necessary to offer a global vision of the *CgI-S* environment. Thus *CgI-M* allows a better understanding of the *ISD* to which *CgI-S* belongs with the aim of:

- visualize *what is* or *how it should be* a solution;
- establishes a guide to specify the structure or behaviour of the domain in relation to an implementation of a possible solution; and
- document the decisions and actions that are carried out.

CgI-M is useful for solving problems with both simple and complex solutions. As a problem becomes more and more complex, it's possible solution will also be, therefore, the *CgI-M model* becomes more important for a simple reason: it seems that when a model is proposed for complex situations, for example, to build the scaffolding of a cognitive architecture, it is due to our inability to deal with complexity in its entirety.

In the meantime, *CgI-M* reduces the complexity of what is being addressed, focusing on only one aspect at a time. It is important to highlight that *CgI-M* intends to formalize, because if informality were allowed, it would generate products that could not clearly address the domain of the solution to be implemented.

In conclusion, and in spite of some solution providers, the more complex the domain and the problem to be addressed, the more it is imperative to use a model. Solution providers are already faced with situations where they have developed simple *CgI-S* without starting from any model and, after a short time, notice that the domain grew in complexity nullifying the effectiveness of the solution to the detriment of the quality of its service and the loss of the client.

Finally, the *CgI-M model* after being used in real cases that its components as a whole can, de facto, respond through a cognitive collision to situations that occur within the domains of informal structure. There is a lot of work to be done on the subject of obtaining and representing common-sense information; for existing frames of representation must evolve and be integrated with other frameworks in order to enhance representation and, consequently, reasoning with common sense information. In general, the results obtained by *CgI-M* suggest that the knowledge obtained from it is highly congruent with that expressed by *ahCN* when validated by the client and the results of the solutions provided by it.

However, it is also clear that it is not possible to explain the complete cognitive process of *ahCN* exclusively in the current terms of the *CgI-M model*. Consequently, the model is open and dynamic for the improvement of its components and to better explain the harmonization and integration of different types of cognitive processes that are supposed to coexist in a perspective of heterogeneous representation, for which additional research and collaboration among those we approach is needed.

In particular, in our opinion, such improvements should be oriented to (i) in which cases the components of the *CgI-M model* play a more relevant role in establishing the scaffolding necessary to develop a particular cognitive solution (ii) or cases where they are not at all evoked by a cognitive system, since the need to react in real time is more urgent and, therefore, (iii) accelerate the activities proposed by the model. Since there is no clear answer to such questioning, these aspects will imply, in our opinion and in congruence with [12], the future research agenda of cognitive psychology and the *cognitive—artificial—systems* research.

6. Conclusions and Future Challenges

This paper communicates the convenience of walking in the direction of an archetype that characterizes the essential aspects of *Cognitive Architecture*, namely, what elements make up the *ahCN* and how they interact with each other, the *cognitive architecture* and the activities, tasks that lead out the *Cognitive Architect*.

It was argued that, based on the results of client studies, these aspects should be addressed to formalize and accelerate the establishment of *Cognitive Architecture* with the limitations and challenges that require the daily tasks of a cognitive process. Such challenges, from a technological perspective, are crucial to being addressed in order to be able to operate cognitive solutions and make decisions in general scenarios exploiting a plethora of integrated reasoning mechanisms. Based on these assumptions, we confirm the convenience of integrating a model to deal, jointly, with the aspects mentioned above.

Finally, there are already several crucial problems of real situations that have been addressed by our model, of which one of them was mentioned, where the cognitive processes are harmonized in the *CgI-M*, interacting with an *ahCN*, and reflected in a cognitive architecture that supports to the *CgI-S* implemented by the *Cg.Ar*. The results obtained suggest that, although the systematic process for knowledge management *KMoS-REload* provided by the *CgI-M* represents an adequate way to integrate different knowledge acquisition and representation mechanisms, it is still not clear if they are sufficient and robust. Therefore, it is still an open question of what and what kind of processes, techniques or elements should be part of a general architectural mechanism and if it is worth implementing them in the processes of the model to operate their conceptual structures. As mentioned above, answers to questions or efforts will require a joint research effort on the part of cognitive psychology and the community of cognitive models and processes, cognitive computation, machine learning, and artificial intelligence.

References

[1] Kamsu-Foguem, B. and Noyes, D. 2013. Graph-based reasoning in collaborative knowledge management for industrial maintenance, in: Computers in Industry, pp. 998–1013.

[2] Santa, M. and Selmin, N. 2016. Learning organization modelling patterns. Knowledge Management Research & Practice, 14(1): 106–125.

[3] Camarinha-Matos, L. and Afsarmanesh, H. 2006. Collaborative networks value creation in a knowledge society. *In*: Proceedings of PROLAMAT'06, Springer, pp. 15–17.

[4] Rosenbloom, P., Demski, A. and Ustun, V. 2015. The sigma cognitive architecture and system: Towards functionally elegant grand unification. Journal of Artificial General Intelligence, 7(1).

[5] Rodas-Osollo, J. and Olmos-Sánchez, K. 2017. Knowledge management for informally structured domains: Challenges and proposals. *In*: Mohiuddin, M. (Ed.). Knowledge Management Strategies and Applications, InTech, Rijeka, 2017, Ch. 5. doi:10.5772/intechopen.70071. URL https://doi.org/10.5772/intechopen.70071.

[6] Bjørner, D. Domains: Their Simulation, Monitoring and Control—A Divertimento of Ideas and Suggestions, Vol. 6570 of Computer Science, Springer, Berlin, Heidelberg, 2011, Ch. Domains: Their Simulation, Monitoring and Control—A Divertimento of Ideas and Suggestions.

[7] Finnie, G. and Sun, Z. 2003. R5 model for case-based reasoning, Knowledge-Based Systems, 16: 59–65.

[8] Recio-García, J., González, C. and Díaz-Agudo, B. 2014. jcolibri2: A framework for building case-based reasoning systems, Science of Computer Programming, 79(1): 126–145.

[9] Ito, J. and Howe, J. Whiplash: How to Survive Our Faster Future, Hachette Book Group USA, 2016.URL https://books.google.com.mx/books?id=HtC6jwEACAAJ.

[10] Sussman, A. and Hollander, J. 2014. Cognitive Architecture: Designing for How We Respond to the Built Environment, Routledge. URL https://books.google.com.mx/books?id=3TV9oAEACAAJ.

[11] Portillo-Pizaña, J., Ortíz-Valdes, S. and Beristain-Hernández, L. 2018. Applications of Conscious Innovation in Organizations, IGI Global. URL https://www.igi-global.com/book/appli...organizations/182358.

[12] Lieto, A., Lebiere, C. and Oltramari, A. 2018. The knowledge level in cognitive architectures: Current limitations and possible developments, Cognitive Systems Research 48: 39–55, cognitive Architectures for Artificial Minds. doi:https://doi.org/10.1016/j.cogsys.2017.05.001.
URL http://www.sciencedirect.com/science/article/pii/ S1389041716302121.

PART II

Applications to Improve a Smart City

CHAPTER-8

From Data Harvesting to Querying for Making Urban Territories Smart

Genoveva Vargas-Solar,[1,5] *Ana-Sagrario Castillo-Camporro,*[2,5,*]
José Luis Zechinelli-Martini[3,5] and *Javier A. Espinosa-Oviedo*[4,5]

This chapter provides a summarized, critical and analytical point of view of the data-centric solutions that are currently applied for addressing urban problems in cities. These solutions lead to the use of urban computing techniques to address their daily life issues. Data-centric solutions have become popular due to the emergence of data science. The chapter describes and discusses the types of urban challenges and how data science in urban computing can face them. Current solutions address a spectrum that goes from data harvesting techniques to decision making support. Finally, the chapter also puts in perspective families of strategies developed in the state of the art for addressing urban problems and exhibits guidelines that can lead to a methodological understanding of these strategies.

1. Introduction

The development of digital technologies in the different disciplines, in which cities operate, either directly or indirectly, is altering expectations among those in charge of the local administration. Every city is a complex ecosystem with subsystems to make it work such as work, food, clothes, residence, offices, entertainment, transport, water, energy, etc. With the growth of cities, there is more chaos and most decisions are politicized, there are no common standards and data is overwhelming. The intelligence is sometimes digital, often analogue, and almost inevitably human.

[1] University Grenoble Alpes, CNRS, Grenoble INP, LIG, France.
[2] Universidad Nacional Autónoma de México, Mexico.
[3] Fundación Universidad de las Américas Puebla, Mexico.
[4] University of Lyon, LIRIS, France.
[5] French Mexican Laboratory of Informatics and Automatic Control.
Emails: Genoveva.vargas-solar@liris.cnrs.fr, sagrariocastillo@comunidad.unam.mx
* Corresponding author: sagrariocastillo@hotmail.com

Urban computing [36] is a world initiative leading to better exploit resources in a city to offer higher-level services to people. It is related to sensing the city's status and acting in new intelligent ways at different levels: people, government, cars, transport, communications, energy, buildings, neighbourhoods, resource storage, etc. A vision of the city of the "future", or even the city of the present, rests on the integration of science and technology through information systems.

Data-centric solutions are in the core of urban computing that aims at understanding events and phenomena emerging in urban territories, predict their behaviour and then use these insights and foresight to make decisions. Data analytics and exploitation techniques are applied in different conditions and using ad hoc methodologies using data collections of different types. Today important urban computing centres in metropolises, have proposed and applied these techniques in these cities for studying real state, tourism, transport, energy, air, happiness, security and wellbeing. The adopted strategies have to do with the type of context in which they work.

This chapter provides a summarized, critical and analytical point of view of the data-centric solutions that are currently applied for addressing urban problems in cities leading the use of urban computing techniques to address their daily life issues. The chapter puts in perspective families of strategies developed in the state of the art for addressing given urban problems and exhibits guidelines that can lead to a methodological understanding of these strategies. Current solutions address a spectrum that goes from data harvesting techniques to decision making support. The chapter describes them and discusses their main characteristics.

Accordingly, the chapter is organised as follows. Section 2 characterises urban data and introduces data harvesting techniques used for collecting urban data. Section 3 discusses approaches and strategies for indexing urban data. Section 4 describes urban data querying. Section 5 summarizes data and knowledge fusion techniques. Finally, Section 6 discusses the research and applied perspectives of urban computing.

2. Data Harvesting Techniques in Urban Computing

Urban computing is an interdisciplinary field which concerns the study and application of computing technology in urban areas. A new research opportunity emerges in the database domain for providing methodologies, algorithms and systems to support data processing and analytics processes for dealing with urban computing. These processes involve harvesting data about the urban environment to help improve the quality of life for people in urban territories, like cities. In this context, academic and industrial contributions have proposed solutions for building networks of data, retrieving, analysing and visualizing them for fulfilling analytics requirements stemming from urban computing studies and projects.

Urban data processing is done using: (i) continuously harvested observations of the geographical position of individuals (that accept sharing their position) over time; (ii) collections of images stemming from cameras observing specific "critical" urban areas, like terminals, airports, public places and government offices; (iii) data produced by social networks and applications like Twitter, Facebook, Waze and similar. Independently of the harvesting strategies and processing purposes, it is important to first characterise urban data. This is done in the next section.

2.1 Urban data

Urban data can be characterized concerning three properties: time, space and objects (occupying urban territories). They are elementary properties that can guide the way urban data can be harvested and then processed for understanding urban phenomena. For urban data, time must be considered from two perspectives, as its mathematical definition as a continuous or discrete linearly ordered set consisting of time instants or time intervals, called *time units* [3]. But, also under a cyclic perspective to consider iterations of seasons, weeks and days. Regarding *space*, it can be represented by [3] different referencing models: coordinate-based models with tuples of numbers representing the distance to certain reference points or axes, division-based models using a geometric or semantic-based division of space, and linear models with relative positions along with linear reference elements, such as streets, rivers and trajectories. Finally, the third urban data property, object, refers to physical and abstract entities having a certain position in space (e.g., vehicles, persons and facilities), temporal properties, for objects existing in a certain period (i.e., event), and spatio-temporal properties, which are objects with a specific position in both space and time.

Besides time, space and object properties, Yixian Zheng et al. [25] identify six types of data that can be harvested and represent the types of entities that can be observed within urban territories according to the urban context they refer to, i.e., human mobility, social network, geographical, environmental, health care and divers.

Human mobility data enables the study of social and community dynamics based on different data sources like traffic, commuting media, mobile devices and geotagged social media data.

Traffic data is produced by sensors installed in vehicles or specific spots around the city (e.g., loop sensors, cameras). These data can include vehicles' positions observed recurrently at given intervals. Using these points (positions), it is then possible to compute trajectories which are spatiotemporally time-stamped and can be associated with instant speed and heading directions. Traffic occupation inroads can be measured with loops that compute, within given time intervals, which vehicles travel across two consecutive loops. Using this information, it is possible to compute travel speed and traffic volume on roads. Ground truth traffic conditions are observed using surveillance cameras that generate a huge volume of images and videos. Extracting information such as traffic volume and flowrate from these images and videos is still challenging. Therefore, in general, these data only provide a way to monitor citywide traffic conditions manually.

People's regular movement data are produced by personalized RFID transportation cards for buses or metro that they tap in station entries to enter/exit the public transportation system. This generates a huge amount of records of passenger trips, where each record includes an anonymous card ID, tap-in/out stops, time, fares for this trip and transportation type (i.e., bus or metro). Commuting data recording people's regular movement in cities can be used to improve public transportation and to analyze citywide human mobility patterns.

Records of exchanges like phone calls, messages, internet, between mobile phones and cell stations collected by telecom operators are data that contain communication information, people's locations based on cell stations. These data offer unprecedented information to study human mobility.

Social Networks Data. Social networks posts (e.g., blogs, tweets) are tagged with geo-information that can help to better understand people's activities, the relations among people and the social structure of specific communities. User-generated texts, photos and videos, contain rich information about people's interests and characteristics, that can be studied from a social perspective. For example, evolving public attention on topics and spreading of anomalous information. The major challenges with geo-tagged social network data lie in their sparsity and uncertainty.

Finally, data refer to points of interest (POI) to depict information of facilities, such as restaurants, shopping malls, parks, airports, schools and hospitals in urban spaces. Each facility is usually described by a name, address, category and a set of geographical coordinates.

Environmental data. Modern urbanization based on technology has led to environmental problems related to energy consumption and pollution. Data can be produced by monitoring systems observing the environment through different variables and observations (e.g., temperature, humidity, sunshine duration and weather conditions), air pollution data, water quality data and satellite remote sensing data, electricity and energy consumption, CO_2 footprints, gas. These data can help to provide insight regarding consumption patterns, on correlations among actions and implications and foresight about the environment.

Divers data. Other data are complementary to urban data, particularly those concerning social and human aspects, such as health care, public utility service, economy, education, manufacturing and sports.

Figure 1 summarizes the urban data types considered in urban computing: environmental monitoring data that concern meteorological data, mobile phone signals used for identifying behaviours, citywide human mobility and commuting data for detecting urban anomalies, city's functional regions and urban planning, geographical data concerning points of interest (POI), land use, traffic data, social networks data, energy data obtained from sensors, and economies regarding city economic dynamics like transaction records of credit cards, stock prices, housing prices and people's income.

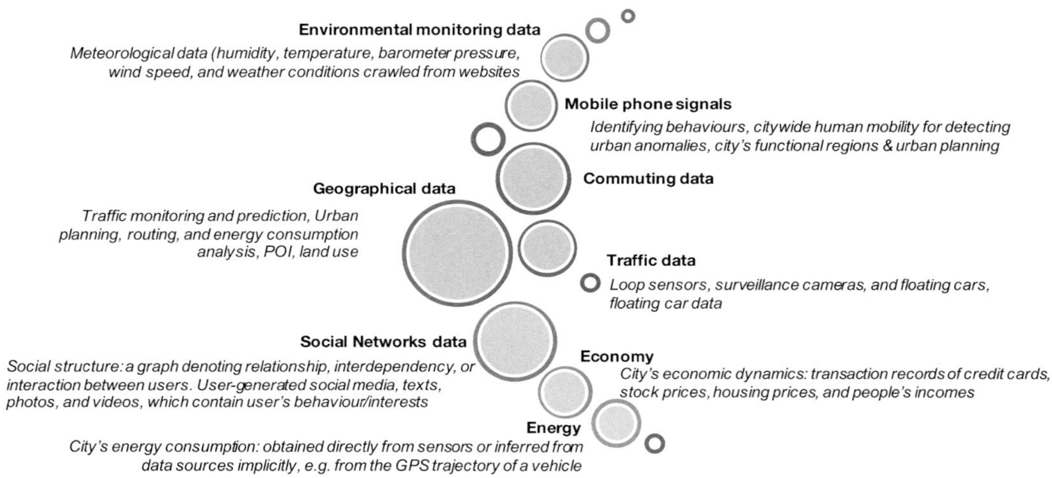

Figure 1: Urban Data Types.

Urban data can be harvested from different sources and using different techniques. These aspects are discussed next.

2.2 Data harvesting techniques

Data acquisition techniques can *unobtrusively* and *continually* collect data on a *citywide* scale. Data harvesting is a non-trivial problem, given the three aspects to consider: *(i) energy consumption and privacy, (ii) loose-controlled and non-uniform distributed sensors, (iii) unstructured, implicit, and noisy data.*

Crowdsensing. The term "crowdsourcing" is defined as the practice of obtaining needed services or content by soliciting contributions from a large group of people. People play the role of urban data consumers, but also participate in the data analysis process through crowdsourcing. Techniques use explicit and implicit crowdsourcing for collecting data that contain information about the way people evolve in public and private places. These data collections can be used as input for learning crowd behaviour and simulating it more accurately and realistically.

The advances of location-acquisition technologies like GPS and Wi-Fi have enabled people to record their location history with a sequence of time-stamped locations, called trajectories. Regarding non-obstructive data harvesting, work has been carried out using cellular networks for user tracking, profiting from call delivery that uses transitions between wireless cells. Geolife[1] is a

[1] https://www.geospatialworld.net/article/geo-life-health-smart-city-gis/

social networking service which aims to understand trajectories, locations and users, and mine the correlation between users and locations in terms of user-generated GPS trajectories. In [17] a new vision has been proposed regarding the smart cities' movement, under the hypothesis that there is the need to study how people psychologically perceive the urban environment, and to capture that quantitatively. Happy Maps uses crowdsourcing and geo-tagged pictures and the associated metadata to build alternative cartography of a city weighted for human emotions. People are more likely to take pictures of historical buildings, distinctive spots and pleasant streets instead of car-infested main roads. On top of that, Happy Maps adopts a routing algorithm that suggests a path between two locations that is the shortest route and maximizes the emotional gain.

2.3 Discussion and synthesis

An important aspect to consider is that data is non-uniformly distributed in geographical and temporal spaces, and it is not always harvested homogeneously according to the technique and the conditions of the observed entities in an urban territory.

Having the entire dataset may be always infeasible in an urban computing system. Some information is transferrable from the partial data to the entire dataset, for example, the travel speed of taxis on roads can be transferred to other vehicles that are also travelling on the same road segment. Some information cannot be transferable, for example, the traffic volume of taxis on a road may be different from private vehicles.

In some locations, when crowdsensing is used, more data can be harvested as required and in other places fewer data than required. In the first case, a down-sampling method, e.g., compressive sensing, could be useful to reduce a system's communication loads. In the last case, in the context of crowdsensing, some incentives that can motivate users to contribute data should be considered. How to configure the incentive for different locations and periods to maximize the quality of the received data (e.g., the coverage or accuracy) for a specific application is yet to explore.

Three types of strategies can be adopted for harvesting data. (i) Traditional sensing and measurement that implies installing sensors dedicated to some applications. (ii) Passive crowdsensing using wireless cellular networks built for mobile communication between individuals to sense city dynamics (e.g., predict traffic conditions and improve urban planning). We described how this technique can be specialised into three strategies:

- Sensing City Dynamics with GPS-Equipped Vehicles: mobile sensors continually probing the traffic flow on road surfaces processed by infrastructures that produce data representing city-wide human mobility patterns.
- *Ticketing Systems of Public Transportation* (e.g., model the city-wide human mobility using transaction records of RFID-based cards swiping).
- *Wireless Communication Systems* (e.g., call detailed records CDR).
- *Social Networking Services* (e.g., geotagged posts/photos, posts on natural disasters analysed for detecting anomalous events and mobility patterns in the city).

(iii) Participatory sensing where people obtain information around them and contribute to formulating collective knowledge to solve a problem (i.e., human as a sensor):

- *Human crowdsensing*: users willingly sense information gathered from sensors embedded in their own devices (e.g., GPS data from a user's mobile phone used to estimate real-time bus arrivals).
- Human crowdsourcing: users are proactively engaged in the act of generating data: reports on accidents, police traps, or any other road hazard (e.g., Waze), citizens turning into cartographers to create open maps of their cities.

3. Managing and Indexing Urban Data

The objective of managing and indexing urban data is to harness a variety of heterogeneous data to quickly answer users' instant queries, e.g., predicting traffic conditions and forecasting air pollution. Three problems are addressed in this context: stream and trajectory data management, graph data management and hybrid indexing structures.

3.1 Stream and trajectory data management

Urban data, often collected recurrently or even continuously (velocity), can lead to huge volumes of data collections that should be archived, organized (indexed) and maintained on persistence supports with efficient associated read and write mechanisms. Indexing and compression techniques are often applied to deal with data velocity and volume properties.

The continuous movement of an object is recorded in an approximate form as discrete samples of location points. A high sampling rate of location points generates accurate trajectories but will result in a massive amount of data, leading to enormous overhead in data storage, communications, and processing. Thus, it is necessary to design data reduction techniques that compress the size of a trajectory while maintaining the utility of the trajectory. There are two major types of data reduction techniques running in batch after the data is collected (e.g., Douglas-Peucker algorithm [7]) or in an online mode as the data is being collected (such as the sliding window algorithm [12,16]). Trajectory reduction techniques are evaluated concerning three metrics: processing time, compression rate, and error measure (i.e., the deviation of an approximate trajectory from its original presentation). Recent research [18], has proposed solutions to the trajectory reduction through a hybrid spatial compression algorithm and error-bounded temporal compression algorithm. Chen et al. [5] propose to simplify a trajectory by considering both the shape skeleton and the semantic meanings of the trajectory [31,32]. For example, when exploring a trajectory (e.g., travel route) shared by a user, the places where she stayed, took photos, changed moving directions significantly would be more significant than other points. Consequently, points with an important semantic meaning should be given a higher weight when choosing representative points for a simplified trajectory.

3.2 Graph data management

Graphs are used to represent urban data, such as road networks, subway systems, social networks, and sensor networks. Graphs are usually associated with a spatial property, resulting in many spatial graphs [36]. For example, the node of a road network has a spatial coordinate and each edge denoting a road segment has a spatial length. Graphs also contain temporal information; for instance, the traffic volume traversing a road segment changes over time, and the travel time between two landmarks is time-dependent: spatio-temporal graphs [36]. Queries like "*find the top-k tourist attractions around a user that are most popular in the past three months*", can be asked on top of graphs.

Hybrid Indexing Structures are intended to organize different data sources; for example, combining POIs, road networks, traffic, and human mobility data simultaneously. Hybrid structures can be used for indexing special regions; for instance, a city partitioned into grids by using a quad-tree-based spatial index (see Figure 2) where each leaf node (grid) of the spatial index maintains two lists storing the POIs and road segments. Then, each road segment ID points to two sorted lists: a list of taxi IDs sorted by their arrival time *ta* at the road segment, and a list of drop-off and pick-up points of passengers sorted by the pick-up time (*tp*) and drop-off time (*td*).

Different kinds of index structures have been proposed to manage different types of data individually. Hybrid indexes can simultaneously manage multiple types of data (e.g., spatial, temporal, and social media) and enable the efficient and effective learning of multiple heterogeneous data sources. In an urban computing scenario, it is usually necessary to harness a variety of data and integrate them into a data-mining model. This calls for hybrid indexing structures that can organize

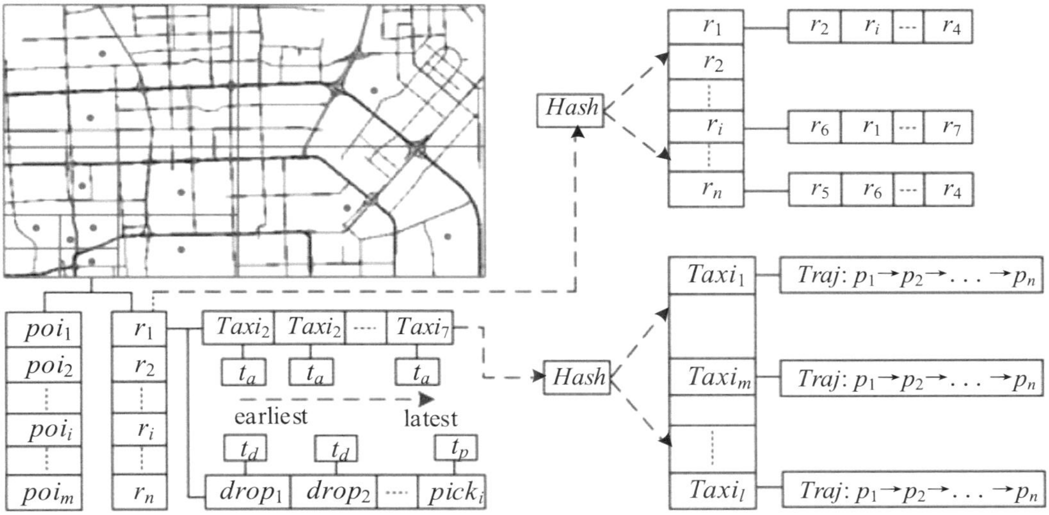

Figure 2: Hybrid index for organizing urban data [36].

different data sources, like hybrid indexing structure, which combines a spatial index, hash tables, sorted lists, and an adjacency list.

4. Querying Urban Data

Querying the actual location of a moving object has been studied extensively in moving object databases using 3DR-Tree [19] and MR-Tree [28]. Yet, sometimes queries must explore historical trajectories satisfying certain criteria, for example, retrieving the trajectories of tourists passing a given region and within a period. This corresponds to a spatiotemporal range query [23,24]), for example, taxi trajectories that pass a crossroad (i.e., a point query), or the trajectories that are similar to a query trajectory [6,20] (i.e., a trajectory query).

Dealing with the uncertainty of a trajectory refers to positioning moving objects while their locations can only be updated at discrete times. The location of a moving object between two updates is uncertain because the time interval between two updates can exceed several minutes or hours. This can, however, save energy consumption and communication bandwidth.

Map matching is to infer the path that a moving object like a vehicle has traversed on a road network based on the sampled trajectory. Map-matching techniques dealing with high-sampling-rate trajectories have already been commercialized in personal navigation devices, while those for low-sampling-rate trajectories [15] are still considered challenging. According to Yuan et al. [30], given a trajectory with a sampling rate around 2 minutes per point, the highest accuracy of a map-matching algorithm is about 70%. When the time interval between consecutive sampling points becomes even longer, existing map-matching algorithms do not work very well any more [36]. Wei et al. [26] proposed to construct the most likely route passing a few sampled points based on many uncertain trajectories.

Krumm et al. and Xue et al. [13,29] propose solutions to predict a user's destination based on partial trajectories. More generally, a user's and other people's historical trajectories as well as other information, such as the land use of a location, can be used in destination prediction models.

Other important problems include observing a certain number of moving objects travelling a common sequence of locations in similar travel time where the locations in a travel sequence are not consecutive for finding *sequential patterns* from trajectories. Other approaches discover a group of objects that move together for a certain time period, under different patterns such as flock

[8,9], convoy [10,11], swarm [14], traveling companion [21,22], and gathering [34,35,36,25]. These "group patterns" can be distinguished based on how the "group" is defined and whether they require the periods to be consecutive. For example, a flock is a group of objects that travel together within a disc of some user-specified size for at least k consecutive timestamps [10]. Li et al. [14] relaxed strict requirements on consecutive periods and proposed the pattern swarm, which is a cluster of objects lasting for at least k (possibly non-consecutive) timestamps.

5. Data and Knowledge Fusion

In urban computing scenarios, it is necessary to exploit a variety of heterogeneous data sources that need to be integrated. Then, it is necessary to fusion knowledge to explore and exploit datasets to extract insight and foresight of urban patterns and phenomena.

Data fusion. There are three major ways to achieve this goal:

- Fuse data sources at a feature level putting together the features extracted from different data sources into one feature vector. Beforehand, and given the heterogeneity of data sources, a certain kind of normalization technique should be applied to this feature vector before feeding it into a data analytics model.
- Use different data at different stages. For instance, first partition an urban region, for example, a city, into disjoint regions by major roads and then use human mobility data to glean the problematic configuration of a city's transportation network [33].
- Feed different datasets into different parts of a model simultaneously given a deep understanding of the data sources and algorithms applied to analyse them.

Building high-quality training datasets is one of the most difficult challenges of machine learning solutions in the real world. Disciplines like data mining, artificial intelligence and deep learning have contributed to building accurate models but, to do so, they require vastly larger volumes of training data. The traditional process for building a training dataset involves three tasks: data collection, data labelling and feature engineering. From the complexity standpoint, data collection is fundamentally trivial as most organizations understand what data sources they have. Feature engineering is getting to the point where it is 70%–80% automated using algorithms. The real effort is in the data labelling stage. New solutions are emerging for combining strong and weak supervision methods to address data labelling.

Knowledge fusion. Data mining and machine-learning models dealing with a single data source have been well explored. However, the methodology that can learn mutually reinforced knowledge from multiple data sources is still missing. The fusion of knowledge does not mean simply putting together a collection of features extracted from different sources but also requires a deep understanding of each data source and the effective usage of different data sources in different parts of a computing framework.

End-to-end urban computing scenarios call for the integration of algorithms of different domains. For instance, data management techniques with machine-learning algorithms must be combined to provide both efficient and effective knowledge discovery ability. Similarly, integrating spatio-temporal data management algorithms with optimization methods. Visualization techniques should be involved in a knowledge discovery process, working with machine-learning and data-mining algorithms.

6. Perspectives of the Role of Data Science for Making Urban Spaces Smart

This chapter discussed and described issues regarding data for enabling urban computing tasks that can lead to the design of smart urban territories. Having a data-centred analysis of the problems and challenges introduced by urban computing exhibits the requirement to study data concerning different perspectives. First, the chapter characterised data produced within urban territories in terms of their mathematical properties (spatio-temporal), concerning the "semantics" of the entities composing urban territories (e.g., points of interest, roads, infrastructure) and also from the mobile entities that populate urban territories, like people, vehicles and the built environment. This variety of data is produced by producers with different characteristics, and approaches today use hardware, software and passive and active participation of people to generate phenomenological observations of urban territories. Finally, the chapter discusses how to create insight and foresight of important situations happening in urban territories, for example, computing trajectories of entities evolving in these territories observed in space and time, and other social foresight of behaviours like popular POIs, the population of regions, etc.

The vision of urban computing—acquisition, integration, and analysis of big data to improve urban systems and life quality— is leading to smarter cities. Urban computing blurs the boundary between databases, machine learning, and visualization and even bridges the gap between different disciplines (e.g., computer sciences and civil engineering). To revolutionize urban sciences and progress, quite a few techniques still need to be explored, such as the hybrid indexing structure for multimode data, the knowledge fusion across heterogeneous data sources, exploratory visualization for urban data, the integration of algorithms of different domains, and intervention-based analysis.

Bibliography

[1] Aigner, W., Miksch, S., Schumann, H. and Tominski, C. 2011. Visualization of time-oriented data. Springer Science & Business Media.

[2] Andrienko, G. and Andrienko, N. 2008. Spatio-temporal aggregation for visual analysis of movements. In Visual Analytics Science and Technology, 2008. VAST'08. IEEE Symposium on pages 51–58. IEEE.

[3] Andrienko, G., Andrienko, N., Bak, P., Keim, D. and Wrobel, S. 2013. Visual analytics of movement. Springer Science & Business Media.

[4] Andrienko, G., Andrienko, N., Hurter, C., Rinzivillo, S. and Wrobel, S. 2011. From movement tracks through events to places: Extracting and characterizing significant places from mobility data. In Visual Analytics Science and Technology (VAST), 2011 IEEE Conference on, pages 161– 170. IEEE.

[5] Chen, Y., Jiang, K., Zheng, Y., Li, C. and Yu, N. 2009. Trajectory simplification method for location-based social networking services. In Proceedings of the 1st ACM GIS Workshop on Location-based Social Networking Services. ACM, 33–40.

[6] Chen, Z., Shen, H.T., Zhou, X., Zheng, Y. and Xie, X. 2010. Searching trajectories by locations: An efficiency study. In ACM SIGMOD International Conference on Management of Data. ACM, 255–266.

[7] Douglas, D. and Peucker, T. 1973. Algorithms for the reduction of the number of points required to represent a line or its caricature. Canadian Cartographer, 10(2): 112–122.

[8] Gudmundsson, J. and Kreveld, M.V. 2006. Computing longest duration flocks in trajectory data. In Proceedings of the 14th International Conference on Advances in Geographical Information Systems. ACM, 35–42.

[9] Gudmundsson, J., Kreveld, M.V. and Speckmann, B. 2004. Efficient detection of motion patterns in spatio-temporal data sets. In the Proceedings of the 12th International Conference on Advances in Geographical Information Systems. ACM, 250–257.

[10] Jeung, H., Yiu, M., Zhou, X., Jensen, C. and Shen, H. 2008a. Discovery of convoys in trajectory databases. Proceedings of the VLDB Endowment, 1(1): 1068–1080.

[11] Jeung, H., Shen, H. and Zhou, X. 2008b. Convoy queries in spatio-temporal databases. In Proceedings of the 24th International Conference on Data Engineering. IEEE, 1457–1459.

[12] Keogh, E., Chu, J., Hart, S.D. and Pazzani, M.J. 2001. An on-line algorithm for segmenting time series. In Proceedings of the International Conference on Data Mining. IEEE, 289–296.

[13] Krumm, J. and Horvitz, E. 2006. Predestination: Inferring destinations from partial trajectories. In Proceedings of the 8th International Conference on Ubiquitous Computing. ACM, 243–260.
[14] Li, Z., Ding, B., Han, J. and Kays, R. 2010. Swarm: Mining relaxed temporal moving object clusters. Proceedings of the VLDB Endowment, 3(1-2): 723–734.
[15] Lou, Y., Zhang, C., Zheng, Y., Xie, X., Wang, W. and Huang, Y. 2009. Map-matching for low-sampling-rate GPS trajectories. In Proceedings of the 17th ACM SIGSPATIAL Conference on Geographical Information Systems. ACM, 352–361.
[16] Maratnia, N. and de By, R.A. 2004. Spatio-temporal compression techniques for moving point objects. In Proceedings of the 9th International Conference on Extending Database Technology. IEEE, 7.
[17] Quercia, Daniele, Rossano Schifanella and Luca Maria Aiello. 2014. The shortest path to happiness: Recommending beautiful, quiet, and happy routes in the city. Proceedings of the 25th ACM conference on Hypertext and social media. ACM.
[18] Song, R., Sun, W., Zheng, B., Zheng, Y., Tu, C. and Li, S. 2014. PRESS: A novel framework of trajectory compression in road networks. In Proceedings of 40th International Conference on Very Large Data Bases.
[19] Theodoridis, Y., Vazirgiannis, M. and Sellis, T.K. 1996. Spatio-temporal indexing for large multimedia applications. In Proceedings of the 3rd International Conference on Multimedia Computing and Systems. IEEE, 441–448.
[20] Tang, L.A., Zheng, Y., Xie, X., Yuan, J., Yu, X. and Han, J. 2011. Retrieving k-nearest neighbouring trajectories by a set of point locations. In Proceedings of the 12th Symposium on Spatial and Temporal Databases. Volume 6849, Springer, 223–241.
[21] Tang, L.A., Zheng, Y., Yuan, J., Han, J., Leung, A., Peng, W.-C., Porta, T.L. and Kaplan, L. 2013. A framework of travelling companion discovery on trajectory data streams. ACM Transaction on Intelligent Systems and Technology.
[22] Tang, L.A., Zheng, Y., Yuan, J., Han, J., Leung, A., Hung, C.C. and Peng, W.C. 2012. Discovery of travelling companions from streaming trajectories. In Proceedings of the 28th IEEE International Conference on Data Engineering. IEEE, 186–197.
[23] Wang, L., Zheng, Y., Xie, X. and Ma, W.Y. 2008. A flexible spatio-temporal indexing scheme for large-scale GPS track retrieval. In Proceedings of the 9th International Conference on Mobile Data Management. IEEE, 1–8.
[24] Wang, F., Chen, W., Wu, F., Zhao, Y., Hong, H., Gu, T., Wang, L., Liang, R. and Bao, H. 2014. A visual reasoning approach for data-driven transport assessment on urban roads. In Visual Analytics Science and Technology (VAST), 2014 IEEE Conference on, pages 103–112. IEEE.
[25] Wang, Z., Lu, M., Yuan, X., Zhang, J. and Wetering, H.v.d. 2013. Visual traffic jam analysis based on trajectory data. Visualization and Computer Graphics, IEEE Transactions on, 19(12): 2159–2168.
[26] Wei, L.Y., Zheng, Y. and Peng, W.C. 2012. Constructing popular routes from uncertain trajectories. In Proceedings of the 18th SIGKDD Conference on Knowledge Discovery and Data Mining. ACM, 195–203.
[27] Wu, Y., Liu, S., Yan, K., Liu, M. and Wu, F. 2014. Opinion flow: Visual analysis of opinion diffusion on social media. Visualization and Computer Graphics, IEEE Transactions on, 20(12): 1763–1772.
[28] Xu, X., Han, J. and Lu, W. 1999. RT-tree: An improved R-tree index structure for spatio-temporal databases. In Proceedings of the 4th International Symposium on Spatial Data Handling, 1040–1049.
[29] Xue, A.Y., Zhang, R., Zheng, Y., Xie, X., Huang, J. and Xu, Z. 2013. Destination prediction by sub-trajectory synthesis and privacy protection against such prediction. In Proceedings of the 29th IEEE International Conference on Data Engineering. IEEE, 254–265.
[30] Yuan, J., Zheng, Y., Zhang, C., Xie, W., Xie, X., Sun, G. and Huang, Y. 2010. T-Drive: Driving directions based on taxi trajectories. In Proceedings of ACM SIGSPATIAL Conference on Advances in Geographical Information Systems. ACM, 99–108.
[31] Zheng, Y., Xie, X. and Ma, W.Y. 2008. Search your life over maps. In Proceedings of the International Workshop on Mobile Information Retrieval, 24–27.
[32] Zheng, Y. and Xie, X. 2010. GeoLife: A collaborative social networking service among user, location and trajectory. IEEE Data Engineering Bulletin, 33(2): 32–40.
[33] Zheng, Y., Liu, Y., Yuan, J. and Xie, X. 2011. Urban computing with taxicabs. In Proceedings of the 13th International Conference on Ubiquitous Computing. ACM, 89–98.
[34] Zheng, Y., Liu, F. and Hsieh, H.P. 2013. U-Air: When urban air quality inference meets big data. In Proceedings of 19th SIGKDD Conference on Knowledge Discovery and Data Mining. ACM, 1436–1444.
[35] Zheng, K., Zheng, Y., Yuan, N.J., Shang, S. and Zhou, X. 2014. Online Discovery of Gathering Patterns over Trajectories. IEEE Transactions on Knowledge Discovery and Engineering.
[36] Zheng, Yu, et al. 2014. Urban computing: concepts, methodologies, and applications." ACM Transactions on Intelligent Systems and Technology (TIST), 5.3: 38

CHAPTER-9

Utilization of Detection Tools in a Human Avalanche that Occurred in a Rugby Stadium, Using Multi-Agent Systems

Tomás Limones,[1,*] *Carmen Reaiche*[2] and *Alberto Ochoa-Zezzatti*[1]

This article aims to make a simulation model of an avalanche that occurred at a Rugby football match due to the panic caused by riots between fanatical fans of the teams that were playing. To carry out this model, the specific Menge simulation tool is used, which helps us to evaluate the behavior of people who consciously or unconsciously affect the contingency procedures established at the place of the event, to define them preventively to reduce deaths and injuries. From the definition of the factors, an algorithm is developed from the combination of the Dijkstra tool and the simulation tool that allows us to find the route to the nearest emergency exit, as well as the number of people who could transit safely. Additionally, Voroni diagrams are used to define perimeter adjacency between people.

1. Introduction

Thousands of deaths have happened in different parts of the world where football is like a religion. The very serious disturbances that occurred after a football match, avalanches caused by panic, riots between fanatical fans, landslides in poor condition, overcapacity, are just a few examples of events that generate deaths in the stadiums. The tragedies have been numerous, and the main causes occur when people enter a panic, which unfortunately causes an imbalance in their thinking, failing to have control over their actions and causing the agglomerations with catastrophic consequences.

Some historical events with the greatest consequence in deaths during football games are described in Table 1:

As can be seen on the Table 1, most of the eventualities presented here have their origin in the disturbances incited by the same fans, causing stampedes wherein, due to closed doors, people become pressed against bars or meshes causing human loss by severe blows and asphyxiation. This type of agglomeration is not exclusive to football. In the article "Innovative data visualization of collisions in a human stampede occurred in a religious event using multiagent systems" [1], the author analyzes this type of phenomenon, but focused on religious events, where large concentrations of people come together. In this example, an analysis is made about the tragedy that occurred in Mecca in 2015, where 2717 people died and 863 were injured as a result of the largest human stampede ever recorded.

[1] Universidad Autónoma de Ciudad Juárez, México.
[2] The University of Adelaide, Australia.
* Corresponding author: al183244@alumnos.uacj.mx

Table 1: Events of greater consequence in deaths during football soccer games.

No.	Place	Year	Causes	Injured	Deaths
1	National Perú Stadium	1964	Fan riots. Stampede towards exit. Closed doors in tunnels	500	328
2	River Plate, Argentina Stadium	1968	Avalanche, exit sector Door 12 Closed	46	71
3	Loujniki, Moscu Stadium	1982	Avalanche Shock. Some wanted to leave and others to enter	61	66
4	Heysel, Bélgica Stadium	1985	fan riots, cause avalanche	600	39
5	Valley Parade, England Stadium	1985	Fire. Closed doors	265	56
6	Hillsborough Stadium	1989	Excess capacity causes avalanche. It did not meet the security requirements		96
7	Mateo Flores, Guatemala Stadium	1996	Over sale due fake tickets and closed stadium doors (The doors opened inwards)	200	83
8	Said Port Stadium	2012	Caused by fans who attacked players and fans with weapons	1000	74

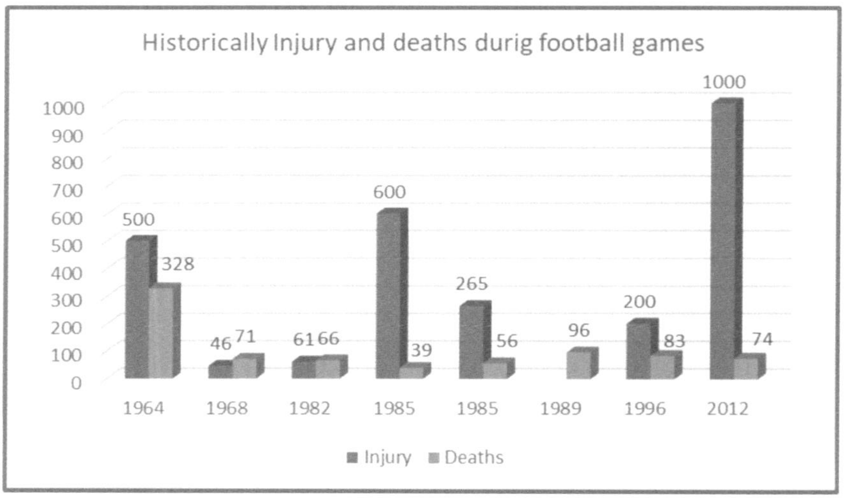

Figure 1: Representative graph of injuries and deaths in the history of soccer.

In the case of this study, the simulation exercise will be carried out in the Rugby Stadium of the Australia Adelaide City, known as Oval Stadium. Its characteristics are described below:

NAME	The Adelaide Oval
DESCRIPTION	It is a multipurpose stadium located in the city of Adelaide, Australia. It is mainly used for the practice of cricket and Australian rules football, as well as soccer and rugby
DIRECTION	War Memorial Dr, North Adelaide SA 5006
CAPACITY	53,500
OPERATION SINCE	1871
PROPERTY	Government of South Australia

The city of Adelaide is in southern Australia and is characterized as a peaceful city, where eventualities due to fighting or aggressions are unusual. Historically, there has been a fight raised on August 25, 2018, where two fans in a Rugby match between Port Adelaide and Essendon AFL started a fight. The fans themselves tried to intervene to avoid this quarrel. It is noted that the actions of these two individuals was an isolated element among a crowd of more than 39,000 fans. The realization of this exercise will be carried out simulating an avalanche in the Oval stadium, provoked by the panic caused by riots among fanatical fans. The result will help us to define the best alternatives of preventive solutions to avoid possible catastrophes.

The Figure 2, shows a distribution graph of the Oval Adelaide stadium.

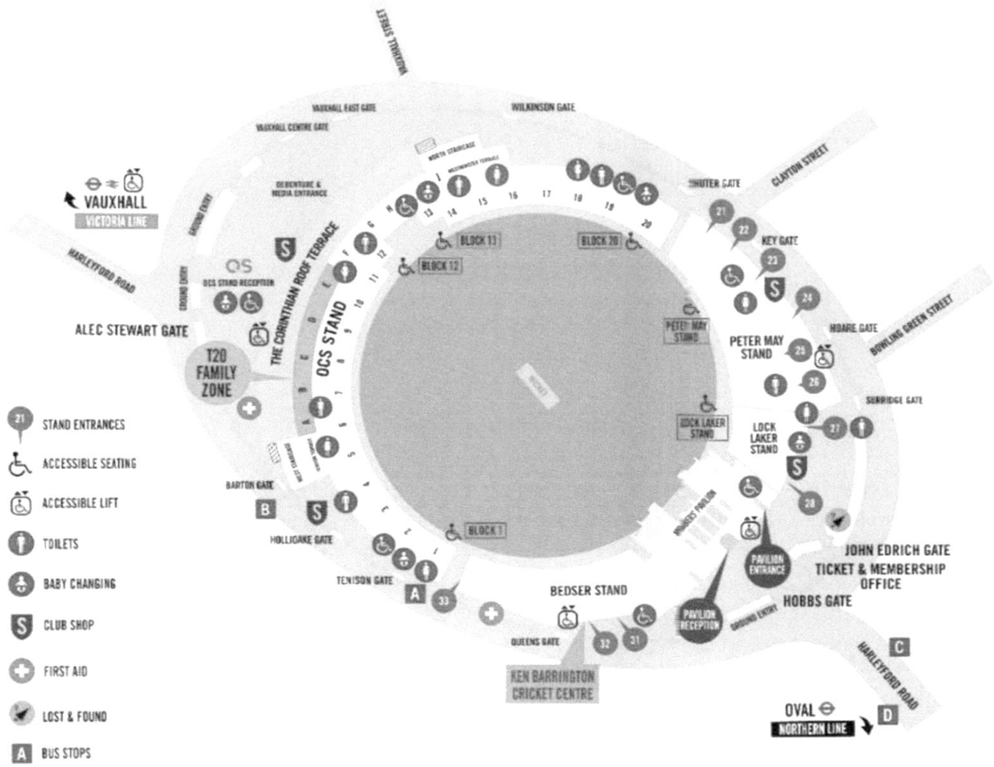

Figure 2: Oval Adelaide stadium of rugby football.

The anthopometry

Anthropometry is considered as the science in charge of studying the physical characteristics and functions of the human body, including linear dimensions, weight, volume, movements, etc., in order to establish differences between individuals, groups and races [2]. This science turns out to be a guideline in the design of the objects and spaces necessary for the environment of the human body and that, therefore, must be determined by their dimensions [3]. By knowing these data, the minimum spaces that human needs to function daily are known, which must be considered in the design of his environment. Some factors that define the physical complexion of the human being are race, sex, diet, age. The reference plane distributes 3 imaginary flat surfaces that cross the body parts and are used as a reference in taking body dimensions (See Figure 3). Sports fans have seen the evolution and development of professional players and how it has been shocking in recent years. A rugby defender of 80 kg, or 160 pounds, which was previously considered enough, now looks less heavy with not enough weight to take the job of a restorer. Dimensional standards and spatial requirements must be constantly adequate. The need to establish standards that guarantee the adaptation of the interior spaces for sports practices to the human dimension and the dynamics of people on the move constitutes, today, a potential threat to the safety of the participants. The lack of this kind of regulation not only involves a serious threat to

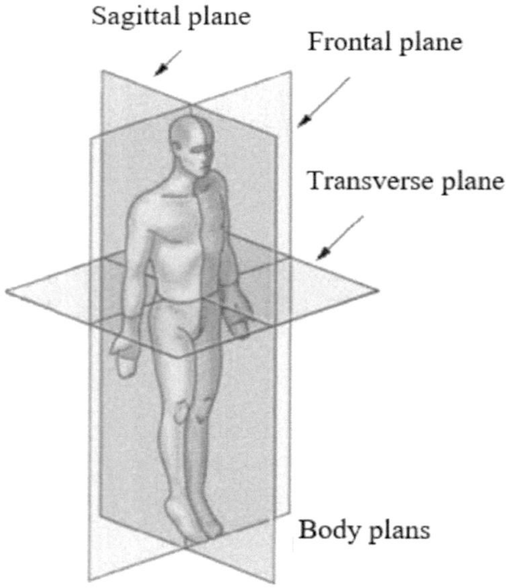

Figure 3: Reference plane.

the physical integrity of the users, but also makes the client and the designer potentially legally liable in the event of an accident with injury or death. The inference of the human body-interior space not only influences the comfort of the first but also in public safety. The size of the body is the fundamental measurement reference for dimensioning the width of doors, corridors and stairs in any environment, whether public or private. Every precaution is little in the use and acceptance of existing methods or empirical rules to establish critical clearances without questioning their anthropometric validity, even for those likely to be part of affected codes and ordinances. In short, certain dimensions and clearances that guarantee public safety must be defined. Public spaces must be designed so as not to hinder their use for people outside a standard, such as children, small people, overweight people. The designs of the different attachments and accessories will also have the reach of these people; the stairs, seats, hallways, open spaces among others.

Horizontal space

Two measures are important to consider in a space of the people movement: (1) Body dimensions and (2) larger people. Slacks should be considered for both measures Figure 4 shows two fundamental projections of the human body, which include the critical dimensions of the 95th percentile. A tolerance of 7.6 cm (3 inches) has been included for width and depth. The final dimension with the tolerance included is 65.5 cm (28.8 inches); The critical anthropometric dimension to be used during a massive agglomeration is the body width. The diagram representing the body ellipse and the lower Table 2 have proven utility in the design of circulation spaces. The latter is an adaptation of a study of the movement and formation of pedestrian queues, prepared by Dr. John Fruin, whose purpose was to set the relative levels of service based on the density of pedestrians. The basic unit is the human body, which is associated with an elliptical shape or ellipse body of 45.6 x 61 cm (18 x 24 inches).

The panic

Panic attacks, also known as crisis of distress, are usually accompanied by various manifestations of somatic nature, such as tachycardia, sweating, tremor, choking sensation, chest tightness, nausea,

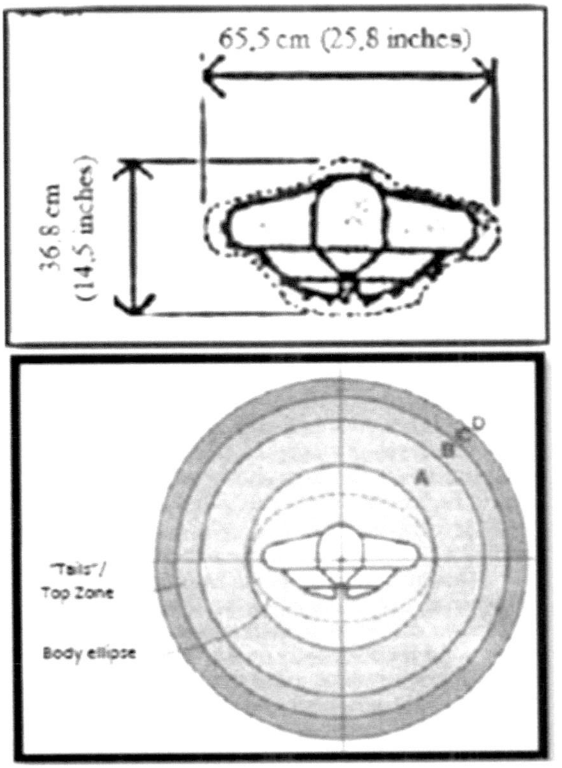

Figure 4: Two fundamental projections of the human figure.

Table 2: Analysis of the circulation space for the human being "density of queues".

Density Analysis In "Queues"						
Denomination	**Description**	**Ratio**		**Surface**		
		Inches	**cm**	**Ft2**	**cm2**	
A-Contact zone	In this area of occupation, body contact is almost inevitable; circulation impossible, movement reduced when walking, shuffling; occupation like an elevator.	12	30.5	3	0.25	
B-Non-contact zone	While it is not necessary to move, body contact can be avoided; possible movement in group form.	18	45.7	7	0.65	
C-Personal zone	The depth of the body separates people; limited lateral circulation by passing people; This area is in the selected space occupation category, experiencing comfort standards.	21	53.3	10	0.95	
D-Circulation zone	It is possible to move in «queue» without disturbing other people.	24	61	13	1.4	

dizziness, fainting, hot flashes, feeling of unreality and loss of control [4]. This can happen when the person experiences the sensation of being near imminent death and has an imperative need to escape from a feared place or situation (aspect congruent with the emotion that the subject is feeling in the perceived imminent danger). The fact of not being able to physically escape the situation of extreme fear in which the affected person is greatly accentuates the symptoms of panic [5]. Taking this in consideration, the relationship with possible triggers of panic attacks can be classified as:

- Unexpected. The beginning of the episode does not match manifest triggers.

- Situationally determined. Attacks occur in the presence or anticipation of a specific stimulus or situation.
- Situationally predisposed. Episodes are more common in specific situations, although they are not completely associated with them.

Panic attacks can originate from different situations, especially in those capable of generating a state of high physiological activation or in the event of a specific stress event. The panic attack is linked to agoraphobia, characterized by an intense anxiety response to situations in which it is difficult to escape or get help [6].

Factors that cause a human stampede

Most of the human stampedes have occurred during religious, sporting and musical events since they are the ones that gather more people. The most common causes occur when people nervously react in moments of panic, whose detonator is fear. This fear can be caused by a fire, an explosion, fear of a terrorist attack, etc. When people want to escape, people from behind push those in front, not knowing that those in the front are being crushed. This stacking thrust force occurs in both forms, vertically and horizontally. The vast majority of deaths are caused by compression asphyxiation and rarely by trampling. The degree of physical strength is the main ally to cling to life. That is why, in most cases, children, elderly people, and women are the most affected ones. For the Honduran human behavior specialist, Teodosio Mejía (2017), one of the reasons is that when people are in a crowd, they "lose their condition of being rational". Mass men place their ego to the collective ego "and that is criminal" because when human beings are frustrated, they begin to despair, and this causes bad decisions to be made [7,10].

Multiple agent systems

A complex system can be defined as a system that presents a large number of interactive components whose aggregate activity is not derivable from the sums of the activity of individual (non-linear) components and typically exhibits hierarchy of self-organization under selective constraints [8]. Multiple agent-based simulations (MABS) offer the possibility of creating an artificial universe in which experiments representing individuals, their behaviors and their interactions can be performed by modeling and simulating complex systems of multiple levels of abstraction. To conceive a MABS (Figure 5), here is the multi-view approach proposed by Michel [9], which distinguishes four main

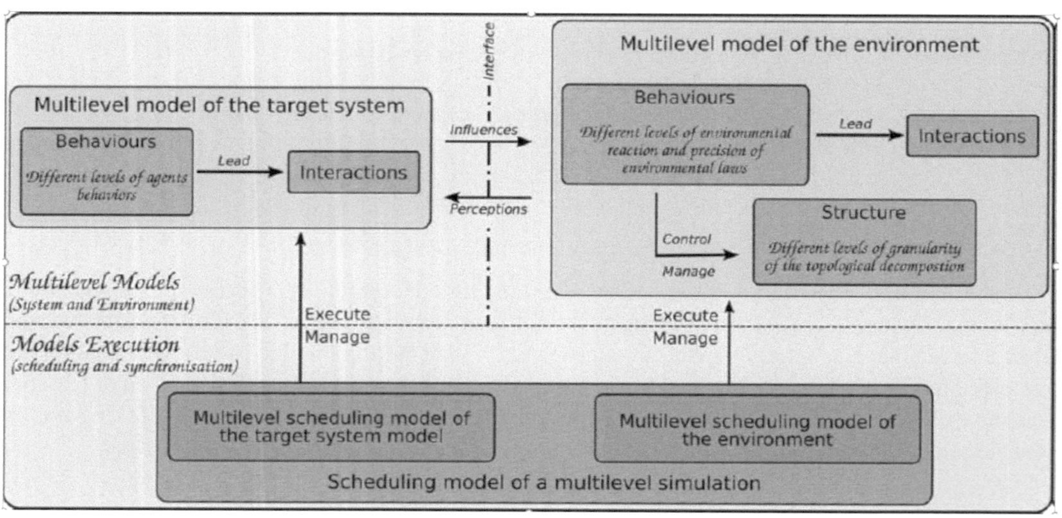

Figure 5: Example of different aspects of a multilevel simulation.

aspects in a MABS: (i) Agent behavior that deals with agent modeling of the deliberative process (their minds). (ii) The environment that defines the different physical objects in the simulated world (the situated environment and the physical body of the agents) and the endogenous dynamics of the environment. (iii) The programming that deals with the modeling of the passage of time and the definition of programming policies is used to execute the behaviors of the agents. (iv) The interaction focuses on the result of the modeling of the actions and interactions between agents at a given time. Our approach broadens these different perspectives to integrate related multilevel aspects.

In the next chart (Table 3) a comparison between some different methodologies for the use of multiagent systems with Mathematical models, are showed:

2. Review of the Problem

Table 1 breaks down a total of 8 events that have represented a greater impact on deaths and injuries in the history of football themselves that passed from 1964 to 2012. The main cause was death due to suffocation for the pressure exerted by the masses, caused by human avalanches. According to the analysis of the table the main reasons for these avalanches were:

- Fan riots.
- Excess of stadium capacity.
- Aggression with a weapon.
- Fire

In most of the events, a factor that influenced these outcomes was the closure of the accesses, since people could not leave because the doors were closed to prevent the entry of people who did not pay for a ticket or because the doors were opened in the opposite direction to the flow of people. Considering the reasons described on all the bibliography reviewed, the following variables that affect the results of a human avalanche can be defined:

1. Anthropometry considering the definition of the minimum horizontal space necessary to ensure integrity between people (4 different areas of circulation).
2. The population in the event (Considered 2 groups according to the sample of the number of people involved (734 people and 170 people).
3. The distribution of spaces in the stadium (Corridors and emergency exits).

Considering the different combinations of these three variables, the movement of people can be simulated using the multiple agent system.

Distribution and spaces in the oval stadium

The architectural design of the Rugby Stadium contemplates a good security system. Four sets of photographs on the architectural distributions of the oval stadium are shown below

The stadium's design includes spacious areas to avoid the crowding of people, as well as security systems, fire areas and areas for people with disabilities. Photographs 1 represents the open spaces in the stadium, both at the entrance and internally. Photographs 2 represents the access stairs to the middle and upper part of the stadium, these being spacious. Photographs 3 represents the exits of the stadium, external, internal and escalators.

For the realization of this simulation model, the specific area of the exits on the internal central part of the stadium will be contemplated (Photographs 4). These exits are placed at the bottom of the stairs, almost at the level of the football field. A greater concentration of people is located in this area due to the seating arrangement. There are 23 rows arranged alphabetically. The probability that

Table 3: Some different methodologies for the use of multiagent systems.

Methodology	Description	Features	Formula
MABS [8]	A forward force (Equation (1)) attracting the holon to its next local goal.	$\vec{F}_{obj} = \beta_{obj} \cdot \overrightarrow{A_i O_{bj}}$	A_i O_{bj} O_{bs} d_{ij} m_i $d = \lVert\overrightarrow{A_i O_{bs}}\rVert$, $\alpha = \angle(\overrightarrow{A_i O_{bj}}, \overrightarrow{A_i O_{bs}})$ $\vec{n} = \begin{pmatrix} 0 & -\text{sign}(\alpha) \\ \text{sign}(\alpha) & 0 \end{pmatrix} \cdot \frac{\overrightarrow{A_i O_{bs}}}{\lvert\overrightarrow{A_i O_b}\rvert}$ $\beta_{obj}, \beta_{ij}, \beta_{obs}$ are constants Position of the holon i. Position Next holon's goal. Position of a given obstacle. Vector from agent i to j. dij the norm of this vector mass of agent i. Distance between agent i and his next objective.
	the force resulting from the repulsions of the other holons (Equation (2)).	$\vec{F}_{rep_{ij}} = \beta_{ij} \cdot \frac{m_i \cdot m_j}{d_{ij}^2} \overrightarrow{d_{ji}}$	
	the force resulting from the repulsions of the environmental obstacles (Equation (3)).	$\vec{F}_{rep_{obs}} = \beta_{obs} \cdot \frac{m_i}{(d \cdot sin(\alpha))^4} \cdot \vec{n}$	
Perception of a normal person. Psychophysical studies [11]	According to scientists [11], human horizontal perception field is divided into three main parts.	The first part is defined by a vision angle of 30 degree. In this field a human can see object in detail.	
		The second part is defined by a vision angle of 100 degree. In this part, a human can see only object forms.	
		In the third part (Its vision angle is 200 degree), only movements of objects are seen. This subdivision determines the certainty degree of the pedestrians' knowledge of each observed object.	
Velocity of normal person. Psychophysical studies [11]	The pedestrian velocity depends on the crowd density. Schnable and define the Kladek function which describes the variation of velocity function of density [11]. The following equation describes the Kladek function.	$V_{F,hi} = V_{F,hf} * [1 - e^{(-* \frac{1}{D} - \frac{1}{D_{max}})}]$	VF;hi is the velocity of the pedestrian located in a crowd with a given density [m=s]. VF;hf is the pedestrian velocity with freedom movement [m=s]. _ Dmax is the crowd density within a pedestrian stop moving.

PEDESTRIAN MODEL [12].	This environment consists of streets, which can be represented by a network consisting of *N* nodes and *L* links, and a set of establishments, consisting of *J* stores, restaurants, etc. A subset *E* of these *N* nodes represents the entry/departure points of the system. Let us assume that the pedestrians can be represented by a multiagent system approach with *I* agents. Each agent *i* is supposed to carry out a set of activities *Ai*	*Perceive: E → P** *Interpret: P* x I x G →A** *UpdStatepandG: P* x I x G →I x G* *act: I x G →A** *UpdEnv: E x A* →E*	The function *perceive* represents the perception process of the agent. It maps the environment *E* to a set of percepts. The interpret function maps the perception (P) of the environment, the current internal sate (I) and the current activity agenda (G) into a list of one or more actions A, for instance The interpret function updates the internal state based on its percepts and the activity agenda; select actions (*act*) based on the updated activity agenda and the *updated* internal state. The function updEnv represents the reaction of the environment to the agents' actions. A new state of the environment is realized.
Pedestrian equation based on the BDI methodology [1].	The pedestrian equation is based on several factors: the desires, beliefs and intentions of the individuals. The pedestrian equation will allow us to determine the speed of everyone according to specific criteria.	Vi(t)= [(v+h+nc)/(a+d)] * f * imc * s	-xi is the velocity of agent i at time t. -To solve Vi (t) determines the position of an agent with respect to time. -v is the average pedestrian speed for all agents. -h represents the height of the simulated person -a represents the age of the person -d is the density of people per square meter -imc is the individual's body mass index -s is the sex of the simulated individual -nc is the level of consciousness of the individual
	The minimum space necessary to consider the calculations of a multitude is equal to the vector	Xi = [xi vi] ^ t	where xi and vi belong to R ^ 2 (radius between two agents)

Photographs 1: Open access spaces, Oval Stadium.

Photographs 2: Access stairs to the Oval stadium and back side of the open stadium.

Photographs 3: Oval stadium exits. Photo 1 and 2 external part. Photo 3 and 4 internal central part. Photos 5 and 6 internal electrical second floor.

'Photo 1' 'Photo 2' 'Photo 3' ' Photo 4' 'Photo 5'

Photographs 4: Exits from the Oval stadium, internal central part.

these two points are a focus of attention for a possible crash problem or agglomeration of people is greater than in the rest of the exits.

3. Methodological Approximation

For the development of this simulation exercise, two pedestrian equations developed in a study by Ochoa et al. (2019) "Innovative data visualization of collisions in a human stampede occurred in a religious event using multiagent systems" will be used as a reference, where it is considered a catastrophic incident with critical levels of concentration of people at a religious event in Mecca. These equations will be used to simulate the movement of people within a stampede and determine the probability of their survival. The equation is based on the BDI methodology, which involves 3 fundamental factors that define the result: (1) Desires, (2) Beliefs, (3) Intentions.

Equipmet

Equipment description used during the simulation trials:
Machine name: DESKTOP-G07PBE6
Operating System: Windows 10 Pro, 64-bits
Language: Spanish (Regional Setting: Spanish)
System Manufacturer: Lenovo
System Model: ThinkPad
Product ID: 8BF29E2C-5A1A-4CA2-92E8-BE228436613D
Processor: Intel (R) Core (TM) i5-2520M CPU @2.50 GHz.
Memory: 10.0 GB
Available OS Memory: 9.89 GB usable RAM
Disk Unit ID: ST320LT007-9ZV142
Hard Disk: 296 GB
Page File: 243 GB used; 53.5 GB available
Windows Dir: C:\WINDOWS
DirectX Version: 10.0.17134.1

Software

Menge: A framework for modular pedestrian simulation for research and development, free code.

Unity: A multiplatform video game engine created by Unity Technologies; a free personal account was used.

Sublime Text 3: A sophisticated text editor to encode, used a free trial evaluation.

Git Bash: An emulator used to run Git from command line, a code software

Layout definition

Taking into consideration the exit door shown in Photo no. 4, where the width of the space of the tunnel that leads to the exit has a dimension of 2.4 meters, begins with the formalization of the layout. For the realization of the layout, the set of seats located on the left side and right side of the tunnel will be considered, making an initial group of 304 people who could leave through this exit door.

The initial distribution is as follows:

1. Number of people located on the left side of the exit tunnel: 80 people.
2. Number of people located above the exit tunnel in three groups: 8, 32, 8 people.
3. Number of people located on the right side of the exit tunnel: 80 people.
4. Number of people located under the exit tunnel in two group: 48 people each. The distribution are 48 people on the left side and 48 people on the right side.

To make the distribution of the layout, the coordinates of the dimensions of the seats as well as stairs are considered, making a distribution of coordinates which have been handled in excel for it to define a preliminary space according to the following Figure 6:

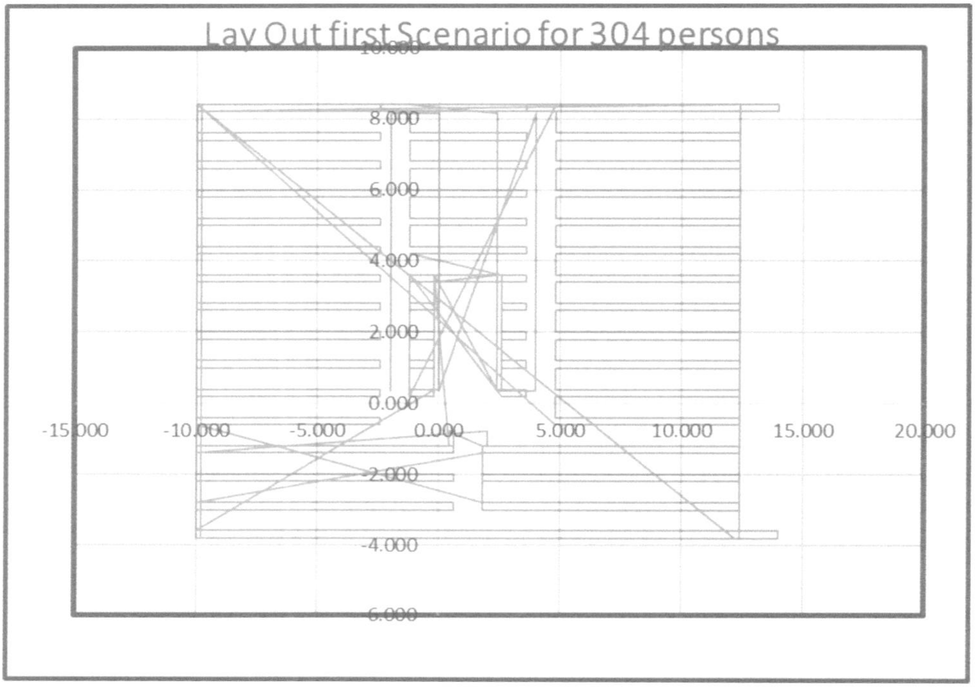

Figure 6: Layout of the scenario exit taking in consideration 304 persons for this evacuation simulation.

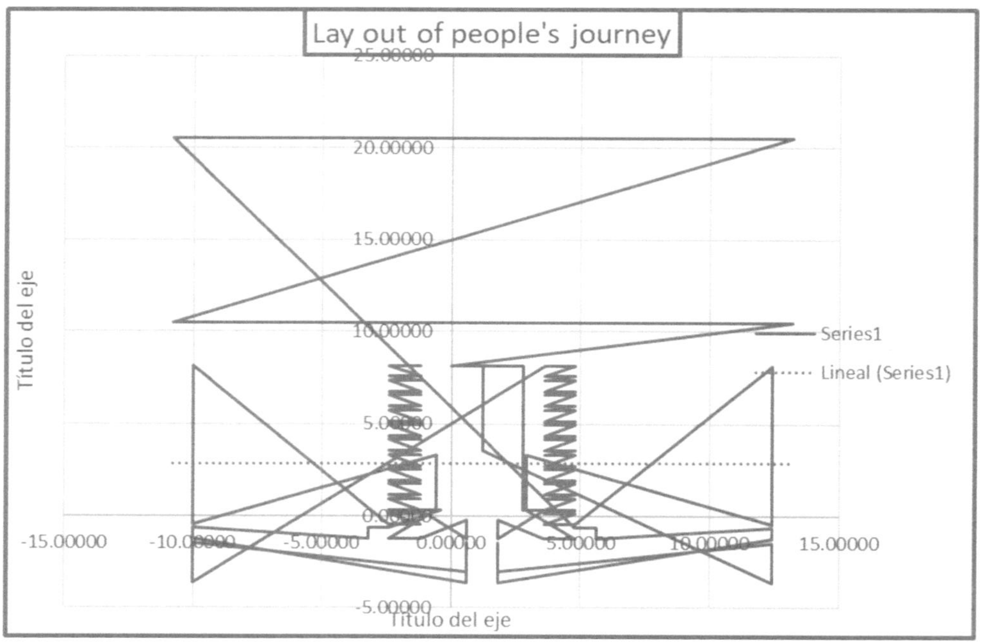

Figure 7: Layout of the journey that is made by the 304 persons.

Performing the first run using the coordinates of the scenario defined, as well as using the pattern that people will follow during the evacuation, the image that is defined during the run of simulation [9] in menge shows, as a result, the following Figures 8 and 9.

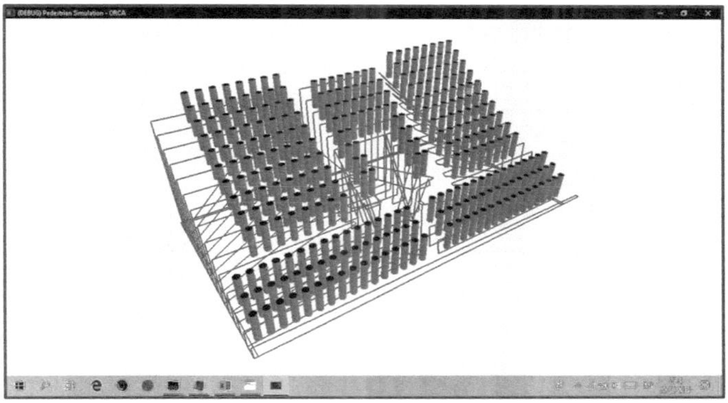

Figure 8: First trial simulation on menge for the 304 people evacuation.

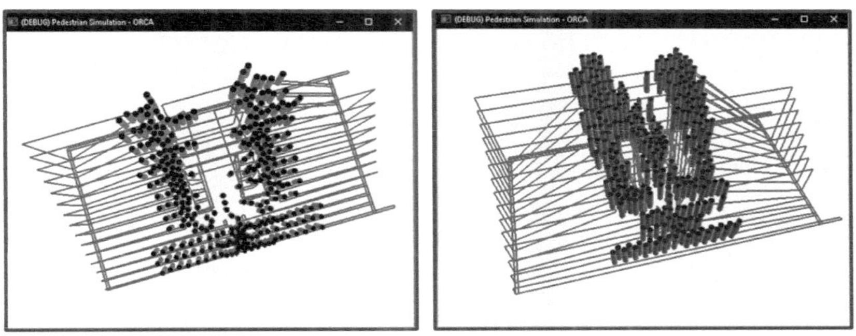

Figure 9: First trial simulation on menge for the 304 people evacuation (simulation advance).

Figure 9 shows how the agglomeration of agents causes a bottleneck at the entrance to the tunnel. This agglomeration is due to the narrow dimension of the roadway to the tunnel.

The elements used during the development of this scenario run are shown in the following Table 4, achieving a total time of 762,994 seconds during the evacuation, a time considered very high for an evacuation process.

Table 4: elements used during the first simulation evacuation trial.

Common

max_angle_vel=	max_neighbors=	obstacleSet=	neighbor_dist=	r =	class=	pref_speed=	max_speed=	max_accel=
90	10	1	5	0.19	1	1.34	2	50

Full frame	scene update	scene draw	buffer swap	simulation time
(avg): 976.929 ms in 762 laps	(avg): 927.179 ms in 763 laps	(avg): 47.0648 ms in 763 laps	(avg): 2.21781 ms in 763 lasp	762.994

4. Looking for Evacuation Time Improvement

To improve the evacuation time defined in the first test run, an experiment should be carried out [11], making changes with the different elements that are used during the development of this run, using the menge simulation software, to optimize the time during the evacuation process.

Table 5 shows the results from after the change elements, with the purpose to define the best condition during the simulation related to the elements that impact in the best time evacuation result.

Table 5: Elements used during the experiment to define the best elements condition.

	max_angle_vel=	max_neighbors=	obstacleSet=	neighbor_dist=	r =	class=	pref_speed=	max_speed=	max_accel=
1	90	2	1	3	0.19	1	1.34	3	50
2	90	2	1	3	0.19	1	1.34	3	70
3	360	2	1	3	0.19	1	1.34	3	70
4	360	2	1	3	0.19	1	1.34	4	70
5	60	2	1	3	0.19	1	1.34	4	70
6	60	2	1	3	0.19	1	1.5	5	80

	Full frame	scene update	scene draw	buffer swap	simulation time
1	(avg): 563.187 md in 357 laps	(avg): 514.112 ms in 358 laps	(avg): 48.0976 m in 390 laps	(avg): 2.25161 ms in 390 laps	35.8
2	(avg): 577.609 ms uin 359 laps	(avg): 526.616 ms in 360 laps	(avg): 48.3162 ms in 360 laps	(avg): 2.27213 ms in 360 laps	36
3	(avg): 572.743 ms in 359 laps	(avg): 522.278 ms in 360 laps	(avg): 47.8306 ms in 360 laps	(avg): 2.3383 ms in 360 laps	36
4	(avg): 551.692 ms in 368 laps	(avg): 554.692 ms in 368 laps	(avg): 47.5738 ms in 369 laps	(avg): 2.26544 ms in 369 laps	36.9
5	(avg): 544.659 ms in 368 laps	(avg): 495.334 ms in 369 laps	(avg): 46.699 ms in 369 laps	(avg): 2.26265 ms in 369 laps	36.9
6	(avg): 560.557 ms in 308 laps	(avg): 510.916 ms in 309 laps	(avg): 46.8925 ms in 309 laps	(avg): 2.29343 ms in 309 laps	30,9001

Elements defined to be changed for simulation time improvement

Performing experimentation tests by making changes to the elements of algorithms agents, like maximum speed angle, maximum neighbor, maximum neighbor distance, pre speed, maximum speed, and maximum acceleration, gives us, as a result, the definition of the best number to be considered during the performance of the simulation test using menge. Some algorithms agents perform better than others [12].

The elements that need to be changed in order to improve the simulation time are:

1. max_angle_ve = from 90 to 60
2. max_neighbors = from 10 to 2
3. neighbor_dist = from 5 to 3
4. pref_speed = from 1.34 to 1.5
5. max_speed = from 2 to 5
6. max_accel = from 50 to 80
7. the entrance to the tunnel dimension from 2.5 to 3.6 with the purpose of increasing the entrance of the exit tunnel.
8. increase the quantity of people to be involved during the evacuation simulation from 304 to 352 people.

The new layout of the scenario for the evacuation simulation is considering improvements on the tunnel entrance, increasing the dimension related to this scenario. The new Layout is shown on the Figure 10.

Performing the multiagent simulation trial with the new data (elements change), the simulation time result is improved, reaching 30.90 s (Table 6).

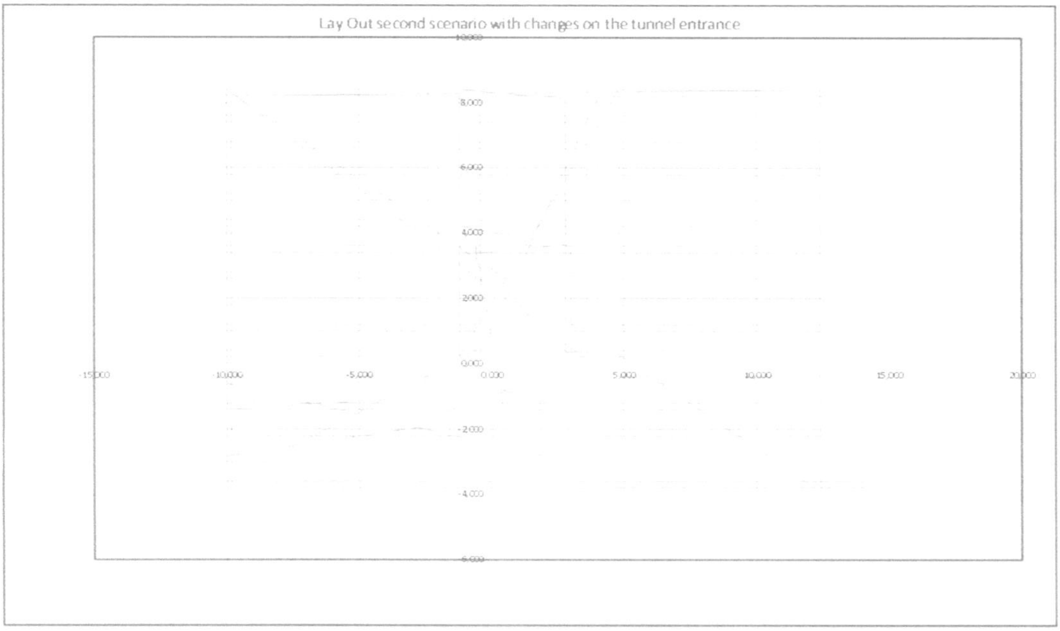

Figure 10: New improved layout considering the change in the tunnel entrance dimension.

Table 6: Elements defined to be use for the best simulation time 30.9.

Common

max_angle_vel=	max_neighbors=	obstacleSet=	neighbor_dist=	r =	class=	pref_speed=	max_speed=	max_accel=
60	2	1	3	0.19	1	1.5	5	80

Full frame	scene update	scene draw	buffer swap	simulation time
(avg): 560.557 ms in 308 laps	(avg): 510.916 ms in 309 laps	(avg): 46.8925 ms in 309 laps	(avg): 2.29343 ms in 309 lasp	30,9001

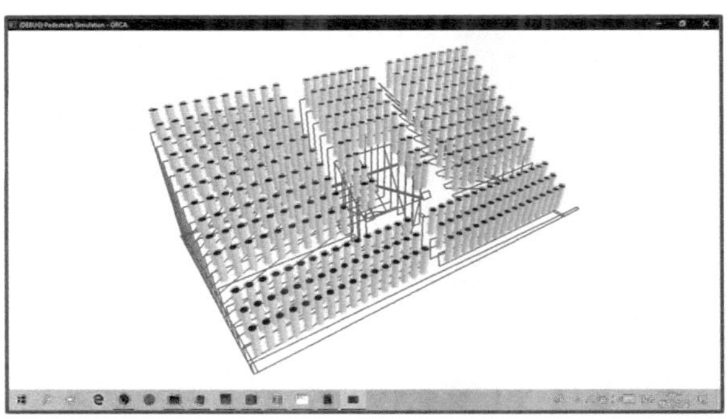

Figure 11: Second trial simulation on menge for the 352 people evacuation.

Figure 12 shows the second run of simulation with the increase to 352 agents as well as the improvements included. It is possible to appreciate the increase in the dimension of the entrance of the tunnel.

To facilitate and appreciate the movements of the agents, after including the improvements proposed during the run of the simulation, Figure 13 shows the agents that are separated in 7 groups and different colors, assigned to each one of them.

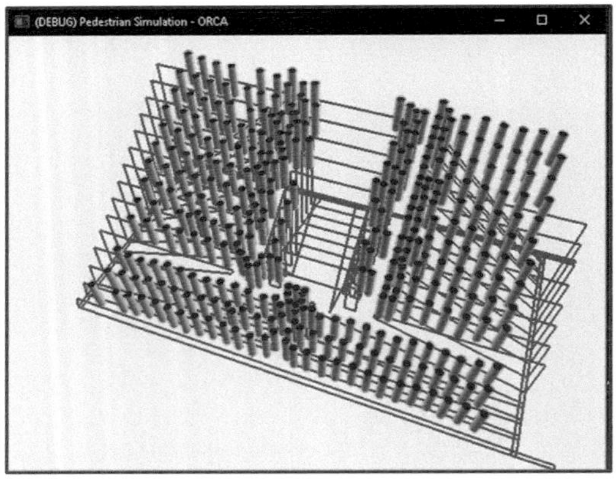

Figure 12: Second trial simulation on menge for the 352 people evacuation (simulation advance).

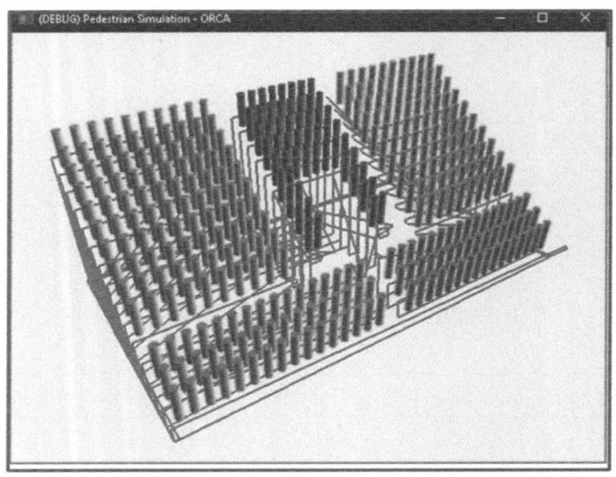

Figure 13: Third trial simulation on menge for the 352 people evacuation defined on 7 groups.

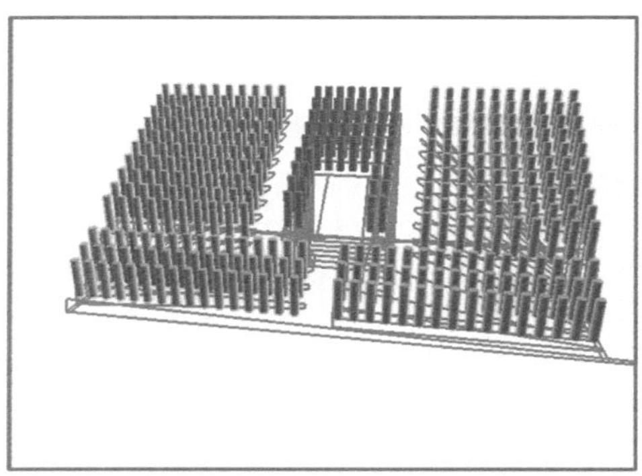

Figure 14: Third trial simulation on menge for the 352 people evacuation for 7 groups (Tunnel view).

Figure 14 shows the increase in the number of agents, which rises from 304 to 352, separating into 7 groups and identifying them are different colors. This will allow us to see the improvement in terms of the decrease in the agglomeration of agents at the entrance of the tunnel.

In Figure 15, can be appreciated and see the improvement in width dimension at the entrance of the tunnel, which greatly facilitates the exit of agents, avoiding collisions between them.

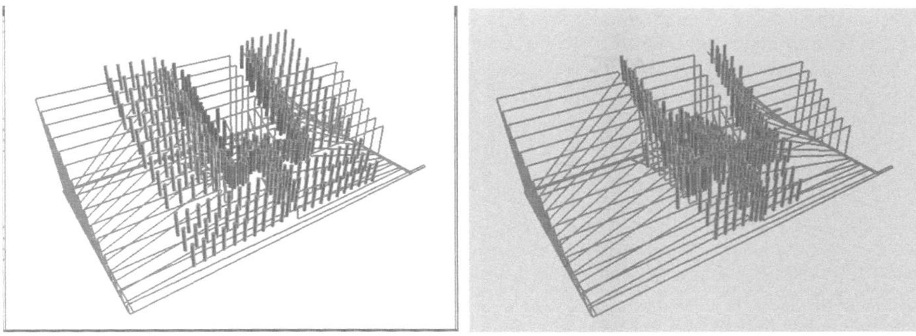

Figure 15: Third trial simulation on menge for the 352 people evacuation for 7 groups (simulation advance).

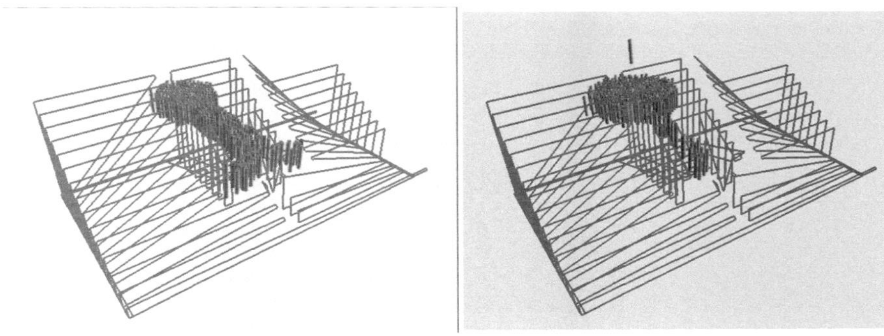

Figure 16: Third trial simulation on menge for the 352 people evacuation for 7 groups (Final simulation).

5. Conclusions

The use of the menge tool for the development of this simulation exercise allows us to perform the simulation with different scenarios, considering the changes in the factors that impact on the outcome, facilitating the alternative evaluation, thereby seeking the preservation and safety of the agents involved. In this exercise, it was demonstrated that the changes of these elements during the development of the simulation allow us to have a clearer view of the potentially catastrophic results that may occur in a real eventuality. These results will give us indicators that can be determined for real decision making, which allows us to generate preventive actions.

For future studies, it is necessary to continue with the development of runs considering changes in the elements that affect the behavior of the agents' travel, to improve the travel at higher speed, without agglomerations or generation of bottlenecks. The simulation must consider the improvements in the use of the software and databases currently available, such as UNITY, Visual Studio, among others, which will allow us to find and make proposals for the solution of potential problems of human avalanches.

Bibliography

[1] Alberto Ochoa-Zezzatti, Roberto Contreras-Massé, José Mejía. 2019. Innovative Data Visualization of Collisions in a Human Stampede Occurred in a Religious Event Using Multiagent Systems.

[2] Antropometría, FACULTAD DE INGENIERÍA INDUSTRIAL, 2011-2. Escuela Colombiana de Ingeniería,https://www.escuelaing.edu.co/uploads/laboratorios/2956_antropometria.pdf.

[3] Rosmery Nariño Lescay, Alicia Alonso Becerra, Anaisa Hernández González, 2016. DOI: https://doi.org/10.24050/reia.v13i26.799.

[4] Lic. Lorena Frangella/Lic. Monica Gramajo, MANUAL PSICOEDUCATIVO PARA EL CONSULTANTE, Fundacion FORO; www.fundaciónforo.com Malasia 857 – CABA.

[5] Jorge Osma, Azucena García-Palacios y Cristina Botella, Anales de Psicología, 2014, vol. 30, no 2 (mayo), 381– 394 http://dx.doi.org/10.6018/analesps.30.2.150741.

[6] https://www.psicologosmadridcapital.com/blog/causas-ataques-panico/.

[7] https://confidencialhn.com/psicologo-explica-el-salvajismo-en-la-estampida-que-dejo-cinco-muertos-en-estadio-capitalino/.

[8] Nicolas Gaud, Stéphane Galland, Franck Gechter, Vincent Hilaire, Abderrafiâa Kouka, 2008. 1569-190X/$ - see front matter _ 2008 Elsevier B.V. All rights reserved; doi:10.1016/j.simpat.2008.08.015.

[9] Michel, F. 2004. Formalism, Tools and Methodological Elements for the Modeling and Simulation of Multi-Agents Systems', Ph.D. Thesis, Montpellier Laboratory of Informatics, Robotics and Microelectronics, Montpellier, France, December 2004.

[10] Michela Milano and Andrea Roli, 'MAGMA: A Multiagent Architecture for Metaheuristics', IEEE TRANSACTIONS ON SYSTEMS, MAN, AND CYBERNETICS—PART B: CYBERNETICS, VOL. 33, NO. 2, APRIL 2004.

[11] Olfa Beltaief, Sameh El Hadouaj, Khaled Ghedira, 2011; Psychophysical studies, DOI: 10.1109/LOGISTIQUA.2011.5939418.

[12] Jan Dijkstra, Joran Jessurun, Bauke de Vries, Harry Timmermans, 2006, Agent Architecture for Simulating Pedestrians in the Built Environment, International Joint Conference on Autonomous Agents and Multiagent Systems; 5 (Hakodate): 2006.05.08-12 (pp. 8-16). New York, NY.

CHAPTER-10

Humanitarian Logistics and the Problem of Floods in a Smart City

Aztlán Bastarrachea-Almodóvar, Quirino Estrada Barbosa, Elva Lilia Reynoso Jardón* and *Javier Molina Salazar*

Floods are natural disasters resulting from various factors, such as poor urban planning, deforestation and climate change, to provide just a couple of examples. The consequences of such disasters are often devastating and bring with them not only losses of millions of dollars but also of human lives. The purpose of this work is to offer a first approximation of people's reactions in the case of an evacuation due to a hypothetical flood in an area located in the Colonia Bellavista of Ciudad Juárez that is adjacent to the Río Bravo, the Acequia Madre, and the Díaz Ordaz viaduct, which are plausible to be subject to overflow or flooding after heavy torrential rains in a scenario where climate change has seriously affected the city's climate.

1. Introduction

According to [1]"A flood is referred to when usually dry areas are invaded by water; there are two possible causes of why this type of disaster occurs, the first reason is related to natural phenomena such as torrential rains and rainy seasons, for the second cause there is talk of human actions that largely induce natural disasters; ...". Among the factors associated with human intervention are deforestation, elimination of wetlands, high CO2 emissions that cause climate variations [2,3], bad urban planning, etc. [1]. On the other hand, floods can be of two types according to [1]: sudden/ abrupt and progressive/slow. In addition, floods may occur in urban or rural areas.

The environment of cities is greatly affected by climate change due to flooding [4]. The authors point out that, in general, public spaces do not adapt well to abrupt changes in the environment and that is why their design must be well worked out to avoid problems in the event of a disaster. One of the main problems affecting urban and rural populations is flooding. Table 1 shows the greatest floods in Europe during the 90's decade and their effects.

The characteristics of Ciudad Juárez, as well as its climate, make it propitious to carry out a study referring to sudden floods, since these are characterized by the precipitation of a large volume of water in a short time, causing a rapid accumulation of water in conurbation areas; this because of the rupture of dams, torrential rains or overflowing of basins or rivers [1]. In addition, according to [2] an increase in torrential rainfall is expected that can cause the type of floods mentioned above is expected. In the case of Ciudad Juárez, this is characterized by the existence of the Río Bravo as well as a desert climate with torrential rains, which have caused severe flooding as in 2013 [6], in addition to the fact that the infrastructure and urban planning of the city are additional also factors

Universidad Autónoma de Ciudad Juárez, Av. Hermanos Escobar, Omega, 32410 Cd Juárez, Chihuahua, México.
* Corresponding author: aztlan.ba@cdjuarez.tecnm.mx

Table 1: Heavy Floods in the EU and Neighboring Countries, 1991–2000, and Their Effects on the Population [5].

Region	Year	Fatalities	Evacuations
Wallis (Switzerland); Northern Italy	2000	36	**¿?**
England and Wales	2000	¿?	**¿?**
Eastern Spain	2000	¿?	**¿?**
Hungary and Romania	2000	11	**¿?**
Bavaria (Germany); Austria; Switzerland	1999	12	**¿?**
Southwest France	1999	¿?	**¿?**
Portugal; Western Spain; Italy	1998	31	**¿?**
Belgium; Netherlands	1998	2	**¿?**
Slovakia, Czech Republic, Poland	1998	Ca.100	**¿?**
Eastern Germany; Czech Republic; Western Poland	1997	114	**195,000**
Southern Spain	1996	25	**200**
Southern, western, and northern Germany; Belgium; 1995 Luxembourg; Netherlands; eastern and northern France	1995	27	**300,000**
Piemonte and Liguria (Italy)	1994	73	**3,400**
Greater Athens (Greece)	1994	12	**2,500**
Southwest Germany; Belgium; Luxembourg; southern Netherlands; eastern and northern France	1993–1994	17	**18,000**
Piemonte and Liguria (Italy); southeast France	1993	13	**2,000**
Essex and Devon (UK); Ireland	1993	4	**1,500**
Vaucluse (France)	1992	42	**8,000**
Sicily (Italy)	1991	16	**2,000**

that lead to flooding in the rainy season. Under this scheme, it is imperative to create scenarios of possible evacuations in case torrential rains can cause flooding in areas that are more prone to overflows and water stagnation.

The objective of this work is to make a first approximation, by using a simulation of the behavior of people who live in an area susceptible to flooding as well as to analyze two possible scenarios where people may be located during the incident. All this assumes that climate change could alter the amount of water that falls in the rainy season and cause an overflow of the Río Bravo as well as floods in the Díaz Ordaz viaduct and the Acequia Madre.

2. Mathematical Models

According to [7] there is a form to estimate the velocity of pedestrians by an Equation. "The pedestrian equation is based on the BDI methodology where the factors used are affected by the desires, beliefs, and intentions of the individuals" [7]. The velocity of an agent is dictated by Equation 1:

$$V_i(t) = [v + h + nc)/(a + d]* f * imc * s \tag{1}$$

Where:

- xi is the velocity of agent i at time t.
- To solve $V_i(t)$ determines the position of an agent with respect to time.
- v is the average pedestrian speed for all agents.

- *h* represents the height of the simulated person
- *a* represents the age of the person
- *d* is the density of people per square meter
- *imc* is the individual's body mass index
- *s* is the sex of the simulated individual
- *nc* is the level of consciousness of the individual

According to [7] the criteria that can be used can be justify:

- Density of people: If the density is higher the mobility decreases.
- Level of consciousness: If the person is in a state of drunkenness or has just awakened his speed will not be the most optimal.
- Age: A person's motor performance is affected by their age since a child and an elder do not have the same performance as a young adult.
- Body Mass Index: The body mass index indicates if the person is overweight, obese, is lacking in weight or is in a normal state.
- Height: The stride of a person is directly proportional to the height of the same, so the height is an important factor.
- Gender: The sex of a person intervenes in the force that will have the same, since a person with more strength can push the rest and advance more quickly.
- The main variables that directly affect the pedestrian movement of an agent are fear (f), body mass index (BMI) and sex (s), which are direct multipliers in the equation.
- The average speed added to a person's height significantly affects the final velocity, however, they are affected by the age (a) of the individual and the density of people (d) at time t in a way inversely proportional to vi. This is reflected in that this last pair of attributes divides the two first mentioned.

Additionally:

The minimum space necessary to consider when a multitude is analyzed is equal to the vector $X_i=[x_i v_i]^t$, where x_i and v_i belong to R^2 (radius between two agents) [8].

In this case, the pedestrian equation was implemented in a simulator that is named "Menge", where 2 types of people were taken, which have different size attributes but equal speeds, this because we assume there is a flood on the streets and the analysis is less complex.

3. Materials

The following are the specifications of the equipment, software and materials used for the implementation of the simulations.

Computer equipment to run the simulation

System information

Machine name: DESKTOP-FJF469O

Operating System: Windows 10 Home Single Language 64-bit

Language: Spanish (Regional Setting: Spanish)

System Manufacturer: Dell Inc.

System Model: Inspiron 15 7000 Gaming

Processor: Intel® Core™ i7-7700HQ CPU @2.80GHz

Memory: 8GB RAM

Available OS Memory: 7.78GB RAM

Software

Menge: A framework for modular pedestrian simulation for research and development, free code [9].

4. Scenarios

The stage is located as shown in the red polygon, as shown in Figure 1. These are houses adjacent to the Río Bravo in Ciudad Juarez's downtown neighborhood. The chosen scenario is interesting because the area is trapped between possible sources of flooding. The Río Bravo is located in the northeastern part while the Acequia Madre is in the southwestern, a natural stream; the Díaz Ordaz viaduct is located in the northwestern part. A satellite view of the stage can be seen in Figure 2.

It is estimated that the people closest to the edges through which water flows, and therefore the first to experience flooding when there is torrential rain, will be the first to react to try to evacuate the area, while the more distant people will do so with some delay, which is why the model estimates that there will be an agglomeration of individuals trying to get out that will cause congestion by the exit routes.

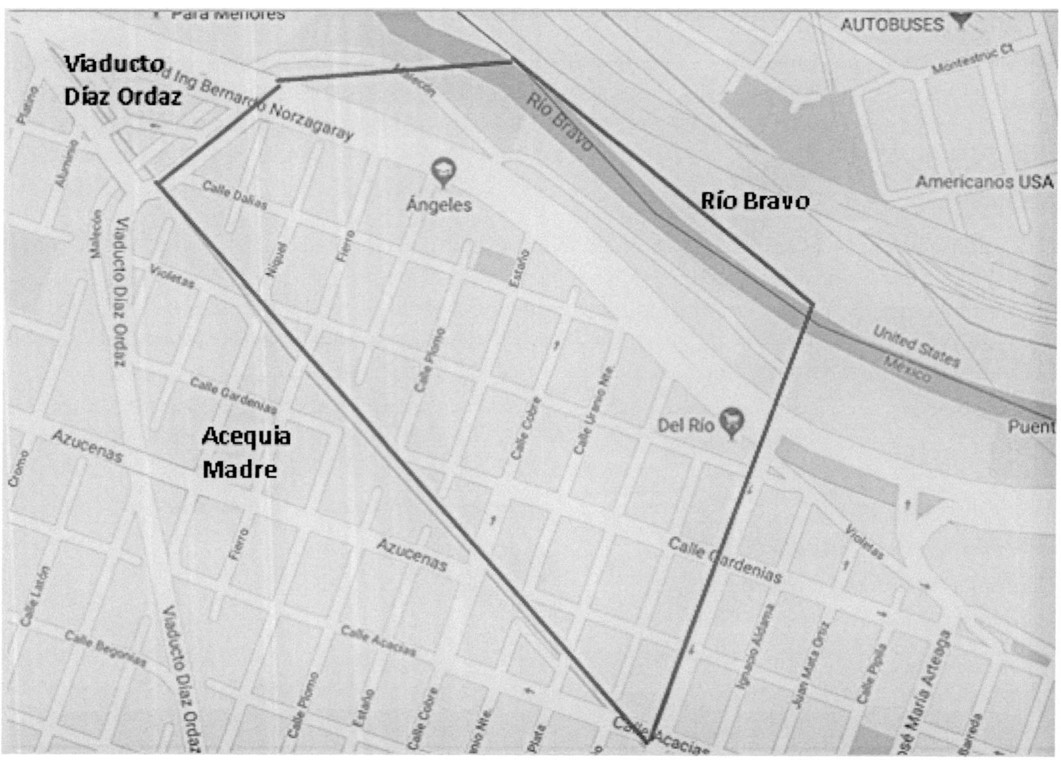

Figure 1: View of the stage in the Colonia Bellavista located in the center of Ciudad Juárez [10]. It can be observed that the group of houses is located between the Río Bravo, the Díaz Ordaz Viaduct and the Acequia Madre.

Figure 2: Satellite view of the stage in the Colonia Bellavista located in the center of Ciudad Juárez. It can be observed that the group of houses is located between the Río Bravo, the Díaz Ordaz Viaduct and the Acequia Madre [10].

5. Simulation

The simulation was performed using Menge software by modifying an open-source example called 4square.xml [9] and adapted to the conditions of the scenario as well as the objectives or goals to be achieved during the simulation of a flood evacuation in the borders surrounding the scenario. The stage was set by making a slight rotation of about 16° as it is shown in Figure 3 to be able to carry out the layout of the streets more easily, but for practical purposes, this is not an issue for the final results.

In one of the scenarios, an evacuation of only people was contemplated, where 610 people intervened in groups of 10 people distributed in different locations of the scenario as shown in Figure 3. An estimate of 122 homes with families of approximately 5 people in equal conditions to mobilize during the disaster was estimated. Besides, it was estimated that the size of the people was the same, with a radius of 1 m for each of them.

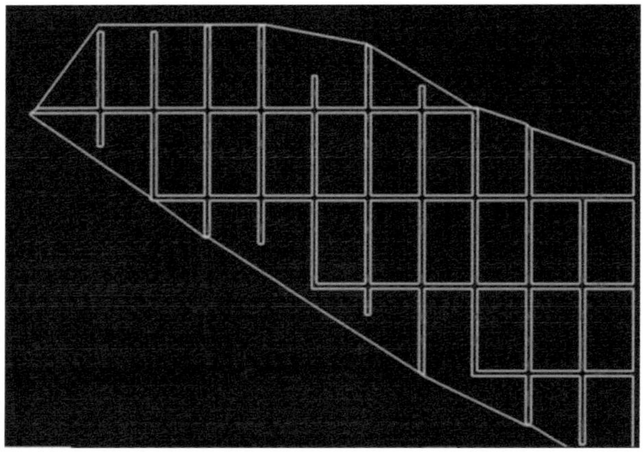

Figure 3: The image shows the distribution of people in groups of 10. Each red dot in the image represents an individual. Only one group of them, the one at the bottom of the stage, has 20 people.

In this part of the simulation, it was contemplated that the only objective of the people was to move from their initial location to a point located in the coordinate (250,0) as shown in Figure 4.

The simulation considers an objective or goal, which must be reached by people. This objective is declared as a point from which displacement vectors which will be a reference of the speed direction to be maintained by the pedestrians are traced. This type of goal is simple, but in terms of simulation, it makes mobility difficult when the pedestrians encounter an obstacle that prevents their mobility in the direction of the displacement vector. That is why, during the simulation, they advance slowly in the "y" direction when there is an obstacle that prevents them from advancing in the "x" direction. Figure 5 shows the evacuation of the pedestrians towards the goal. The simulation takes place over a time of about 200 seconds (the program time is marked as 400 cycles) and is the maximum simulation time, so in order to facilitate the simulation, the movement speed of most pedestrians was chosen to be increased to 8 m/s, almost 5 times faster than the normal speed of a person who is moving freely [11].

According to actual dimensions and without considering obstacles in the path, a person running at an average speed of 1.5 m/s should complete the path indicated in Figure 6, with a length of 375 m, in 250 seconds or 4.16 min. However, it must be considered that the speed of the pedestrians will be affected by the environment, surely flooded, which would imply a reduction of their speed to

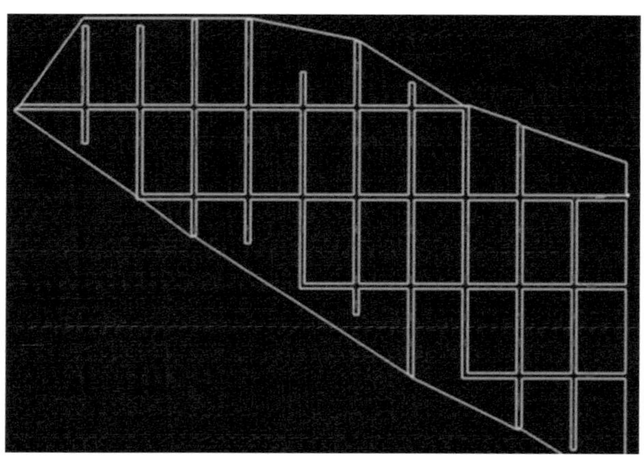

Figure 4: The blue dot represents the coordinates of the goal that people have to reach during the evacuation, which is located at the coordinate (250,0) according to the frame of reference.

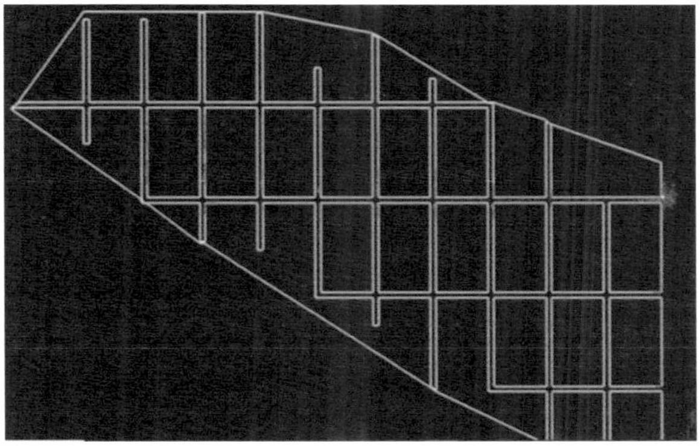

Figure 5: Displacement of the pedestrians who congregate at the evacuation point at (250.0).

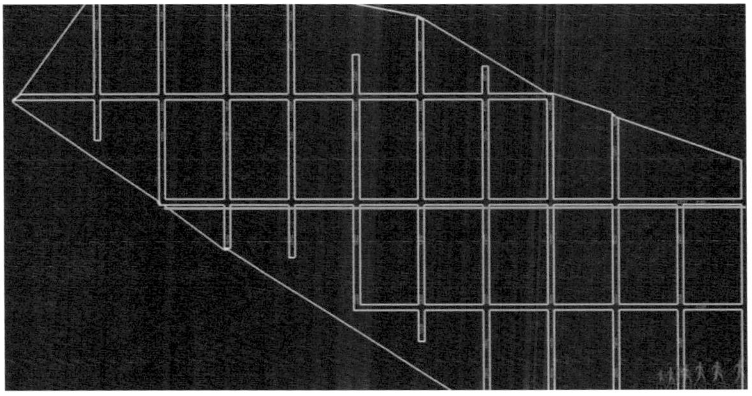

Figure 6: The segment marked in green has a length of 375 meters.

about 0.5 m/s, which would imply that by moving in a straight line it would take not about 4 minutes but a little more than 12 minutes to complete the journey.

Another interesting situation to analyze is one in which people do not escape to an evacuation point, but gradually gather in the geometric center of the stage. Figure 7 shows this point in blue, which is located at the coordinate (41, -2). This situation is less realistic than the previous one because one would expect to escape from the flood sources located at the top, bottom and left of the scenario, however, it can be thought that, in the confusion, people decide to go in the opposite

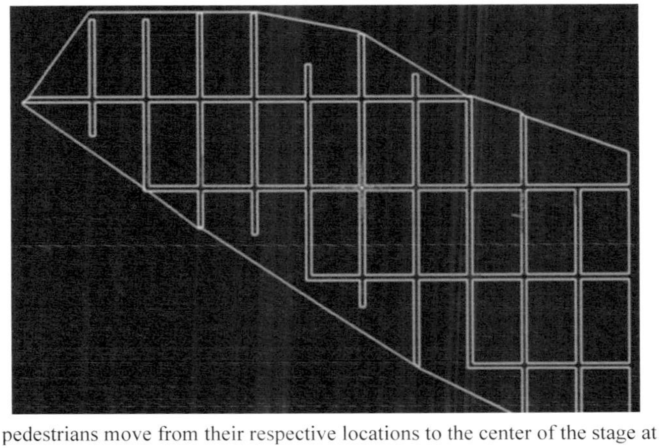

Figure 7: The pedestrians move from their respective locations to the center of the stage at point (41, -2).

direction from the nearest flood source. However, the behavior of the people located to the right of the scenario would not have to move and agglomerate with all the others.

5.1 A more complex scheme

Despite the results, it should be noted that these were obtained under the consideration that people evacuate the area on foot and do not move in vehicles. It is for this reason that a simulation that contemplates vehicles, with the capacity to move more people without them being exposed to rain or accidents, should also be considered. A vehicle was considered for each group of people, i.e., a total of 61 vehicles, each with an estimated radius of 2 m and moving at similar speeds to the people on foot. Figure 8 represents this new scheme, where the blue dots are the vehicles.

The simulation reflects, in the case where evacuation to the point (250.0) is treated, how people and vehicles as a whole make movement slower as they become obstacles during travel due to their interactions. This behavior can be seen in Figure 9 and Figure 10. In this scenario, the vehicles have the same speed when they move, but in reality, their speed will be affected by the flooded environment, as well as by various obstacles. Also, the versatility of vehicles is much less than that of people, so evacuation times, in general, will be severely affected by the presence of vehicles.

On the other hand, if one considers the case where people conglomerate in the center of the stage (41,-2), see Figure 11, it is observed that people and vehicles occupy more and more space and their interactions make movement difficult. Figure 12 shows this behavior.

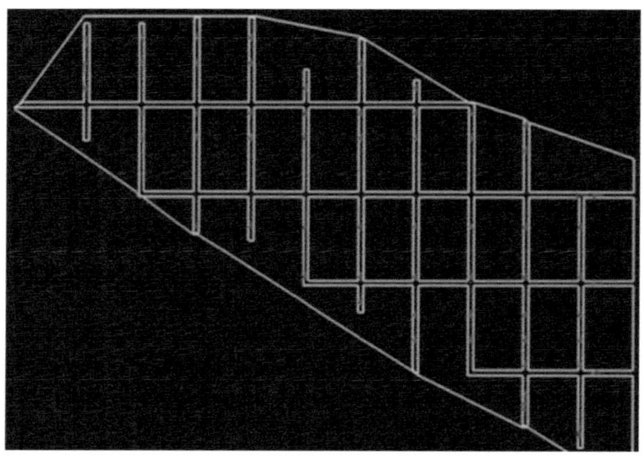

Figure 8: Vehicles (blue dots) are located between people groups.

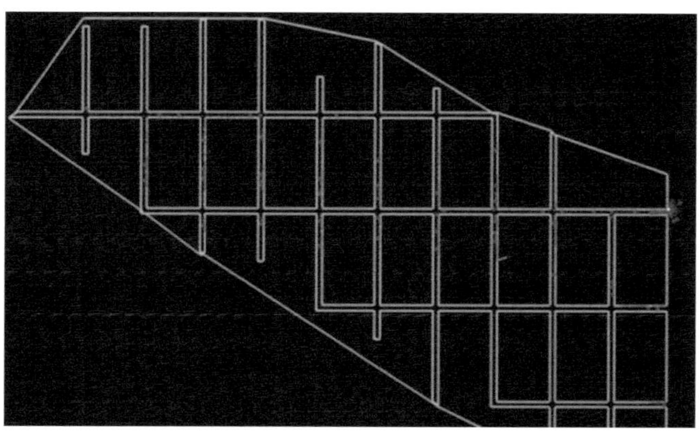

Figure 9: Evacuation simulation with vehicles and people. Vehicles are marked as blue dots while people are red dots.

Figure 10: Closer view of the simulation. Vehicles (blue dots) interact with people (red dots) and serve as obstacles during displacement.

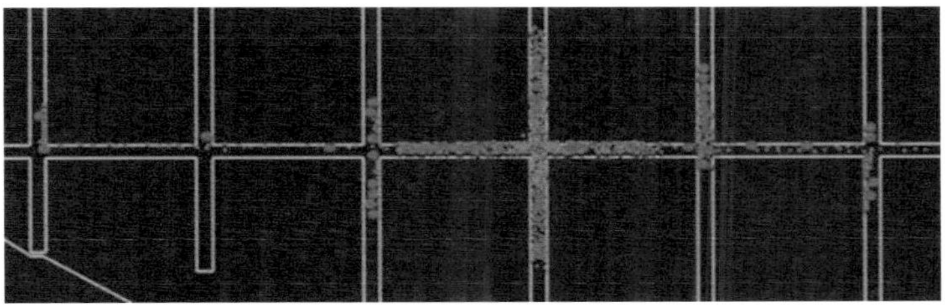

Figure 11: The pedestrians move from their respective locations to the center of the stage at point (41, -2), blue dots represent vehicles while red dots represent people.

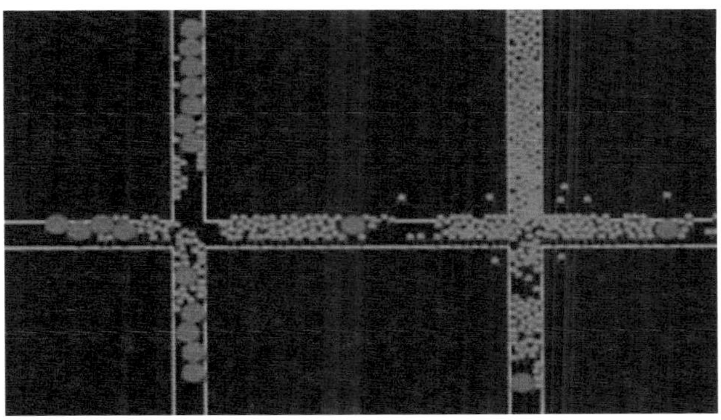

Figure 12: Interactions between pedestrians and vehicles cause congestions that block movement.

It is important to indicate that the *CgI-M model* is open, constantly revised, enriched, updated, and that it is currently implemented as a *modus operandi* of a *Cognitive Architects team* to build *Cognitive Solutions*. Subsequent subsections give a review of the parts of the model.

6. Conclusions and Future Work

This simulation exercise should not be considered as a real predictive case of the behavior of a group of people in the middle of a disaster, such as a flood, but as a first approximation to estimate behavior and response times in the event of an evacuation.

It should be noted that the real scenario has multiple quantitative variables that have not been considered in this paper, such as the age of people, sex, physical complexion, etc., due to the complexity of the mathematical model that would be required in order to address the behavior of pedestrians and vehicles involved in the simulation.

On the other hand, there is no way to quantify the qualitative factors of people in a scenario like this and put them into an equation to estimate behavior. These factors can be the shock caused by the situation, fear of leaving home and possessions, and even disbelief at the consequences of the overflow of the Río Bravo and the imminent flooding in the area.

Besides the above, these types of models can help in urban planning at the time of building lots of houses, especially in areas susceptible to floods or overflows. In this sense, as in this hypothetical case, it could be observed that in the face of the flooding of the Río Bravo, the Acequia Madre and the flooding of the Díaz Ordaz viaduct, there are only a few evacuation routes for the people who live in the area. Also, the evacuation time will depend on several factors, but in the best-case scenario, as analyzed in Figure 6, it would take just over 4 minutes without considering severe obstacles or flooding that impede mobility, i.e., moving at a constant speed of approximately 1.5 m/sec. However, this speed could be reduced by two-thirds due to terrain conditions and obstacles, so that time could easily be tripled to 12 minutes, which would put people's lives at risk. In the case of the more complex scheme, which involves the presence of vehicles, these play an important role in the mobility of individuals since they function as obstacles on the stage because they are less versatile than people, which will result in a much longer evacuation time than if it were only people. That is why this simulation represents a first step in the elaboration of protocols and evacuation routes in case this type of flooding occurs in the future.

As future work, the idea would be to modify the simulation to establish possible emergency routes in case of floods, which would be used by people depending on their location in the area. Besides, the simulation can be improved by adapting menge to new platforms, such as Unity, which can be used to create more realistic scenarios and objects.

References

[1] Reyes Rubiano, L. 2015. Localización de instalaciones y ruteo de personal especializado en logística humanitaria post-desastre - caso inundaciones. Univ. La Sabana.

[2] Mpacts, I., Daptation, A. and Ulnerability, V. 2002. Climate change 2001: Impacts, Adaptation, and Vulnerability, 39(6).

[3] Schaller, N. et al. 2016. Human influence on climate in the 2014 southern England winter floods and their impacts. Nat. Clim. Chang., 6(6): 627–634.

[4] Silva, M.M. and Costa, J.P. 2018. Urban floods and climate change adaptation: The potential of public space design when accommodating natural processes. Water (Switzerland), 10(2).

[5] Bronstert, A. 2003. Floods and climate change: interactions and impacts. Risk Anal., 23(3): 545–557(13) ST-Floods and Climate Change: Inter.

[6] González Herrera, M.R. and Lerma Legarreta, J.M. 2016. Planificación Y Preparación Para La Gestión Sustentable De Riesgos Y Crisis En El Turismo Mexicano. Estudio Piloto En Ciudad Juárez, Chihuahua. Eur. Sci. Journal, ESJ, 12(5): 42.

[7] Ochoa Zezzatti, A., Contreras-Masse, R. and Mejia, J. 2019. Innovative Data Visualization of Collisions in a Human Stampede Occurred in a Religious Event using Multiagent Systems, no. Figure 1, pp. 62–67.

[8] Curtis, S., Best, A. and Manocha, D. Menge: a modular framework for simulating crowd movement. Collect. Dyn., 1: 1–40.

[9] Curtis, S., Best, A. and Manocha, D. MENGE, 2013. [Online]. Available: http://gamma.cs.unc.edu/Menge/developers.html. [Accessed: 29-Oct-2019].

[10] Google Maps. [Online]. Available: https://www.google.com.mx/maps/@31.7483341,-106.4920319,17.66z. [Accessed: 28-Oct-2019].

[11] Fruin, J.J. 1971. Designing for pedestrians: A level of service concept. Highw. Res. Rec., 355: 1–15, 1971.

CHAPTER-11

Simulating Crowds at a College School in Juarez, Mexico

A Humanitarian Logistics Approach

Dora Ivette Rivero-Caraveo,[1,*] *Jaqueline Ortiz-Velez*[2] and *Irving Bruno López-Santos*[2]

1. Introduction

Due to the frequency of natural disasters and political problems, interest in humanitarian logistics among academics and politicians has been increasing. In the literature, studies that analyze trends in humanitarian logistics were found to focus more on how to deal with the consequences of a disaster than on its prevention [1]. Simulation can be useful to pose different scenarios and be able to make decisions about strategies to help avoid stampedes when a natural or man-made disaster occurs. This helps to define a preventive strategy in the eventuality of a disaster. As a case study, we present a model to simulate crowds, based on a building of a college in Ciudad Juárez, Mexico: the Instituto Tecnológico de Ciudad Juárez (ITCJ).

Ciudad Juárez is in the northern border area of Mexico. It is a city that has had a population growth due to a migratory process of great impact, receiving a significant number of people from the center and south of the country in search of better opportunities, which has resulted in many cases in the settlement of areas not appropriate for urban development, a situation that has been aggravated as the natural environment has changed negatively [2]. Recently, the migratory flow has also come from countries in Central and South America in the form of caravans seeking asylum in the United States.

In 2016, the municipal government of Ciudad Juárez made an list of natural and anthropogenic risks. As for geological risks, the document mentions that in 2014 there were some earthquakes that measured up to 5.3 on the Richter scale, it is mentioned that: the province has a tectonic activity is an internally active zone and will have seismic activity sooner or later [2].

The ITCJ, founded on October 3rd 1964, is ranked number 11 among the National System of Technological Institutes [3]. The institution is located at 1340 Tecnológico Ave, in the northern part of the city. Figure 1 shows the satellite location obtained through Google Maps. Ciudad Juárez has a total of 27 higher education schools, and the ITCJ ranks third with a total of 6510 students enrolled [2]. To date, the institution offers 12 bachelor's degrees, three master's degrees, a doctorate and an open and distance education program [3].

[1] Universidad Autónoma de Ciudad Juárez.
[2] Instituto Tecnológico de Cd. Juárez.
* Coresponding author: al183255@alumnos.uacj.mx

Figure 1: Satellite view of the ITCJ obtained through Google Maps [4].

Over the years, the institution has grown and new buildings have been built, with the Ramón Rivera Lara being the oldest and most emblematic. Figure 2 shows photographs of the building. To model the simulation, classrooms were taken into account in the aforementioned building, since it is the oldest building in the institute and it is where the majority of students are concentrated.

This work presents a model and simulation based on the Ramón Rivera Lara building of the ITCJ. The objective is to evaluate the feasibility of using the Menge framework to simulate the evacuation of students and teachers in the event of a disaster. In the context of humanitarian logistics, simulations help to plan strategies before the occurrence of a natural or anthropogenic disaster. As future work, the aim is to model the whole institution and contrast the simulation against the simulacrums that are carried out in the school. Finally, to develop an informatic tool based on Menge and Unity so that decision-makers can evaluate different distributions in the classrooms and through the simulation, to be able to evaluate if it is possible to obtain a more efficient one that minimizes the risks in case of a disaster.

Figure 2: Ramón Rivera Lara Building [5].

2. Related Work

This section presents a brief review of the literature from previous work related to the topic presented. It is divided into three subsections: humanitarian logistics, crowd simulation, and mathematical models.

2.1 Humanitarian logistics

Humanitarian logistics is a process by which the flow and storage of goods and information is planned, implemented and controlled from a point of origin to the point where the emergency occurred [6].

Three phases are identified in the life cycle of a disaster: pre-disaster (preparedness phase), post-disaster (response phase), and finally, the recovery phase [7]. In the initial phase of the life cycle mentioned above, risk preparedness and prevention plans are established; in this regard, simulation can be a tool to evaluate prevention plans.

In the pre-disaster or preparedness phase, it is important to identify different scenarios, specific risks, and complexity; simulations help to assess the risks in different scenarios [8]. The next section discusses crowd simulation, both literature, and tools.

2.2 Crowd simulation

Crowd simulation is a fundamental problem of video games and artificial intelligence; recently, it has also been used for other serious applications, such as evacuation studies [9]. For the previous one, this sort of simulation can contribute to the planning phase in the humanitarian logistics life cycle, specifically to elaborate evacuation plans in case of a possible natural or man-made risk.

These types of simulations apply to humanitarian logistics in four types of situations: trampling and crushing at religious events, trampling, crushing and sinking of ships, crushing at concerts and in bars, and contingency situations due to natural disasters, such as earthquakes, floods, fires, etc., that cause destruction to man-made structures [10].

For the simulation presented in this paper, we used Menge, which is an open platform based on C++. Menge is based on the needs of pedestrians and breaks down the problem into three sub-problems: target selection (where people will move), computational plan and adaptation plan. This platform has the advantage that it does not require advanced knowledge of programming and multi-agent systems for its use [11–13]. It provides documentation and examples in order to be able to adapt it to different contexts.

2.3 Mathematical models

According to [14], there is a way to estimate the velocity of pedestrians using an Equation. The velocity of an agent is dictated by Equation 1:

$$V_i(t) = [(v + h + nc)/(a + d]* f * imc * s \quad (1)$$

Where:

- xi is the velocity of agent *i* at time t.
- To solve *Vi*(*t*) determines the position of an agent with respect to time.
- *v* is the average pedestrian speed for all agents.
- *h* represents the height of the simulated person
- *a* represents the age of the person
- *d* is the density of people per square meter
- *imc* is the individual's body mass index
- *s* is the sex of the simulated individual
- *nc* is the level of consciousness of the individual

According to [14] the criteria that can be used can justify:

- Density of people: If the density is higher the mobility decreases.
- Level of consciousness: If the person is in a state of drunkenness or has just awakened his speed will not be optimal.
- Age: A person's motor performance is affected by their age since a child and an elder do not have the same performance as a young adult.
- Body Mass Index: The body mass index indicates if the person is overweight, obese, is lacking in weight or is in a normal state.
- Height: The stride of a person is directly proportional to the height of the same, so the height is an important factor.
- Gender: The sex of a person intervenes in the force that will have the same since a person with more strength can push the rest and advance more quickly.
- The main variables that directly affect the pedestrian movement of an agent are fear (f), body mass index (BMI) and sex (s), which are direct multipliers in the equation.
- The average speed added to a person's height significantly affects the final velocity, however, they are affected by the age (a) of the individual and the density of people (d) at time t in a way inversely proportional to vi. This is reflected in that this last pair of attributes divides the two first mentioned.

3. Materials

The following describes the hardware and software used to perform the simulation.

3.1 Hardware materials

A Lenovo Laptop was used for the simulation and the characteristics of the device are shown in Figure 3.

Device specifications

ideapad 330-15ARR

Device name	LAPTOP-42B8IO2E
Processor	AMD Ryzen 3 2200U with Radeon Vega Mobile Gfx 2.50 GHz
Installed RAM	8.00 GB (7.55 GB usable)
Device ID	91E29D50-3E98-41D1-9606-1DCEAE58FE88
Product ID	00325-81128-68145-AAOEM
System type	64-bit operating system, x64-based processor
Pen and touch	No pen or touch input is available for this display

Figure 3: Specifications of the device where the simulation was run.

3.2 Software

As far as software is concerned, the materials used are listed below.

- **Operative System.** Windows 10 Home.
- **Operative System type.** 64-bit operating system, x64 based processor.
- **IDE.** Microsoft Visual Studio community 2019. In order to compile y generate menge.exe application.
- **Text Editor.** Visual Studio Code, version 1.38.1.
- **Menge software.** A framework for modular pedestrian simulation for research and development, free code [13].
- **Windows Command Prompt.** Used to run simulations.

4. Methodology

To model the rooms and the section of the Ramón Rivera Lara building, first, the architectural plans of the building were analyzed. Figure 4 shows the upper view of the ground floor of the building.

To establish the coordinates of the agents and the obstacles, measurements were taken of four halls adjacent to the ground floor. First, a single room was simulated and later, the simulation was made with the four rooms. To establish the speed of the pedestrians, the characteristics of the morning shift students were analyzed in the classes from 8:00 to 9:00 AM and Equation 1, which can be viewed in the previous section, was applied.

5. Simulation

The simulation was divided into two stages. First, a single classroom was simulated using the methodology described in the previous section. Subsequently, four adjacent classrooms were used.

5.1 Simulation of a single classroom

For this simulation, a project XML file was defined, which is shown in Figure 5. Inside the project folder, three XML files require Menge: scene, behavior, and view, as well as a file called graph.txt which contains the trajectories of multiple agents. Figure 6 shows the project folder with the four files mentioned.

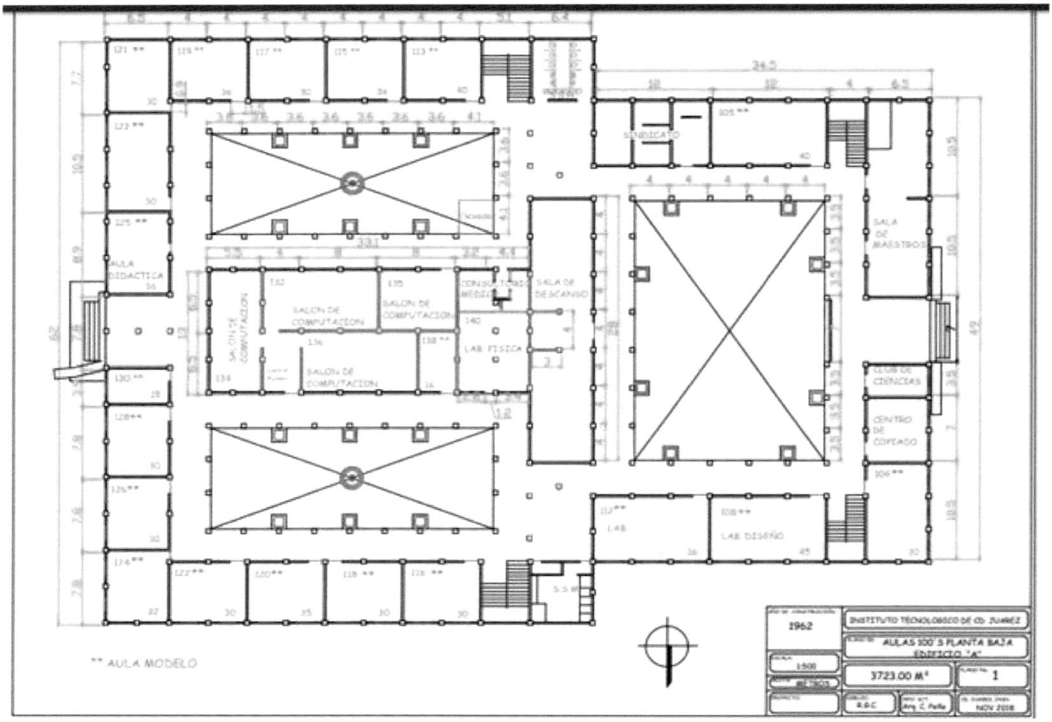

Figure 4: Ground floor of the Ramón Rivera Lara building.

```
c: > Users > diriv > Documents > software > Menge-master > Menge-master > examples > core > Salon5.xml
<?xml version="1.0"?>

<Project
    scene="Salon5/sem6S.xml"
    behavior="Salon5/sem6B.xml"
    view="Salon5/sem6V.xml"
    model="orca"
    dumpPath="images/test"
/>
```

Figure 5: Project XML File to simulate one classroom.

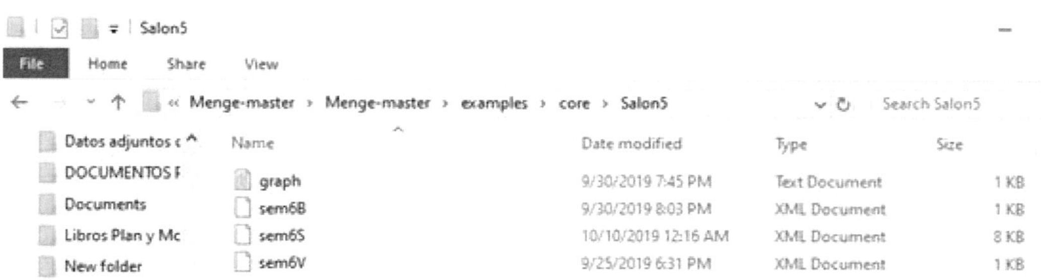

Figure 6: Project folder with scene, behavior and view XML files, as well as the graph file.

In the graph.txt file, the paths of the different agents were defined; Figure 7 shows some of the paths defined in that file. It is worth mentioning that the darkest blue agent represents the teacher, while the students are represented in light blue.

One of the files that most require configuration is the scene file, since there are agents and obstacles declared. As the number of agents increases, this file grows proportionally. Figures 7 and 8 show some sections of this file.

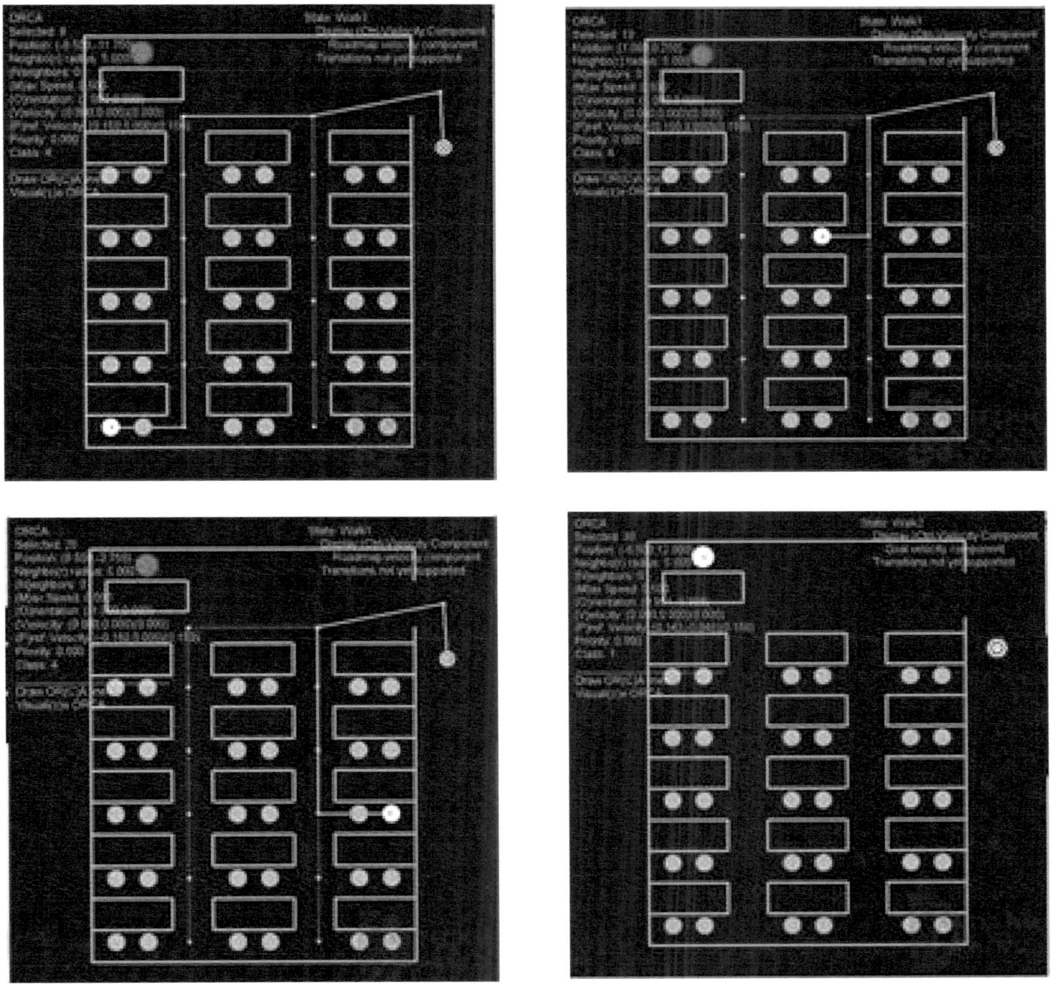

Figure 7: Some of the trajectories towards the target of the agents.

To run the simulation, we must run the menge.exe and send as a parameter the project we want to run (XML file of the project) which in this case is Salon5.xlm. Figure 9 shows an example of how to run the simulation. Figure 10 shows the simulation in its initial, intermediate and final stages.

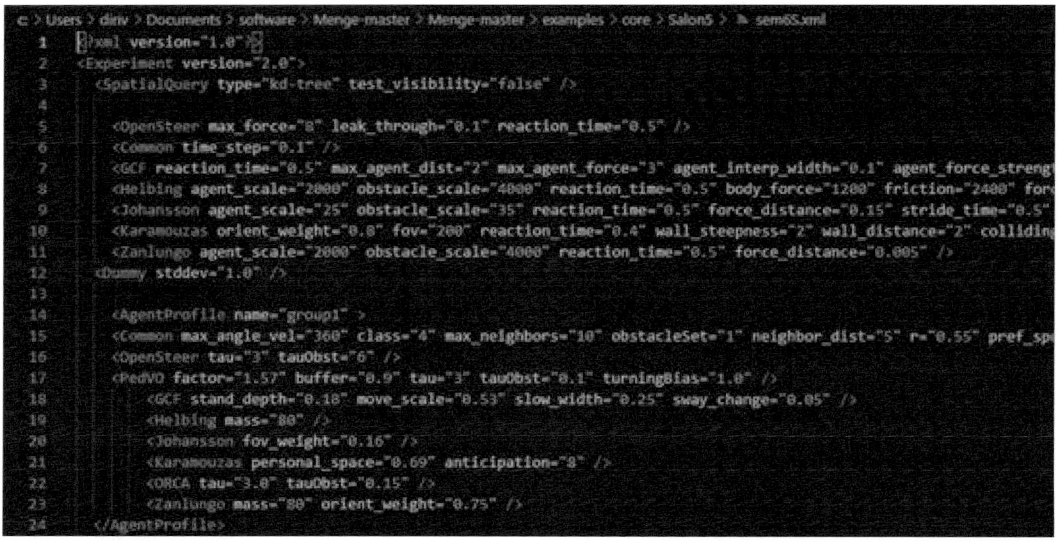

```
c > Users > diriv > Documents > software > Menge-master > Menge-master > examples > core > Salon5 > sem6S.xml
<?xml version="1.0"?>
<Experiment version="2.0">
  <SpatialQuery type="kd-tree" test_visibility="false" />

    <OpenSteer max_force="8" leak_through="0.1" reaction_time="0.5" />
    <Common time_step="0.1" />
    <GCF reaction_time="0.5" max_agent_dist="2" max_agent_force="3" agent_interp_width="0.1" agent_force_streng
    <Helbing agent_scale="2000" obstacle_scale="4000" reaction_time="0.5" body_force="1200" friction="2400" for
    <Johansson agent_scale="25" obstacle_scale="35" reaction_time="0.5" force_distance="0.15" stride_time="0.5"
    <Karamouzas orient_weight="0.8" fov="200" reaction_time="0.4" wall_steepness="2" wall_distance="2" collidin
    <Zanlungo agent_scale="2000" obstacle_scale="4000" reaction_time="0.5" force_distance="0.005" />
  <Dummy stddev="1.0" />

    <AgentProfile name="group1" >
    <Common max_angle_vel="360" class="4" max_neighbors="10" obstacleSet="1" neighbor_dist="5" r="0.55" pref_sp
    <OpenSteer tau="3" tauObst="6" />
    <PedVO factor="1.57" buffer="0.9" tau="3" tauObst="0.1" turningBias="1.0" />
        <GCF stand_depth="0.18" move_scale="0.53" slow_width="0.25" sway_change="0.05" />
        <Helbing mass="80" />
        <Johansson fov_weight="0.16" />
        <Karamouzas personal_space="0.69" anticipation="8" />
        <ORCA tau="3.0" tauObst="0.15" />
        <Zanlungo mass="80" orient_weight="0.75" />
  </AgentProfile>
```

Figure 8: Scene XML file segment 1.

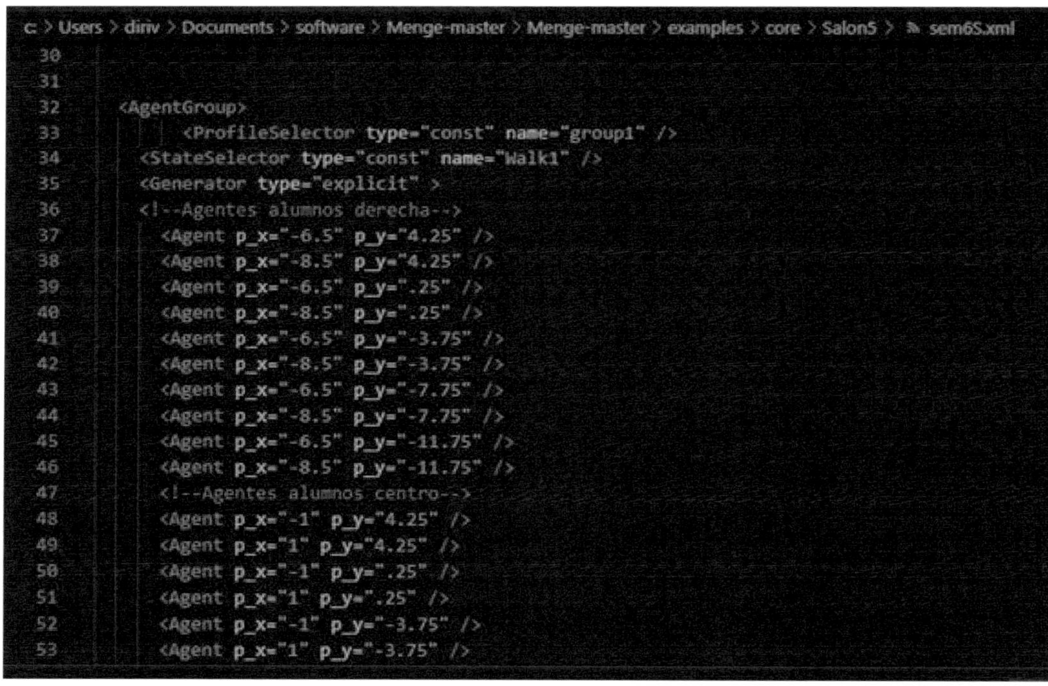

```
c > Users > diriv > Documents > software > Menge-master > Menge-master > examples > core > Salon5 > sem6S.xml


    <AgentGroup>
          <ProfileSelector type="const" name="group1" />
      <StateSelector type="const" name="Walk1" />
      <Generator type="explicit" >
      <!--Agentes alumnos derecha-->
        <Agent p_x="-6.5" p_y="4.25" />
        <Agent p_x="-8.5" p_y="4.25" />
        <Agent p_x="-6.5" p_y=".25" />
        <Agent p_x="-8.5" p_y=".25" />
        <Agent p_x="-6.5" p_y="-3.75" />
        <Agent p_x="-8.5" p_y="-3.75" />
        <Agent p_x="-6.5" p_y="-7.75" />
        <Agent p_x="-8.5" p_y="-7.75" />
        <Agent p_x="-6.5" p_y="-11.75" />
        <Agent p_x="-8.5" p_y="-11.75" />
        <!--Agentes alumnos centro-->
        <Agent p_x="-1" p_y="4.25" />
        <Agent p_x="1" p_y="4.25" />
        <Agent p_x="-1" p_y=".25" />
        <Agent p_x="1" p_y=".25" />
        <Agent p_x="-1" p_y="-3.75" />
        <Agent p_x="1" p_y="-3.75" />
```

Figure 9: Scene XML file segment 2.

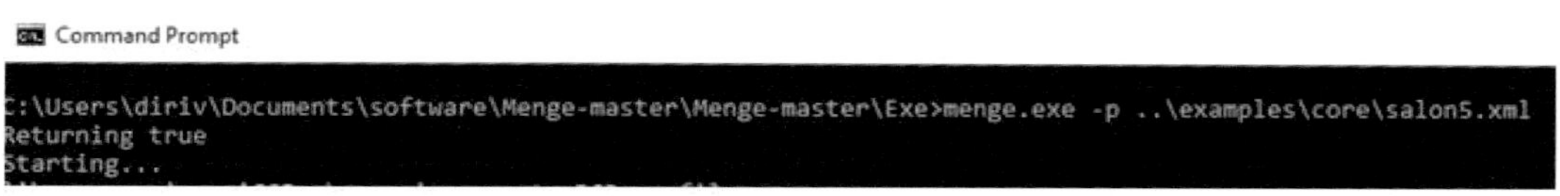

```
Command Prompt

:\Users\diriv\Documents\software\Menge-master\Menge-master\Exe>menge.exe -p ..\examples\core\salon5.xml
Returning true
Starting...
```

Figure 10: Running simulation.

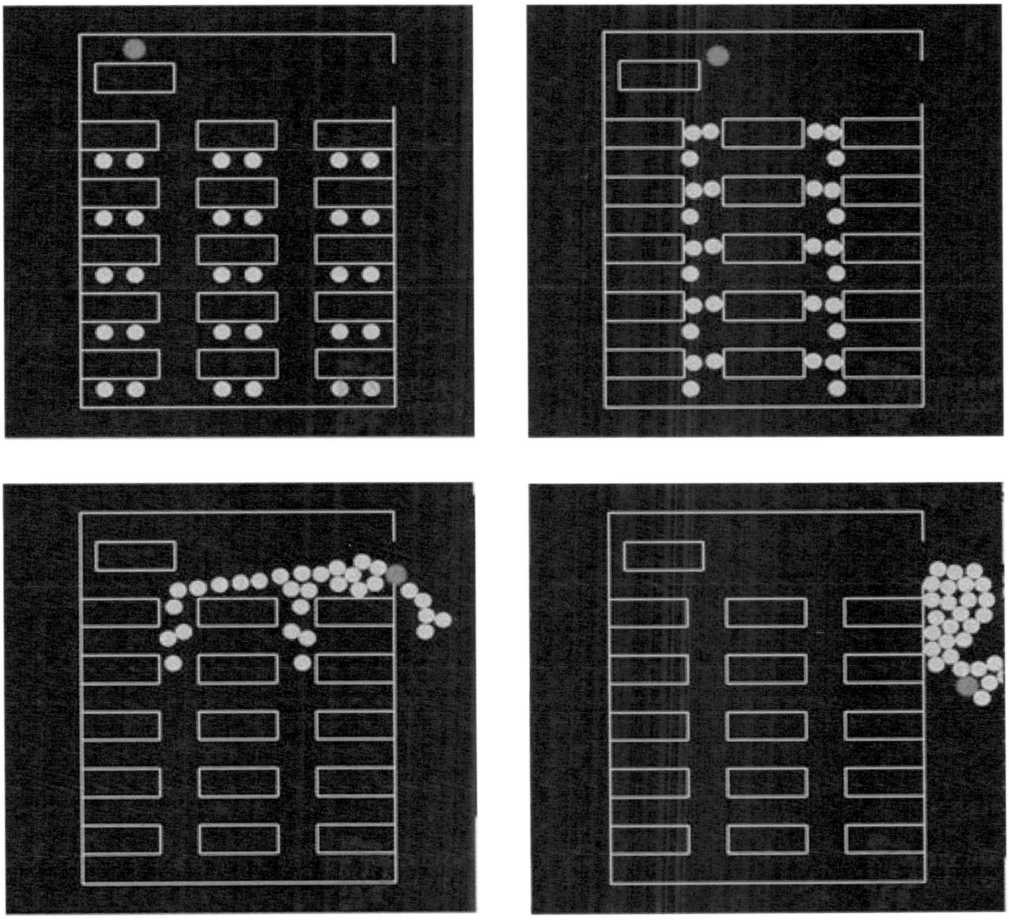

Figure 11: Simulation of a classroom.

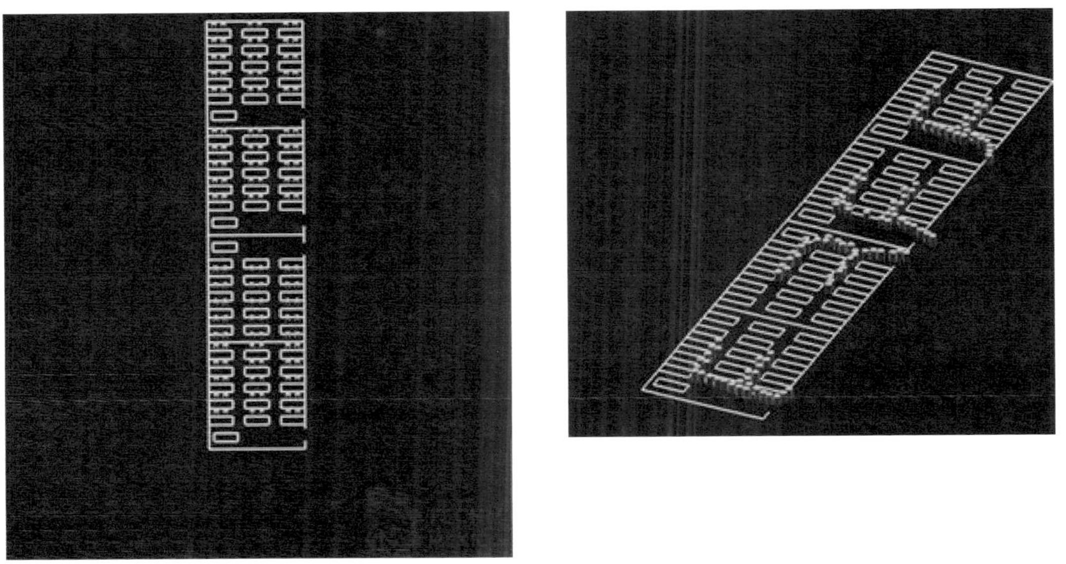

Figure 12: Simulation of four classrooms.

5.2 Simulation of four adjacent rooms

To run this simulation, the procedure mentioned in Section 5.1 was used. The structure of the files is similar, simply scaled. The figure shows the simulation of the four adjacent rooms.

6. Conclusions and Future Work

Crowd simulation can be a first approach to analyzing possible evacuation plans in an emergency. It can help to detect bottlenecks in the event of a mass evacuation, so the distribution of halls can be improved, minimizing such bottlenecks. One of the disadvantages is that this model does not take into account the stress and panic behaviors that a disaster situation can induce.

As future research, the plan is to simulate the entire Ramón Rivera building, including the upper floor, as well as other ITCJ buildings. It is also planned to make a computer tool based on Menge and Unity for a more realistic simulation. This tool is intended to be used so that people who have no knowledge of the use of Menge can move some parameters in a way and change the simulation scenarios.

References

[1] Chiappetta Jabbour, C.J., Sobreiro, V.A., Lopes de Sousa Jabbour, A.B., de Souza Campos, L.M., Mariano, E.B. and Renwick, D.W.S. 2017. An analysis of the literature on humanitarian logistics and supply chain management: paving the way for future studies. Ann. Oper. Res., pp. 1–19.

[2] Instituto Municipal de Investigación y Planeación. Atlas de riesgos naturales y atlas de riesgos antropogénicos. Ciudad Juárez, Chihuahua. 2016. [Online]. Available: https://www.imip.org.mx/atlasderiesgos/.

[3] ITCJ - Nosotros. 2019. [Online]. Available: http://www.itcj.edu.mx/nosotros.

[4] Google Maps. 2019. [Online]. Available: https://www.google.com/maps/place/Instituto+Tecnológico+de+Ciudad+Juárez/@31.7211545,-106.4251575,1122m/data=!3m1!1e3!4m5!3m4!1s0x86e75dc249fd3e4b:0x58a769357165487b!8m2!3d31.7213256!4d-106.4238612.

[5] Aguilar, F. 2017. 'Liebres', de fiesta, El Diario de Juárez.

[6] van der Laan, E., van Dalen, J., Rohrmoser, M. and Simpson, R. 2016. Demand forecasting and order planning for humanitarian logistics: An empirical assessment. J. Oper. Manag., 45: 114–122.

[7] Özdamar, L. and Ertem, M.A. 2015. Models, solutions and enabling technologies in humanitarian logistics. Eur. J. Oper. Res., 244(1): 55–65.

[8] Souza, J.C. and de C. Brombilla, D. 2014. Humanitarian logistics principles for emergency evacuation of places with many people. Procedia - Soc. Behav. Sci., 162, no. Panam: 24–33.

[9] Van Toll, W., Jaklin, N. and Geraerts, R. 2015. Towards Believable Crowds: A Generic Multi-Level Framework for Agent Navigation. Ict.Open.

[10] Ochoa, A., Rudomin, I., Vargas-Solar, G., Espinosa-Oviedo, J.A., Pérez, H. and Zechinelli-Martini, J.L. 2017. Humanitarian logistics and cultural diversity within crowd simulation. Comput. y Sist., 21(1): 7–21.

[11] Simonov, A., Lebin, A., Shcherbak, B., Zagarskikh, A. and Karsakov, A. 2018. Multi-agent crowd simulation on large areas with utility-based behavior models: Sochi Olympic Park Station use case. Procedia Comput. Sci., 136: 453–462.

[12] Curtis, S., Best, A. and Manocha, D. 2016. Menge: A modular framework for simulating crowd movement. Collect. Dyn., 1: 1–40.

[13] Curtis, S., Best, A. and Manocha, D. 2013. MENGE.

[14] Ochoa Zezzatti, A., Contreras-Masse, R. and Mejia, J. 2019. Innovative Data Visualization of Collisions in a Human Stampede Occurred in a Religious Event using Multiagent Systems, no. Figure 1: 62–67.

CHAPTER-12

Perspectives of State Management in Smart Cities

Zhang Jieqiong and *Jesús García-Mancha**

1. Introduction

The development of technology in public management will be the first step towards the transformation of large cities, the use of big data, technologies for industrial internet within the cloud, new accounting tools, budget management, and others. The transformation towards the development of the "Start cities" in countries like the Russian Federation, the People's Republic of China and Mexico and two societies in Africa as emerging powers of development is the first-level priority, they are usually commissioned to develop innovations in the areas of artificial intelligence, mass data processing, intranet, and computer security, that is why great efforts are being made in legislative matters by prioritizing laws whose main objective is the inclusion of a digital economy through the use of technologies, as it will cover absolutely everything in the development of trade, infrastructure, urban development, public transport, payment of taxes, etc.

In the case of the Russian Federation, cities in full industrial development have a greater advantage over larger metropolitan cities such as Moscow and St. Petersburg, although these cities are large metropolises, their congestion, and limited growth space could hinder the use of new state management systems, for example in the use of new public transport systems and urban development, compared to emerging cities such as Kazan, Ekaterinburg, Rostov-on-Don or Sochi, which is the best-planned city in the federation, the "Intelligent transport" how introduction to the transport systems in these cities and future metropolises will be able to minimize the potential and future problems well in advance, in the particular case of the city of Kazan in the Republic of Tatarstan is contemplated the creation of a model of development of public transport, which will bring together public and private specialists from the sector of construction, insurance, civil protection, transport and communications and the automotive sector. As for the improvement of the quality of transport and road services, one vital point cannot be forgotten: road safety. Road accidents in these big cities are increasing year by year, mainly due to an imbalance between their infrastructure and the needs of citizens and the state; a strong investment is needed in the construction of new avenues, the maintenance of the existing ones, pedestrian crossings, ring roads around metropolitan areas to avoid traffic congestion [1]. In the implementation of such measures, a special role is played by the introduction of technical means of regulation with the use of electronic control systems, automation, telemetry, traffic control and television to control roads in a large area or throughout the city. The

Kazan Federal University, Republic of Tatarstan, Russian Federation.
Email: 1457479864@qq.com
* Corresponding author: jesus0291gm@gmail.com

construction of new communication and road transport centers is not enough in itself. The role of the intellectual component in the organization of the operation of the street and road networks is increasing. The concept of "Smart cities" in recent years in Mexico and Latin America has ceased to be considered a fantasy, due to the rise of interest in this topic, unlike the large metropolises of the Russian Federation mentioned above, the large cities in Mexico do have vital space to develop at an even faster level to increase the quality of life of its citizens, however, the challenge in Mexican cities is not the budget or government development plans, is the absence of political projects and a marked lack of automation strategies and legislation. In Mexico City, the main problem is transportation, since mobility in a city of more than 20 million inhabitants, in addition to the people who travel every day from the State of Mexico, is an alarming priority, as can be seen in Figure 1.

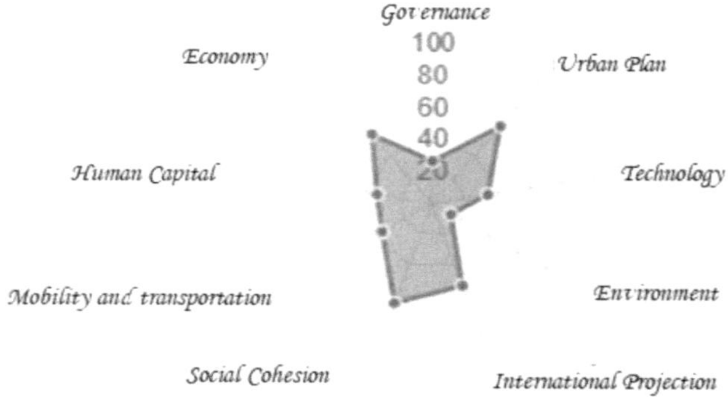

Figure 1: Data based on portal Cities in Motion Índex of the Escuela de Negocios IESE [2].

Mexico City is ranked 118th in the world in the use of technology in state management, and it is clear that there has been little development in the area of mobility and transport on a par with the use of technology and little legislation in this area. However, the combination of public and private initiatives is increasing day by day, and the population spends an average of 45 days a year using transportation [3]. Querétaro on the other hand already has a legislation developed since 2013 focused mainly on online tools, all public service information will be managed and connected to the internet 100% in the city by 2021, will be used in services such as garbage collection, payment of electricity, gas and water services, transportation services, traffic reports, while in the industrial sector will promote the use of sustainable development. In these times of innovation, humanity has entered an urban era, never in the whole history of mankind, half the population of the planet lives in cities, life is more connected than ever, connectivity is not measured in distance, it is measured in data consumption, data that is used as big data, cloud storage, etc. The functioning of government institutions in terms of information management improves performance and its development allowing later regional and municipal governance. There is much discussion about how and how much information should be collected from citizens. Intelligent cities are now an experiment for new public management to ensure the proper use of data, the quality of life of citizens and their rights, seeking a rapprochement between the citizen and the state. Other possible risks derived from the management of information and data collected in the intelligent city would be, on the one hand, the generation of information bubbles that thanks to big data and algorithms deform reality and only show us information according to our respective preferences and, on the other hand, the consolidation of the phenomenon of the so-called "post-truth", which consists of the possibility of lying in the public debate, especially on the Internet, without relevant consequences and without the supporters of those who have lied reacting even though the lie is even judicially proven. The "post-truth", in short, is built on a "certain indifference to the facts" [4]. In countries where corruption indexes are high, the panorama of misuse of citizens' information is one of the biggest challenges for the conception of smart cities, that is why emphasis is placed on the development of general public law, that is why transparency systems must have legal tools in the data collection and storage sets opened by state administrations, over time there will be a serious problem of duplication of data and many other issues associated with the digitization of documents that are currently written on paper, which will be very complicated to develop the digital transformation of state institutions, it will be necessary the participation of the general population and organizations to solve this problem, citizens will have to take responsibility for uploading their data legally if they want to be part of the process in question. For example, when a child is born, its parents are obliged to register it at the local civil registry office to obtain a birth certificate, the secretary of foreign affairs is obliged to obtain a passport, the institute of social security is obliged to obtain a certificate of insurance, and the institute of education is obliged to obtain a certificate of preschool education, at the time of the child's birth all his information will be captured in state databases in a digital data "Cloud", where the data of the citizens will be stored, the interaction between the citizens and the state will change forever, the state will not only be limited to provide basic services but also to manage all these data and more complex but at the same time more dynamic and fast life situations, by using tools and algorithms developed with high quality. The direction and priority of the development of these tools is directly focused on the improvement of bureaucratic, economic and social processes and the improvement of the quality of life of the cities, all the state information managed from a centralized system as an objective to improve the governance system also giving rise to new public services via electronic forming an intelligent system of "self-government". Not only is the use of citizen data proposed in the "Intelligent state management", but the systems will also contain information on the population, territories, properties, urban planning, social development, public space, budgeting, etc. This will mean a considerable increase in state income in a faster and more reliable way, for this reason, the following conceptual diagram is proposed, which explains in a general way the potential of digitalization in the management of public information in the framework of state management. Figure 2.

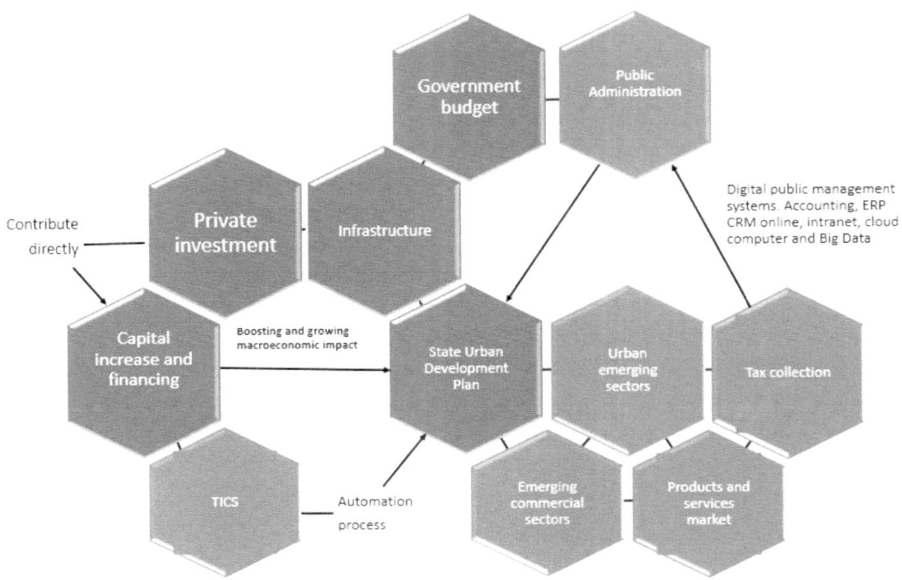

Figure 2: The evolution in the state management to optimize a Smart City.

The evolution in the state management has many questions before the automation of the processes, as it is already known the fear to the disappearance of traditional jobs is not well seen by anybody, it is not necessary to think about the disappearance of state jobs but of the evolution of these, will job disappear? Of course, they will, but at the same time, new ones will also arise, as has always happened throughout the history of the world. Rational management of resources and automation in public management will seek to eliminate excessive costs, diversion of public resources, duplication of jobs, money laundering, etc. It will free up personnel resources that can be used in the improvement of other services to make the bureaucratic system more efficient. Constant monitoring 24 hours a day 7 days a week throughout the year will provide an audit of the resources to detect corrupt processes by detecting possible irregularities thanks to the implementation of new algorithms when creating public contracts, state concessions, recruitment of personnel to avoid conflicts of interest.

1.1 Public health and education

The new management tools in the field of health will seek to adapt programs to provide better services by replacing obsolete, slow, cumbersome information management systems to make them less bureaucratic. With the popularization of applications and their use in the bureaucratic process, technologies for voice interaction, computer vision and cognitive computing, and cloud computing have gradually matured. Several applications of artificial intelligence have become possible in the field of medicine. Saving the time doctors spend on disease diagnosis, long waiting hours and patient mobility, improving diagnosis and greatly reducing care times and costs for the state. For another example, images are an important diagnostic foundation, but how to obtain high-quality medical imaging data is a difficult problem in the medical field, and deep learning can improve the quality of medical images, and deep learning technology using artificial intelligence can be extracted from medical images. Useful information to help physicians make accurate judgments.

1.2 Urban planning

When we think about cities we think about buildings, streets, buildings, noise but we should think about people, cities are fundamentally for people, the buildings in the city and public spaces promoting more community among people, the Smart Cities is just a concept in which we add the management of aspects related to the environment such as water, electricity, transport, etc., with aspects linked to social programs of education, health and aspects associated with the administration, good governance when it comes to being efficient, obviously it is a model that depends on the use of new technologies so what is happening is that several words are always on everyone's lips when we talk about these smart cities, concepts such as sustainability, efficiency, effectiveness, innovation, and investment. Because this is a business with a lot of money in the development model of smart cities, why? Because of course there are many interests, we are not talking about money invested just because we want to solve a problem of resource efficiency, these resources have much to do with the capabilities with the development in the future thanks to the objectives of the smart cities, not only should be a mechanism to meet our challenges to make our economy more competitive and more efficient in the future, ICT in the thread of an intelligent city are from public administration to energy consumption to the use of urban transport all this has to do with the management of the Big data, the information they give us from the state archives, the crossing of information allows us to generate new solutions and applications to live in a much more efficient city, we are in a process of transformation not so much of the model of city but of the economic model, new opportunities for our industry and technology, to look for a different model of the city, an intelligent city cannot exist without an intelligent government, it is necessary to develop a model of education, creativity and innovation that are the motors to find a way so much if the city is a development of ICT as if the city is a development of the common good and the search of procedures that make us the life more "intelligent" the debate of the model of participation is needed, because at the end the great question is that most of the people want to be participant of all this project. The extended city model is considered an obsolete model, little dense and peripheral generates high costs, for this reason, it is important to study it since it generates an inequality in the quality with which the government provides public goods to the citizens, that is why the unplanned human settlements generate that the cost of the public services rises. Public policies and programs should be implemented to encourage the concentration of larger population centers, improving public services in rural areas, with urban planning complaints. This will promote the cooperation of different levels of government and the participation of civil society in the organization of the city, taking care of the economic and environmental order, Through the construction of buildings and urban settlements near major centers to generate jobs and avoid unnecessary expenses, each time you generate a construction of industrial, housing or any other is important to take into account the environmental factor according to the contact in which it is counted to generate a more sustainable building in this way it is possible to plan to organize and use the resources for each space or time.

2. Automation of State Systems in China and use of AI

In China, over time, artificial intelligence is gradually being commercialized and is undergoing profound development in different fields. The project continues to be favored by major AI organizations. In the last five years, investment in artificial intelligence in China has grown exponentially. Since the first years of AI development, in 2015, the total investment reached 45 billion yuan in researching its development and new uses in state management alone, and it continued to increase in frequency in 2016 and 2017. In the first half of 2019, China's artificial intelligence sector raised a total of more than 47.8 billion yuan, and achieved remarkable results [5]. With the advent of the "AI+" era, innovations have been unleashed in hardware and software services such as mobile phones, the Internet of Things, cars and chips, and features such as face recognition and virtual reality have continued to expand. With the deep understanding of artificial intelligence technology by the business and investment communities, investment in artificial intelligence is becoming more rational. While the amount of investment in human wages and energy is decreasing, the amount of investment is increasing year by year. For example, the Shanghai government has provided tax incentives, capital subsidies, talent introduction and has optimized government processes to optimize the business environment, attracting a large amount of investment and funding for its public administration, artificial intelligence companies and talents, its scientific research force is outstanding. Promote the scale effect of upstream and downstream enterprises in the chain of artificial intelligence industry, and increase the strength of urban artificial intelligence industry. The top-level cities represented by Shanghai and Beijing have long been at the top of the ladder in terms of number of talents, number of enterprises, capital environment and scientific research capabilities. The number of artificial intelligence enterprises in Shanghai and Beijing cities has exceeded 600 all through private funding and state control. Among them, Shanghai has established business laboratories with technology giants Tencent, Microsoft and unicorn artificial intelligence merchant Tang and Squirrel AI who are currently working to develop AI uses in smart cities through public and private support and funding. Artificial intelligence empowers the financial industry to build a high-performance ecosystem with a broader range of capabilities, improves the financial efficiency of financial firms and transforms the entire process of internal company operations. Traditional financial institutions and technology companies have jointly promoted the deep penetration of artificial intelligence in the financial industry and state bureaucracy, restructured service architecture, improved service efficiency and provided customized services to long-distance customers, while reducing financial risks. Among the types of application of artificial intelligence technology in the field of education, adaptive artificial intelligence learning is the most widely used in all aspects of learning. In addition, due to China's large population, scarce educational resources and favorable factors such as the importance of education, it is expected that adaptive intelligent learning systems will be applied in recent years and will be able to reach even the most remote rural areas, to ensure that the entire population has access to education publicly, free of charge and universally through the use of distance education based on new educational models, for example the first Chinese textbook on artificial intelligence, aimed at rural secondary school students, was published earlier this year [6]. The construction of digital government affairs depends mainly on top-down promotion so it is very important that the state is the first one to make use of the new tools that are available in terms of technologies since the main beneficiaries will be the citizens who in turn will provide large corporations with better qualified human capital to be able to understand and make optimal use of new technologies regardless of their age, the objective of the digitalization of government affairs is to accelerate the intelligent transformation of government. The requirements of building digital government in different places can become very diversified, so companies must provide customized solutions in view of the country's cultural diversity. The technology requirements in the country's major metropolises will not be the same as those in rural areas or in small or developing cities in the west of the nation. Barriers to entry in the field of public safety have been lifted. The automotive industry, dominated by driverless technology, will mark the beginning of innovation in the industrial chain. The production, channeling and sales models of traditional automotive companies will be replaced by emerging business models. The boundaries of the industry between emerging driverless technology companies and traditional automotive companies will be broken. With the rise of the car-sharing concept. Driverless technology carpools will replace the traditional concept of private cars. With the development of specifications and standards for the unmanned industry, emerging industries such as the safer and faster cars and at the same time will be able to solve 2 of the most serious problems of large cities in China and the world, the problem of traffic due to excessive vehicle fleet and pollution emitted by them significantly lowering travel times and carbon dioxide rates in cities, in addition to reducing health problems caused by pollution that in the end also represent a high cost to the state. For this reason, the potential for the application of artificial intelligence in the field of intelligent car manufacturing in large cities should not be underestimated. At present, the decrease in costs is greater than ever, and it is therefore possible to invest in this area as it is a guarantee of success for the future, even though high-quality data resources are not fully available or fully develop Through the use of algorithms that allow communications to connect their devices to an internet network, decision support systems can be made, to process large amounts of data for user support, control systems that also process data and allow "to manage" in real time such as intelligent lighting for energy saving or the traffic light network for full traffic flow to eliminate traffic problems in addition to obtaining real time data. The development and use of intelligent vehicle traffic management is an obligatory aspect in Smart Cities, which is not only limited to vehicle data, but also by using data obtained from the infrastructure to connect to the internet and process this data. The most used is the use of video cameras and different types of sensors, magnetic, infrared, radar, acoustic and of course the devices that travel inside the vehicles that are circulating. Through the simulation in real time to be able to predict the traffic at a certain time but the accuracy of the data will depend on the quality of the tools and their use, with the simulators it is possible to learn and understand the traffic in the Smart Cities the accomplishment of maintenance to the public roads, the pedestrianization, intelligent traffic lights etc. As previously mentioned, logistics companies will benefit and increase due to the demand for these intelligent systems. In the area of vehicle safety, the possibility of issuing fines in real time for the violation of traffic laws, such as ignoring traffic signs, parking in prohibited places will be detected by video surveillance systems that allow the identification of the vehicle by recording the license plates or the proportion of emergency vehicles in case they are necessary in the event of a breakdown of a vehicle that could compromise the flow of traffic.

3. The Learning of Government through the Entry of AI

The analysis of AI investment trends is mainly divided into the following points:

- Investors are looking for readily available AI application scenarios. In recent years, investment and financing data show that corporate services, robotics, medical and healthcare, industrial solutions, building blocks, and financial sectors are higher in investment frequency and amount of financing than other industries. From a company perspective, the world's most important equipment, financial strength, and technology genes are more likely to be favored by investors in the secondary market. From the industry perspective, the new retail, driverless, medical, and adaptive education that is easy to land indicates more opportunities, so companies in these areas have more investment opportunities. The investment market has begun to favor the underlying new technology companies. Unlike the previous investment preferences of applied artificial intelligence companies, the investment market has gradually started to focus on start-ups with underlying artificial intelligence technologies. The underlying technology is more popular, and due to the high ceiling, these companies are more competitive in the market. The development of the underlying artificial intelligence technology in China continues to lag behind that of the United States, and the underlying technology is an important support for the development of artificial intelligence, with the further development of artificial intelligence in China, investment in the underlying technology will continue to grow.

- The proportion of companies that have won rounds A and B remains the highest, and strategic investments have gradually increased. Currently, more than 1,300 AI companies across the country have received venture capital investments. Among them, the proportion of A-round investment frequency started to gradually decrease. Investors continue to be very enthusiastic about round A, and it is currently the most frequent round of investment. Strategic investments started to explode in 2017. With the gradual maturity of the artificial intelligence market segment, leading companies, mainly the Internet giants, have turned their attention to strategic investments that seek long-term cooperation and development. This also indicates that strategic cooperation between the artificial intelligence industry and the capital level industry has started to increase. The giants are investing in artificial intelligence upstream and downstream of business-related industries. At the height of the development of artificial intelligence, Internet giants with a keen sense of smell have also initiated their strategic design. Technology giants like Alibaba, Tencent, Baidu, and JD.com have invested in various sectors of artificial intelligence, supported by technology investment funds backed by the Ministry of Science and Technology, the National Science Holding of the Chinese Academy of Sciences, the Local Finance Bureau and the Economic and Information Commission. In terms of fields, the projects in which investment institutions decide to invest are all before and after their future strategic industrial design, and these investment projects also promote the implementation of national strategies for the development of artificial intelligence. For example, Alibaba's investment is mainly focused on security and basic components. Representative companies that have won investments include Shangtang, MegTV, and Cambrian Technology. Tencent's investment focus is mainly in the areas of health, education and intelligent cars. Representative companies include Weilai Automobile and Carbon Cloud Smart. Baidu's investment focus is primarily in the areas of automotive, retail and smart homes. JD.com's investment focus is on areas such as automotive, finance and smart homes. The customer transformation and market strategy of the new retail platform Tmall.com, an online sales platform operated by the Alibaba Group. In the age of the internet, as traditional retail modes are concerned with the difficulties of finding sustainability, artificial intelligence technologies have been gaining popularity in the Chinese retail market. In addition to unmanned stores, new emerging innovations such as unmanned delivery vehicles and artificial intelligence customer support have also been launched or planned in China. The National Science Department, which is based on the Chinese Academy of Sciences system, is involved in artificial intelligence technologies and applications such as chips, medical treatment, and education. With the transformation and integration of digitization in various industries, artificial intelligence will become a necessity for giants in many fields such as automotive, medical and health care, education, finance, and intelligent manufacturing.

4. Government and Smart Security

The main purpose of intelligent security is: by transforming the unstructured image information in video surveillance into structured data that can be understood by computers, with the help of data processing, "mass video data" is transformed into "effective intelligence" to perform the security industry's intelligent upgrade from "seeing clearly" to "understanding and analyzing". Intelligent security needs to use machine learning to implement feature extraction, target recognition and power, organize into text information that can be understood by computers and people according to standard video content. This can drive significant improvements in image recognition and rating accuracy. The intelligent security industry chain includes primarily construction and maintenance engineering secondly, hardware and systems manufacturers, representing companies such as the listed intelligent security companies Hikvision, Dahua, etc. Thirdly, the software and algorithm companies in the artificial intelligence start-up companies, four large companies that need facial recognition (Shangtang Technology, Queng Technology, Yuncong Technology, Yitu Technology) and other technologies such as facial recognition, online identity verification, intelligent monitoring, and image recognition [7]. Because of the above mentioned, China already has a highly developed monitoring system unlike other countries like Mexico, which could take that leap and implement their intelligent security systems, the improvement of security processes will ensure the safety of the Smart Cities, public security agencies will constantly improve their working mechanisms by adopting various measures for the standardization of law enforcement and the protection of human rights, which in turn is one of the factors in carrying out security procedures. Inevitably, all municipal entities will at some point employ smart city technologies to optimize resource management and improve the lives of people living in the community. However, how they manage security will be the determining factor in the success of their efforts. This allows the analysis of moving crowds in urban areas, airports, train stations, shopping centers, and sports stadiums, among others. It is also used for forensic analysis, due to the capacity of intensive search of subjects in video recordings, for the location of suspects or automatic classification. However, the advancement of technology and the application of scenarios are a gradual process that should be taken as a priority in emerging countries, the delay in its transformation will pose a great threat to the economy and the efficiency of state processes, it is advisable to make a strong investment by the government to keep up with the technology that is currently evolving at a rapid pace as it is constantly changing day by day.

5. Conclusions and Future Research

Living in an intelligent city will be an inspiration to everyone in the future, for governments, investors, citizens in general, to see groups of all ages use technology for the common good, the fact that everyone lives together as a united society that is connected beyond their creed, color or origin is wonderful, so will the cities of the future, cities that educate the talents and innovators of the future a home of advanced invention, a city where technologies coexist in an urban center, a city where people work together to create a new kind of city, the city of the 21st century, an intelligent city that works for everyone, a city that is not afraid of the challenges of the future. In addition, as a result, 12 societies in Africa will exceed 100 million inhabitants in this century and their capitals will become the next metropolis around the world, a perfect laboratory to implement the efficient management and process optimization of a future Smart City, as is shown in Figure 3.

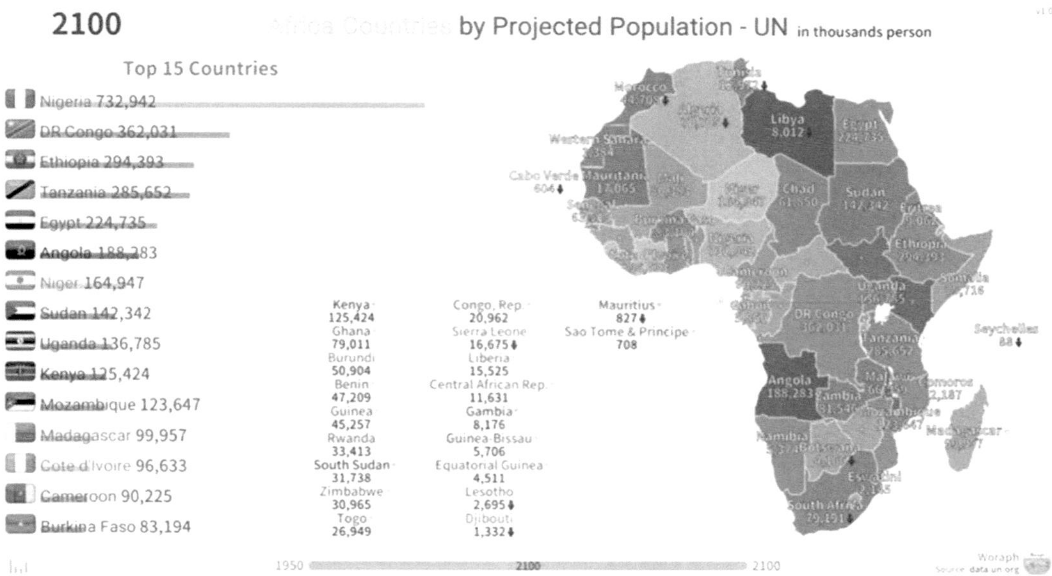

Figure 3: Future population of 59 societies in Africa.

References

[1] Mikhailova, N. //Innovative forms and mechanisms of forming the concept of efficient municipal management// Bulletin of Volgograd stare university//#3.// p. 127–134.

[2] Cities in Motion Index de la Escuela de Negocios IESE. //2018 url: https://citiesinmotion.iese.edu/indicecim/.

[3] Moovit insights/Data and statistics on the use of public transport in Mexico City//2019//url:https://moovitapp.com/insights/es/Moovit_Insights_%C3%8Dndice_de_Transporte_P%C3%BAblico-822.

[4] Rosado, J. and Diaz, R. 2017. //Latin America facing the challenge of the Smart Cities// d+i desarrollando ideas//2017// p. 1–4.

[5] Li, K. 2019. //Global Artificial Intelligence Industry Development// 2019// url: https://xueqiu.com/9508834377/137204731.

[6] How does artificial intelligence develop in various fields? // 2018// URL: https://www.ofweek.com/ai/2018-10/ART-201700-8470-30276953.html.

[7] China Artificial Intelligence Industry White Paper// 2019// URL: https://www2.deloitte.com/cn/en/pages/technology-media-and-telecommunications/articles/global-ai-development-white-paper.html.

[8] Korshunova, E. //Educational Potential of Smart City Management: Analysis of Civil Service Training Standards// Business community power//2017.

[9] Toppeta, D. //The Smart City Vision: How Innovation and ICT Can Build Smart, "Livable"//Sustainable Cities// 2010//. URL: http://www.inta-aivn.org/images/cc/Urbanism/background%20documents/Toppeta_Report_005_2010.pdf.

PART III

Industry 4.0, Logistics 4.0 and Smart Manufacturing

CHAPTER-13

On the Order Picking Policies in Warehouses

Algorithms and their Behavior

Ricardo Arriola, Fernando Ramos, Gilberto Rivera, Rogelio Florencia, Vicente García* and *Patricia Sánchez-Solis*

This chapter explores the relationship among different routing policies for order picking and the features of the problem (describing both warehouse layout and orders), the results obtained by simulation show that some policies are especially sensitive to the presence of certain conditions that are likely to be present in real-world cases.

Moreover, the routing policies are represented—for the first time in the literature as far as our knowledge—on structured algorithms. This contribution can facilitate their implementation because the features of the policies are modeled by formal mathematical structures, laying the foundations to standardize the way they operate.

1. Introduction

A warehouse is a fundamental part of a company, and its performance can impact the entire supply chain [1].

Order picking is a problem that is present in all companies. It has received special focus from research areas related to planning and logistics. This fact is a consequence of several studies that identify order picking as the activity that demands more resources inside the warehouses, reaching up to 55% of the operational cost of the entire warehouse [2].

This activity has a strong impact on production lines, so companies with complex warehouses have areas dedicated to improving their product collection processes.

Universidad Autónoma de Ciudad Juárez, Av. Hermanos Escobar, Omega, 32410 Cd Juárez, Chihuahua, México.
* Corresponding author: gilberto.rivera@uacj.mx

There are optimization models to support the resolution of this problem in the generation of product-picking routes; however, being considered as an NP-complete problem is not feasible to solve the models when working at medium and large scale due to the high cost that this represents. Thus, it is possible to apply some heuristics to get an approximate solution in real cases.

Although several studies in the literature (e.g., [3]) show that these procedures are far from finding solutions close to the optimal one; these heuristics are still applied to real problems due to the simplicity and the way they relax the problem, granting a good balance between the quality of the solution and the ease of implementation.

The picking routes are dependent on the structure of the warehouse and the properties of the orders, so studies have stated [4] that the input elements and the performance of the routes obtained are highly related. For example, a greater number of cross aisles facilitates movements inside the warehouse, so that the distance of the route tends to decrease.

Throughout this chapter, we are going to define the algorithms for five of these heuristics and deepen on the study of which of them are more sensible to the characteristics describing the layout of a warehouse.

2. Background

Warehouses are an important part of supply chain in a factory, and the main activities inside of it, like reception (receive and collect all product data), storage (move products to their locations), pick up (pick products from their storage location), packing (prepare for being transported), and shipping (place product in the transport medium). On this last step, the warehouse operation ends.

2.1 Warehouse Layout

The chief features of the warehouse are: the central depot, where the picker starts its route and finish it, also usually is where the picker gets the order. The picker has to walk by the—the second element— picking aisles. A picking aisle is the space between two racks and facilitates the picker to pick product from the shelf, and the aisles have the following characteristics: length (distance between front aisle and rear aisle), and distance between aisles (distance that exists from the center of one aisle to center of next aisle); based on the distance, we can classify the aisles as short or long, in this case, we are going to use short aisles, which means that we can get the products from the shelves without making lateral displacements over the aisle. Cross aisles are perpendicular to picking aisles and are used to travel from one aisle to another, picking node is the location in the shelf where you can get the product, shelf is the space in the rack where products are stored, block is the area between the cross aisle and picking aisle, and sub-aisle is the area between the picking aisles inside the block if the blocks have picking aisles. Finally, picker is the person, tool or machine that picks up the products.

Figure 1 represents an example of a layout of a warehouse with five cross aisles, four blocks, six picking aisles, and a central depot.

Also, we are going to clarify the key concepts and briefly explain how the order picking routing policies work.

2.2 Steiner STP

In the literature, it is verified that TSP (Travel Salesman Problem) is on the classification of NP-hard problems [5]; likewise, TSP and Order Picking have a close relationship. Unfortunately, the optimal solution may require intolerable run times for high-scale problems. Considering an approximate solution might be more convenient as this provides an favourable relationship between time and cost.

Optimization algorithms based on local searches focus on achieving a good quality solution in a reasonable time to get a minimum or maximum value and avoid being stuck in a local optimum.

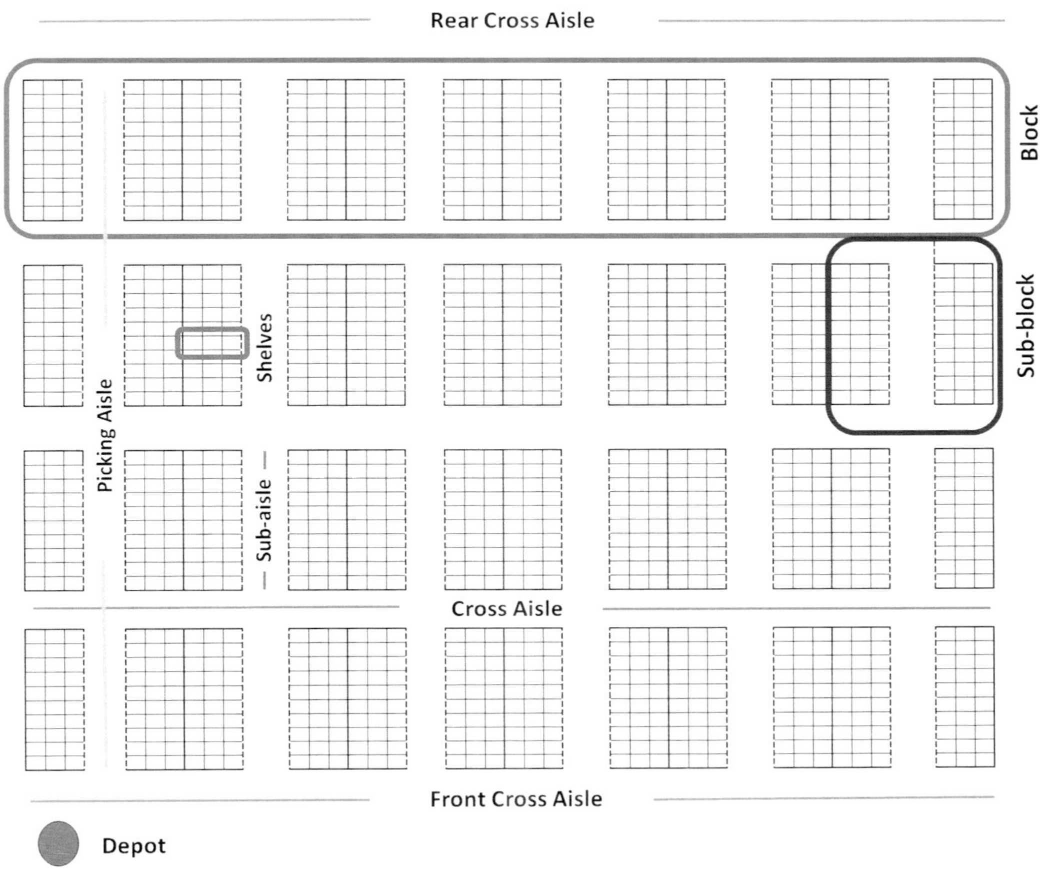

Figure 1: Warehouse layout elements.

It is necessary to start from a solution and, by applying operators, calculate solutions better than the initial solution. Normally, this strategy is applied to NP-hard problems, where heuristic functions are used to eliminate non-promising routes [6].

The solution for this project is represented as SPRP (Single-Picker Routing Problem), which consists of finding the shortest path that includes all the products to pick up [7]. This problem could be represented as a special case for TSP and can be applied as such to solve the initial SPRP problem. The objective is minimizing the distance and the time travel of the picker, either a human or machine, so it becomes a TSP. TSP consists of a salesman and a set of cities. The salesman must visit each of the cities, starting from a specific location (for example, native city), and come back to the same city. The challenge of this problem is that the salesman wishes to minimize the total duration of his/her trip.

SPRP could be modeled as TSP, where the vertices of the correspondent graph are defined by the location of the available products inside of the warehouse and the location of the depot, as presented in Figure 2.

This graph shows all the vertices and not only the picking ones, so the SPRP was modelled as a Steiner TSP (which is a variant of the classical TSP) that is defined as follows:

Let $G = (V, E)$ be a graph with a set of vertices V and a set of edges E. Let P be a subset of V. The elements of $V \backslash P$ are Steiner points. On a Steiner route, each vertex of P is visited only once. Steiner points should not be visited multiple times. However, a Steiner route could travel through

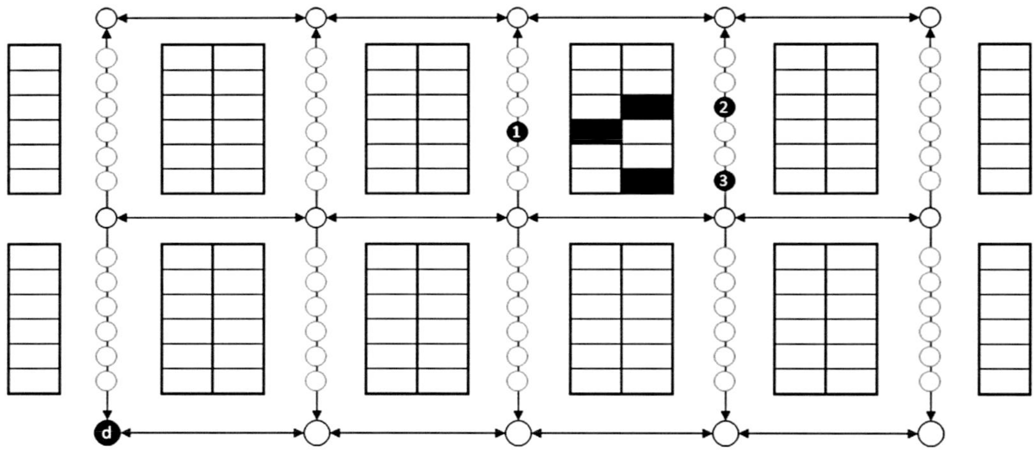

Figure 2: Example of a warehouse structure as a set of V vertices.

some vertices and edges more than one time. In conclusion, the TSP of Steiner consists of finding a route of Steiner with the minimum distance [8].

Figure 3 shows an example of a warehouse layout with different parameters and variables.

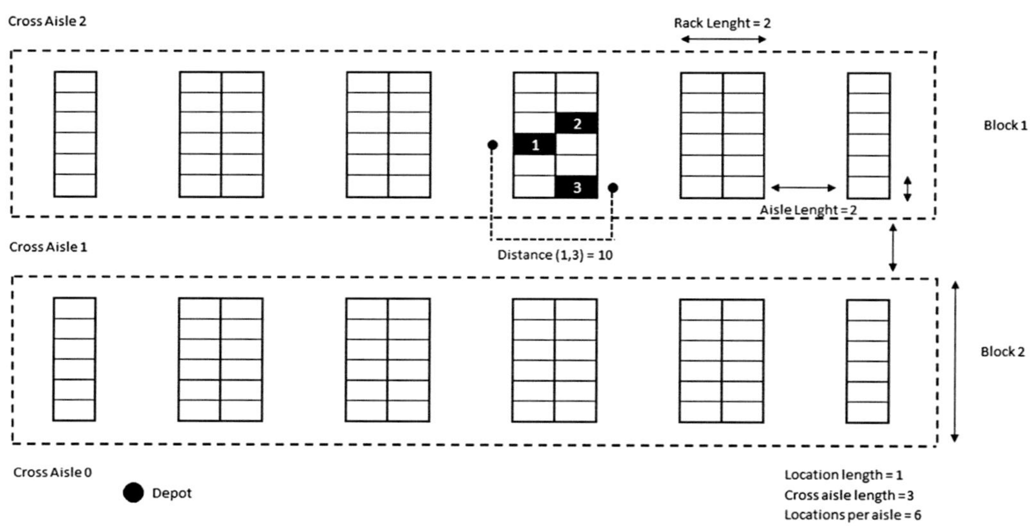

Figure 3: Example of a warehouse structure with different parameters.

In Figure 4, the black-filled vertices are the picking vertices and the initial vertex, also known as the depot. This set of vertices is the set P, a subset of all the vertices V. This subset will form a Steiner graph, and the vertices formed at the intersections of the cross aisles and the picking aisles we will call Steiner points.

Once the graph is obtained, the objective is to find a Hamiltonian circuit with the minimum cost. The initial and finish point of this circuit will always be the depot.

Also, it is important to know that there are six different ways to travel through a picking aisle [9].

Figure 5 describes each one over one example of a unique block, one front cross aisle, and one rear cross aisle.

Picker enters by the front cross aisle, crosses it completely, picks up all required products, and finishes leaving the aisle by the rear cross aisle.

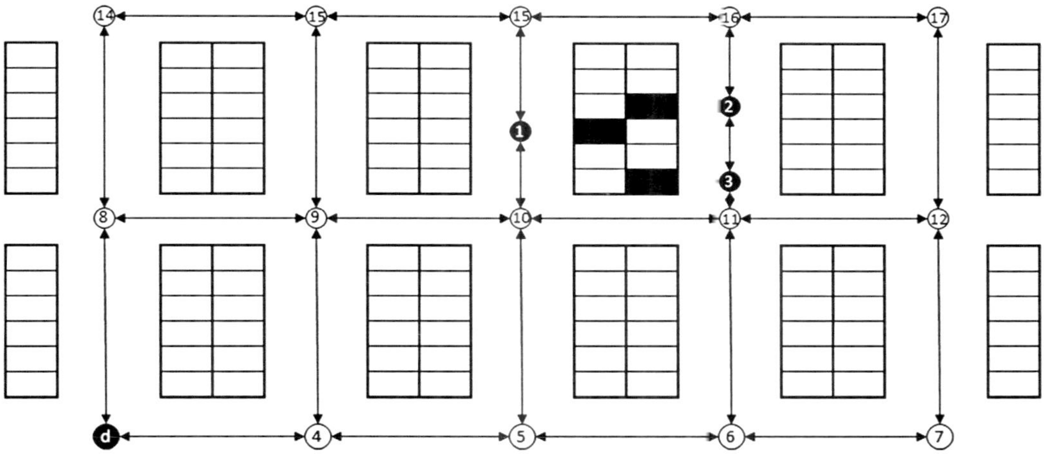

Figure 4: Example of a Steiner graph.

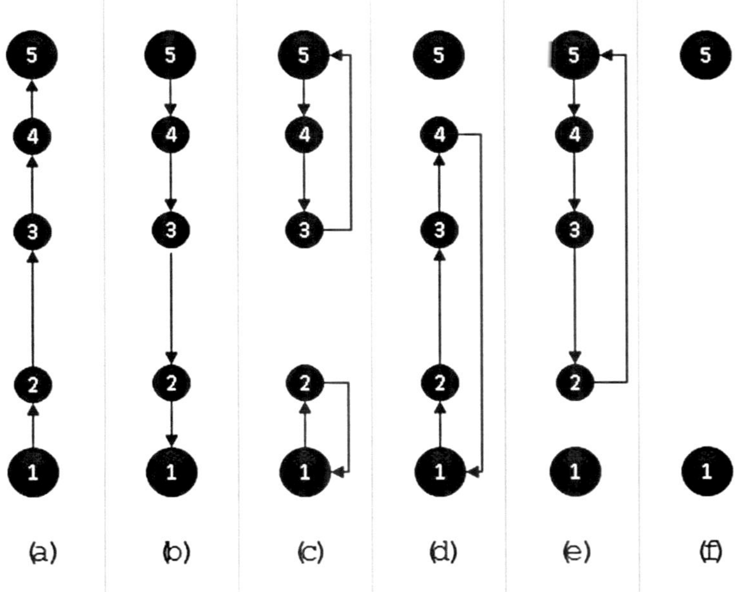

Figure 5: Six ways to travel edges through the picking aisles.

Picker enters by the rear cross aisle, crosses it completely, picks up all required products, and finishes leaving the aisle by the front cross aisle.

Picker enters and leaves twice through the aisle, enters once through the front cross aisle and once more through the rear cross aisle, picker enters and leaves by the same place. The picker will make its return defined by the largest gap, which is the largest distance between two adjacent picking vertices or the picking vertex and cross aisle.

Picker enters through the front cross aisle, and its return point is the picking vertex farthest from the front aisle.

Picker enters through the rear cross aisle, and its return point is the picking vertex farthest from the rear aisle.

The picker doesn't need to travel through the aisle because there are no picking vertices inside.

These ways to travel are combined, generating different routing policies, which are highly popular in practice.

3. Routing Policies

The routing policies determine the collection sequence of the SKUs (Stock-keeping unit) [10]. The objective of the routing policies is minimizing the distance traveled by the picker using simple heuristics [3].

To achieve this, it is necessary to consider the following features of the warehouse layout and the product orders, which can influence the final performance of each policy: quantity of products in the order, picker capacity, aisle length, and the number of aisles.

Five of these heuristics are described below.

3.1 S-Shape

The picker must start by entirely crossing the aisle (with at least one product) that is at the left or right end (depending on which is the closest one to the depot) until reaching the rear cross aisle of the warehouse. Then, the sub-aisles that belong to the farthest block of the depot are visited one by one until they end up at the opposite end of the warehouse. The only case where it is not necessary to cross a sub-aisle completely is when it is the last one in the block. In this case, after picking up the last product, the picker returns to the cross aisle from where it entered the sub-aisle. When changing blocks, the picker visits the closest sub-aisle to the last visited sub-aisle of the previous block. After picking up all the products, the picker must return to the depot [11]. Figure 6 shows an example.

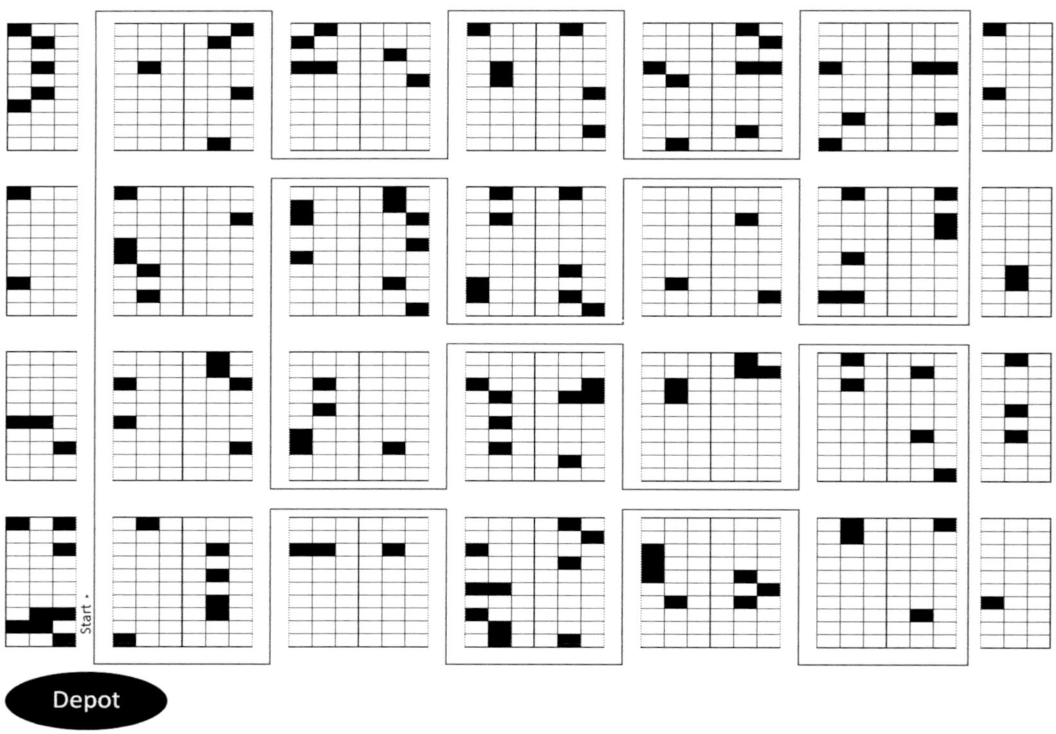

Figure 6: An example of a route applying S Shape.

3.2 Largest gap

This routing policy consists of identifying which is the largest gap in each sub-aisle and then avoiding crossing it [12]. This gap can be either between the rear cross aisle and the first product to be picked, between adjacent products or between the last product of the sub-aisle to the front cross aisle. All products that are before this gap will be picked up from the rear cross aisle and afterwards the picker must return to the cross aisle where it departed and do the same until all the sub-aisles of the block are explored, then, pick up all the remaining products from the front cross aisle.

The sub-aisles that are completely crossed are those belonging to the first visited picking aisle and the last one of each block (this way, it passes from one cross aisle to another).

When the picker picks up all the products, it must return to the depot. Figure 7 shows an example.

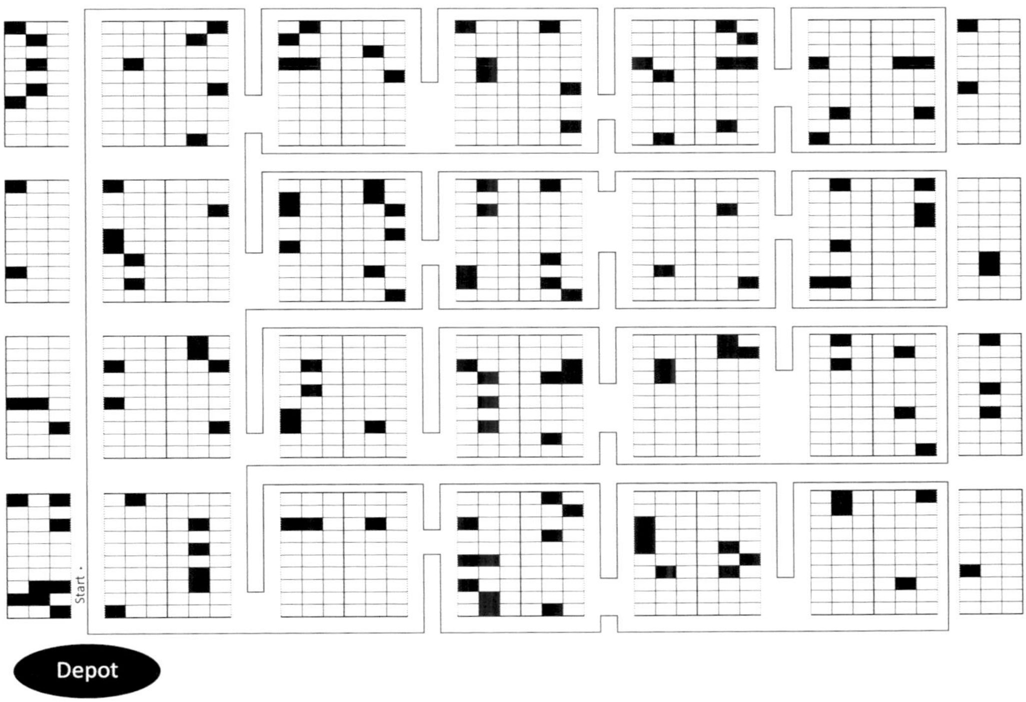

Figure 7: An example of a route applying Largest Gap

3.3 Midpoint

This routing policy is similar to Largest Gap; the main difference is that the picker identifies the product to pick closest to the center of each sub-aisle, which is considered the travel limit [13]; at first, the products that are in the upper half of the sub-aisle are picked up from the rear cross aisle, then, after picking up all the upper half products of the entire block, continues picking up the remaining products from the front cross aisle. If the product is exactly in the center, the picker takes it from either of the two cross aisles. In the end, the picker must return to the depot. An example is represented in Figure 8.

3.4 Return

When applying this routing policy, the picker enters and leaves the sub-aisle from the same cross aisle; this means that, after picking up the last product of the sub-aisle, the picker must return to the cross aisle [14].

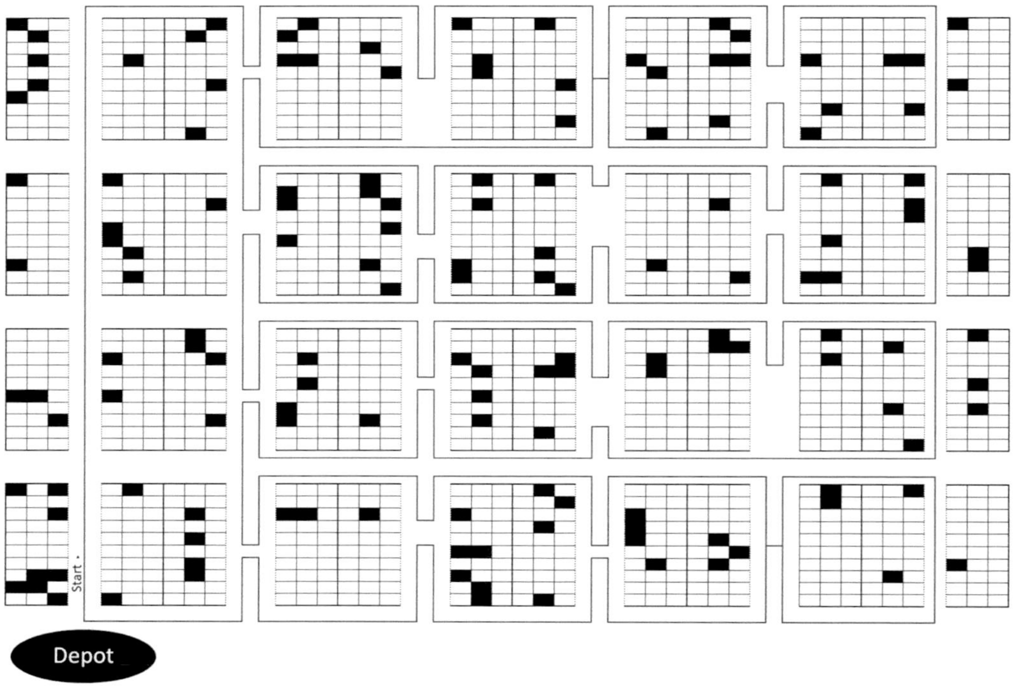

Figure 8: An example of a route applying Midpoint.

In the case that the warehouse configuration contains more than one block, the picker visits all the sub-aisles of the two adjacent blocks to the cross aisle alternately. After that, the picker moves to the next cross aisle that is adjacent to two unexplored blocks. The picker must return to the depot once all the products have been picked. This route is shown in Figure 9.

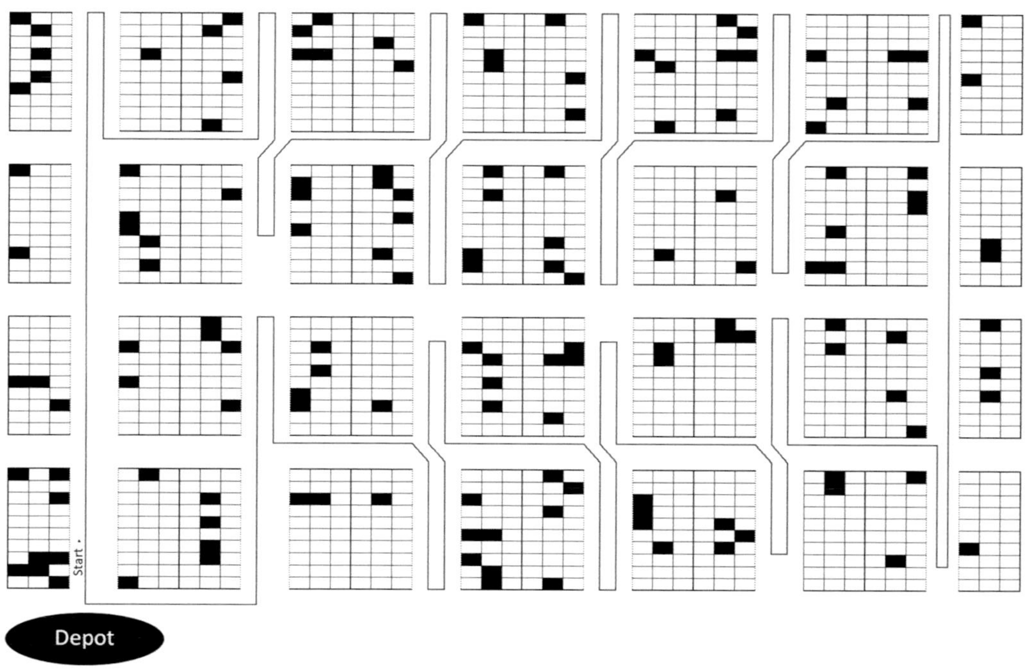

Figure 9: An example of route applying Return.

3.5 Combined

This routing policy is considered a combination of S Shape and Largest Gap policies. When all products are picked up completely, the picker must decide between (1) continuing through the front cross aisle, or (2) returning to the rear cross aisle [14]. This decision is made according to the shortest distance to the next product to pick up. An example of this route is shown in Figure 10.

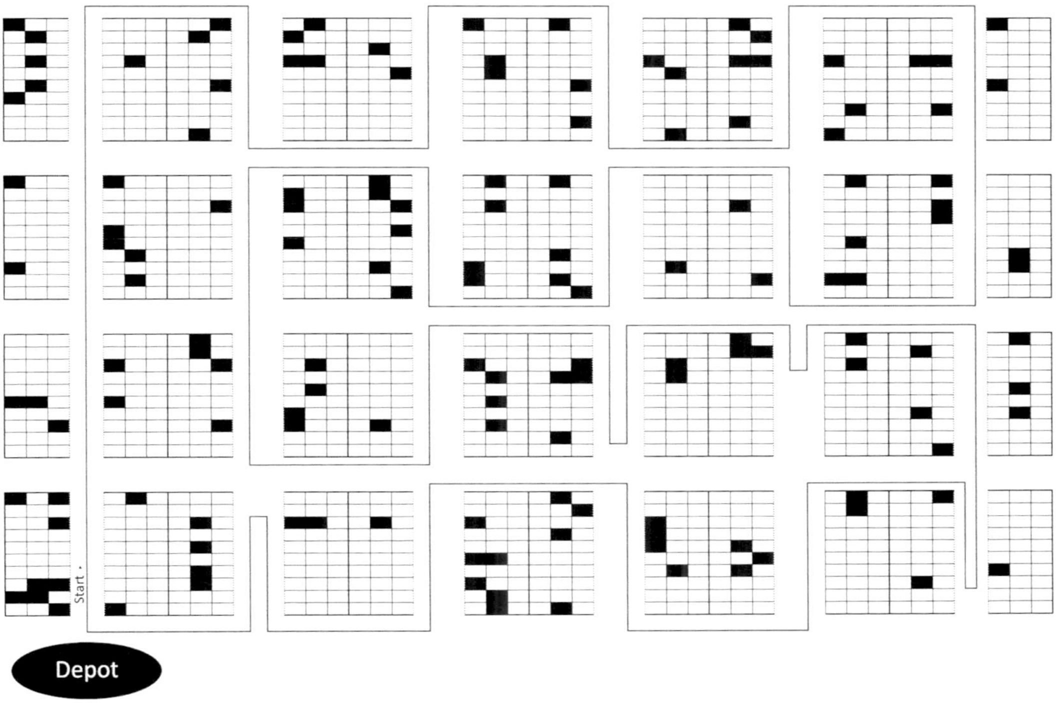

Figure 10: An example of a route applying Combined.

4. Development of Routing Policies

The necessary elements for the development and application of the algorithms for each heuristic will be defined below.

4.1 S-Shape

A fundamental part of the implementation of this heuristic is to define the order in which the picker will visit the sub-aisles, an example of the correct order according to its characteristics is shown in Figure 11.

Also, to help obtain the order, it is necessary to assign "auxiliary coordinates" to each sub-aisle. Figure 12 represents an example.

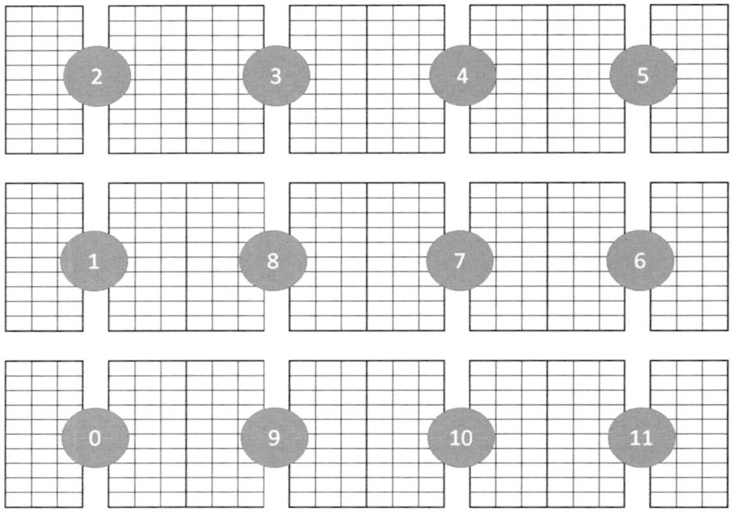

Figure 11: An example of sub-aisles order to visit for S-Shape.

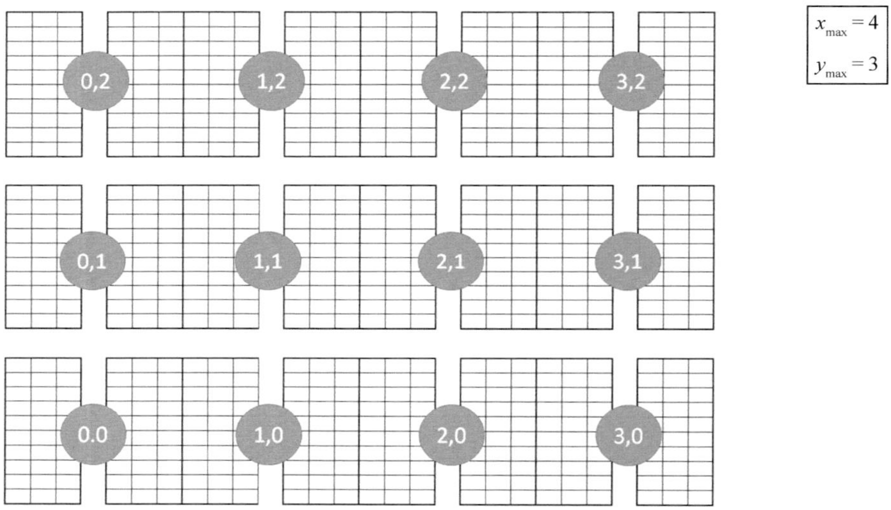

Figure 12: An example of auxiliary coordinates for each sub-aisle.

The following equation returns the picking order of each sub aisle, given x and y coordinates:

$$f(x,y)=\begin{cases} y & \text{if } x=0, \\ y_{\max}+(x_{\max}-1)(y_{\max}-(y+1))+x-1 & \text{if } y_{\max}-(y+1) \bmod 2=0, \\ y_{\max}+\big((x_{\max}-1)(y_{\max}-(y+1))\big)+(x_{\max}-(x+1)) & \text{otherwise}. \end{cases} \quad (1)$$

where:

$y_{\max}$ is the quantity of sub-aisles per aisle,

$x_{\max}$ is the quantity of sub-aisles per block,

x is the current picking aisle, and

y is the current block.

Once the order is defined, the next step is to obtain the direction in which the picker will pick products from each sub-aisle.

Algorithm 1 describes the procedure used to obtain the final path.

Algorithm 1. S Shape

```
Input: Sub-aisles s with at least one product to pick up, visit order of sub-aisles
Output: Final path C
1 Begin
2    While there is unexplored s with product Do
3       Select next s according to the visit order
4       If s is part of the first picking aisle
5          Add products in s in ascending order to C
6       Else if s-1 was explored in an ascending direction
7          Add products in s in descending order to C
8       Else if s-1 was explored in a descending direction
9          Add products in s in ascending order to C
10      If the current block was explored completely
11         Add products in s in descending order to C
12   While end
13   Return C
14 End
```

Where all the sub-aisles that are part of the first picking aisle are explored and added to the final path ascendingly (lines 4–5); then, the direction must alternate between each sub-block (lines 6–9) until the picker has visited the block completely; in that case, the first sub-block the picker visits in the new block is always traversed in a descending direction (lines 10–11).

4.2 Largest gap

The order in which the picker will visit the sub-aisles in this heuristic is different in comparison to S-Shape. In Largest Gap, each explored block starts and ends from the same sub-aisle. Figure 13 shows an example.

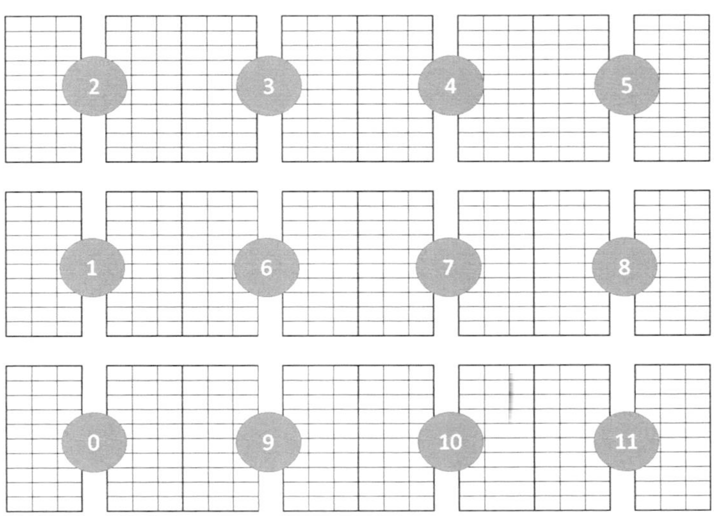

Figure 13: An example of sub-aisles visit order generated by Largest Gap.

The application of this equation only depends on whether the sub-block is on the first picking aisle; in any other case, the blocks are explored from left to right. The following equation represents this:

$$f(x,y)=\begin{cases} y & \text{if } x=0, \\ y_{max}+(x_{max}-1)(y_{max}-(y+1))+x-1 & \text{otherwise}. \end{cases} \quad (2)$$

Once the order in which the picker will visit the sub-blocks is defined, the generation of the final route begins. This method is described in Algorithm 2.

Algorithm 2. Largest Gap

```
Input: Sub-aisles s with at least one product to pick up, visit order of sub-aisles
Output: Final path C
1  Begin
2      While there is unexplored s with product Do
3          Select next s according to the visit order
4          If s is part of the first picking aisle
5              Add products to C in an ascending direction
6          Else
7              Valuate elements in s
8                  Calculate the distance between the current element and the next one
9                      If the current distance is higher than the current limit
10                         The new limit is the current element.
11             Traverse elements in s
12                 If the current element is higher than the limit
13                     Add the current element to C
14                 Else
15                     Add the current element to pending
16         If the current block was explored completely
17             Add elements in pending to C.
18     While end
19     Return C
20 End
```

Where the direction in which the picker traverses the sub-aisles that are in the first picking aisle is ascending, adding the elements to the final path (lines 4–5). Then, the largest gap of each sub-aisle is calculated by the distance between each of the elements that it contains; the new limit is defined by the element that detects the highest travel distance (lines 8–10), the next step is to add the elements that are above this limit to the final path (lines 11–12) and the elements that are below are stored in a *stack* (14–15); once all the elements of the block—that are above this limit—have already been added to the final path, lines 16–17 insert the pending elements in LIFO order (Last In, First Out).

4.3 Midpoint

It is a heuristic similar to Largest Gap. They share the order in which the picker traverses the sub-aisles (Figure 13).
Hereunder, the algorithm designed for the generation of the final path for Midpoint:

Algorithm 3. Midpoint

Input: Sub-aisles *s* with at least one product to pick up, visit order of sub-aisles, locations per sub-block *LS*, locations per aisle *LA*.
Output: Final path *C*

```
1  Start
2      While there is unexplored s with product Do
3          Select next s according to the visit order
4          If s is part of the first picking aisle
5              Add products to C in an ascending direction
6          Define the Midpoint value of the current block: LP-(LS/2)
7          Traverse elements on s
8              If the current element is higher or equal to Midpoint
9                  Add element to C
10             Else
11                 Add element to pending
12         If the current block was explored completely
13             Add elements on pending to C
14             Update Midpoint value of the next block: Midpoint-LS
15     While end
16     Return C
17 End
```

Where the sub-aisles that are in the first picking aisle must be traversed in an ascending direction while adding all the elements to the final path. (Lines 4–5). After, to obtain the midpoint and take it as a limit (line 6), starting from the farthest block to the depot, this value is obtained as follows:

$$Mp = LA - \frac{LS}{2} \tag{3}$$

Where:

Mp is the midpoint,

LA represents the locations per picking aisle, and

LS represents the locations per sub-aisle.

This value is defined as a limit until there is a block change (lines 12 and 14), where it must be updated as follows:

$$Mp = Mp - LS \tag{4}$$

If the element is above the midpoint, it will be added directly to the final path, storing the rest on the pending elements (lines 8–11). The elements of the last sub-block of the block must be added completely in a descending way; so, it passes to the lower cross aisle and then starts adding the pending elements to the final path (line 13).

4.4 Return

The first step in the implementation of this routing policy is to define the order in which the picker will visit the sub-aisles; Figure 14 shows an example of the correct order according to the previously described properties.

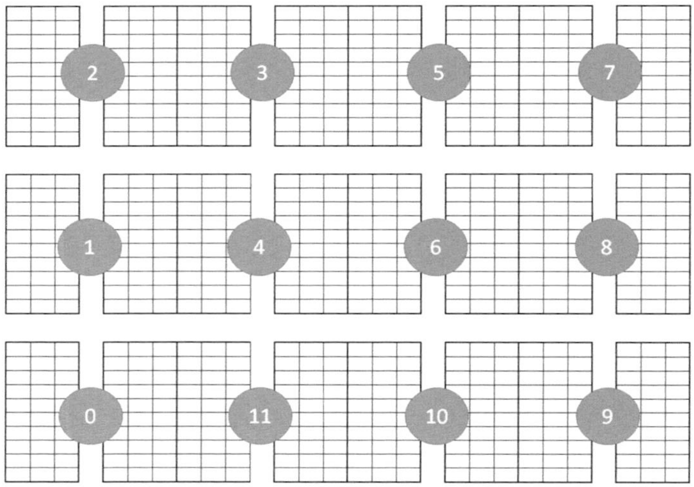

Figure 14: An example of sub-aisles order to visit for Return.

The following equations return the previously mentioned values:

$$g(x,y)=\begin{cases} y & \text{if } x=0, \\ h_1(x,y) & \text{if}(y_{\max}-(y+1)) \bmod 4<2, \\ h_2(x,y) & \text{otherwise}. \end{cases} \tag{5}$$

$$h_1(x,y)=\begin{cases} y_{\max}+(x_{\max}-1)(y_{\max}-(y+1) & \\ h_1(x-1,y)+1 & \text{if}(y_{\max}-(y+1)) \bmod 2=0, \\ & x=1 \\ h_1(x-1,y)+2 & \text{if}(y_{\max}-(y+1)) \bmod 2=0, \\ & y=0 \\ h_1(x,y+1)+1 & \text{if}(y_{\max}-(y+1)) \bmod 2=0, \\ & \text{otherwise} \end{cases} \tag{6}$$

$$h_2(x,y)=\begin{cases} y_{\max}+(x_{\max}-1)(y_{\max}-(y+1)) & \\ h_2(x+1,y)+1 & \text{if}(y_{\max}-(y+1)) \bmod 2=0, \\ & x=x_{\max}-1 \\ h_2(x+1,y)+2 & \text{if}(y_{\max}-(y+1)) \bmod 2=0, \\ & y=0 \\ h_2(x,y+1)+1 & \text{if}(y_{\max}-(y+1)) \bmod 2=0, \\ & \text{otherwise} \end{cases} \tag{7}$$

In this policy, the patterns of the order of travel are more complex than those seen before, so it was necessary to define a series of equations which consist of a function $g(x, y)$ on which you can obtain direct results and functions $h_1(x, y)$ and $h_2(x, y)$ where it is necessary to call them recursively to reach the desired result. For the application of these equations, it is necessary to use the auxiliary coordinates exemplified in Figure 12.

Once having the sub-aisles visit order, the process to generate the final path is shown in Algorithm 4:

Algorithm 4. Return

```
Input: Sub-aisles s with at least products to pick up, visit order of sub-aisles, locations per sub-block LS, locations per aisle LA.
Output: Final path C
1  Start
2      While there is unexplored s with product Do
3          Select next s according to the visit order
4          If s is part of the first picking aisle
5              Add products to C in an ascending direction
6          If the quantity of the blocks is an even number
7              If s is part of a block with an even coordinate
8                  Add elements in s to C in an ascending direction
9              Else
10                 Add elements in s to C in a descending direction
11         Else
12             If s is part of a block with a even coordinate
13                 Add elements in s to C in a descending direction
14             Else
15                 Add elements in s to C in an ascending direction
16            If s is part of the block on y=0
17               Add elements on s to C in an ascending direction
18            If the current block was explored completely
19               Add elements in s+1 to C in an ascending direction
20     While end
21     Return C
22 End
```

The elements found in the first picking aisle are added to the final path in an ascending direction (lines 4–5). Subsequently, it is important to define whether the number of blocks in the warehouse is even or odd (line 6). This fact influences because the characteristics of the routing policy states that the picker must alternate between the sub-blocks of two blocks, this implies that, in cases where the warehouse configuration has an odd number of blocks, the sub-blocks belonging to the last block to explore (closest to the depot) must be explored continuously (without alternating) (line 16). When the total number of blocks is even, all sub-aisles belonging to a block are crossed ascendingly, and in the odd ones in a descending direction (lines 6–10). In the opposite case (warehouses with an odd number of blocks), the sub-aisles that belong to the odd-number blocks are traversed in an ascending direction, and the sub-aisles that belong to the even-number blocks are traversed in a descending direction (lines 11–15). On the last block to be explored, all sub-aisles are visited ascendingly (line 16). In both cases, at the end of every block, the elements of the first sub-block of the next block (lines 18–19) are added to the final path in a descending direction.

4.5 Combined

The order of how to traverse the sub-aisles is similar to S Shape (represented in Figure 11), but there are cases where it can vary because of the characteristics of this routing policy; when picking up the last element of every block, it is important to define which is the sub-aisle of the next block with product that has the smaller distance, if this sub-aisle is on the left end, the path of the block turns in the direction from left to right; otherwise (if the sub-block with the nearest product is in the extreme right sub-aisle), the opposite direction must be taken. The following equation represents the above behavior:

$$f(x,y)=\begin{cases} y & \text{if } x=0, \\ y_{\max}+(x_{\max}-1)(y_{\max}-(y+1))+x-1 & \text{if } d_1<d_2, \\ y_{\max}+(x_{\max}-1)(y_{\max}-(y+1))+(x_{\max}-(x+1)) & \text{otherwise.} \end{cases} \tag{8}$$

Where d_1 is the distance between the last element of the block and the sub-block with the leftmost product and d_2 is the distance between the last element of the block and the sub-block with the product on the rightest location. Once having the order in which the picker will visit the sub-aisles, the final route is generated as presented in Algorithm 5.

Algorithm 5. Combined

```
Input: Sub-aisles s with at least one product to pick up, visit order of sub-aisles.
Output: Final path C
1  Start
2      While there is unexplored s with product Do
3          Select next s according to the visit order
4          If s is part of the first picking aisle
6              Add elements on s to C in an ascending direction
7          Capture the last element on s
8             Calculate d_1: the difference between the last element on s and the first one on s+1 from the rear cross aisle.
9             Calculate d_2: the difference between the last element on s and the first one on s+1 from the front cross aisle.
10         If d_2 is greater than d_1
11             Add elements on s to C in an ascending direction.
12         Else
13             Add elements on s to C in a descending direction.
14         If the current block was completely explored
15             Add elements in s+1 to C in an ascending direction
16     While end
17     Return C
18 End
```

Where the elements that belong to the first picking aisle are added to the final path in an ascending direction (lines 4–5), from this point, the distance from the last element of each sub-aisle to the first element of the next sub-aisle accessing from the front and rear cross aisle must be evaluated (lines 7–9). In the case that the distance to the first element from rear cross aisle is slower, the elements on the next sub-aisle have to be added in a descending direction to the final path; in the opposite case, the elements on the next sub-aisle are added in an ascending direction (lines 10–13).

4.6 Efficiency measurement

To measure the effectiveness of the result, it is necessary to calculate the distance matrix between all product nodes and artificial nodes by applying Dijkstra. Subsequently, a sum of all the distances between consecutive nodes on C is made, represented by the following equation:

$$d(C) = \sum_{i=1}^{p-1} D(x_i, x_{i+1})$$

where:

$C = \langle x_1, x_2, x_3, \ldots, x_p \rangle$ is the sequence of elements to evaluate,

p is the number of vertices that forms a circuit, and

D is the distance matrix.

5. Experimental Results

In this section, the results obtained by this project are shown and interpreted.

A total of 125 different warehouse layouts were processed and defined according to the combination of the following values:

Number of products capacities per sub-block: 18, 30, 42, 54 and 60

Number of picking aisles: 3, 6, 9, 12 and 15

Number of cross aisles: 1, 3, 5, 7 and 9

In each layout, five routing policies were processed and applied to five different orders with 4, 7, 10, 12, and 15 percent of the full capacity of the products in the warehouse.

The results give a total of 3125 different instances (625 results per routing policy). This information was processed by SAS 9.4 software to generate a correlation analysis.

The rank is one of the different variables to consider, it represents the position obtained by each routing policy in comparison to the other ones. The first position is assigned to the one that gets the lower distance and the fifth the highest. The locations per sub-block with or without product, the quantity of picking aisles, the number of cross aisles, and the percent of total locations of the warehouse that contains products to pick up were other variables used.

5.1 Insights

Let us remember that, as the correlation gets closer to 1 or –1, the correlation is greater. Because this is a minimization problem, a correlation with the performance is better if it is negative. So, the main insights are:

- S Shape tends to be sensitive to the number of locations by sub-aisles (negative correlation of 0.41824) and the number of products in the warehouse (negative correlation of 0.24840) (Figures 15a and 16a).
- For Largest Gap, the correlation coefficient that stands out is the number of picking aisles, obtaining a positive 0.37988 (Figures 15b and 16b). Figure 15b demonstrated the tendency generated by this result, where the level of efficiency compared with the other policies decreases as this value becomes greatest.
- For Midpoint, the two most relevant variables are the number of aisles and the number of products to pick up, both with a positive correlation of 0.33772 and 0.26812, respectively (Figures 15c and 16c).
- The variable with the most effect over the performance of Return is the number of locations by aisle, where it gets a positive correlation of 0.58660. The more locations by aisle, the more competitiveness Return obtains. Figures 15d and 16d show that this policy gets better results as the number of aisles increases.
- Regarding the results of Combined, the number of locations by sub-aisle seems to be an important feature, with a negative coefficient of 0.48762, considerably higher compared to the other variables (Figures 15e and 16e show).
- Combined has better results in warehouses where the number of locations by aisle is the greatest variable, while the most degraded is Return.
- A high number of aisles tends to affect the performance of the five policies, but the policy with the most unfavorable results is Largest Gap. Being S-Shape and Combined the least affected.
- In the case of cross aisles, there was no improvement in the performance of any studied policy. The policies where its effectiveness decreases are Return and Combined. While in S Shape is just a little bit affected.
- S-Shape is the most benefited policy in warehouses where the number of products to be picked up increases.

To be able to develop this project, it was necessary to process orders, get five different results, and compare them over different methods. A benchmark of instances was synthetically created, and the performance in this wide range of different conditions was measured.

a) S shape

Pearson Correlation Coefficients, N = 625 Prob > |r| under H0: Rho=0				
	ProdPerSubBlock	Aisles	CrossAisles	OrderSize
Rank	-0.41824 <.0001	-0.12243 0.0024	0.04785 0.2360	-0.34840 <.0001

b) Largest Gap

Pearson Correlation Coefficients, N = 625 Prob > |r| under H0: Rho=0				
	ProdPerSubBlock	Aisles	CrossAisles	OrderSize
Rank	-0.13199 0.0010	0.37988 <.0001	0.14355 0.0004	0.16394 <.0001

c) Midpoint

Pearson Correlation Coefficients, N = 625 Prob > |r| under H0: Rho=0				
	ProdPerSubBlock	Aisles	CrossAisles	OrderSize
Rank	0.01836 0.6496	0.33772 <.0001	0.11537 0.0042	0.26812 <.0001

d) Return

Pearson Correlation Coefficients, N = 625 Prob > |r| under H0: Rho=0				
	ProdPerSubBlock	Aisles	CrossAisles	OrderSize
Rank	0.58660 <.0001	-0.22508 <.0001	-0.18367 <.0001	0.08751 0.0300

e) Combined

Pearson Correlation Coefficients, N = 625 Prob > |r| under H0: Rho=0				
	ProdPerSubBlock	Aisles	CrossAisles	OrderSize
Rank	-0.48762 <.0001	0.11624 0.0039	0.21212 <.0001	-0.03980 0.3245

Figure 15: Pearson correlation coefficient.

The main purpose of this project is to explain and get knowledge on policies and to reduce traveled distances in order picking processes in warehouses, offering an encouraging panorama for the construction of more complex routing policies.

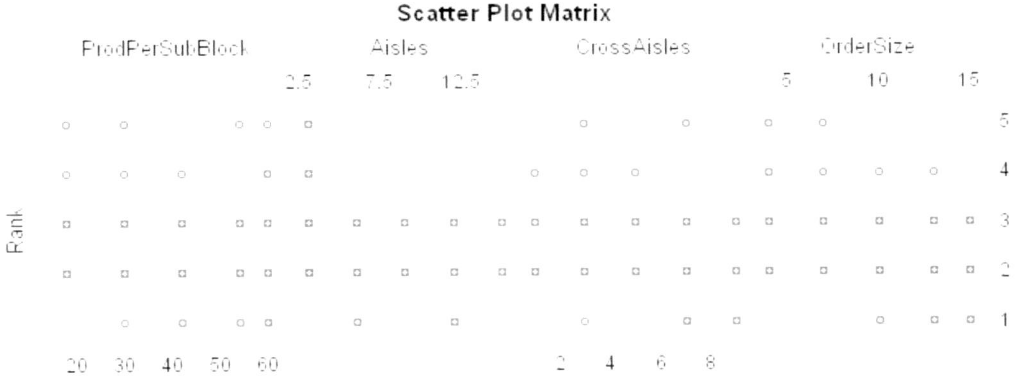

a) S shape

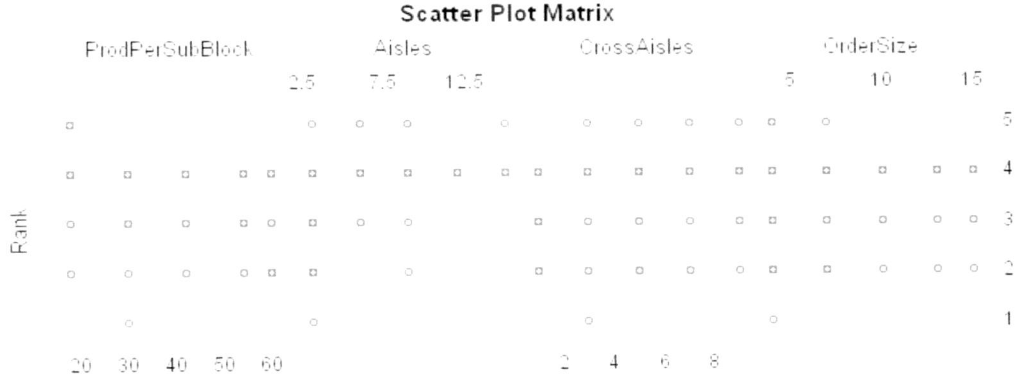

b) Largest Gap

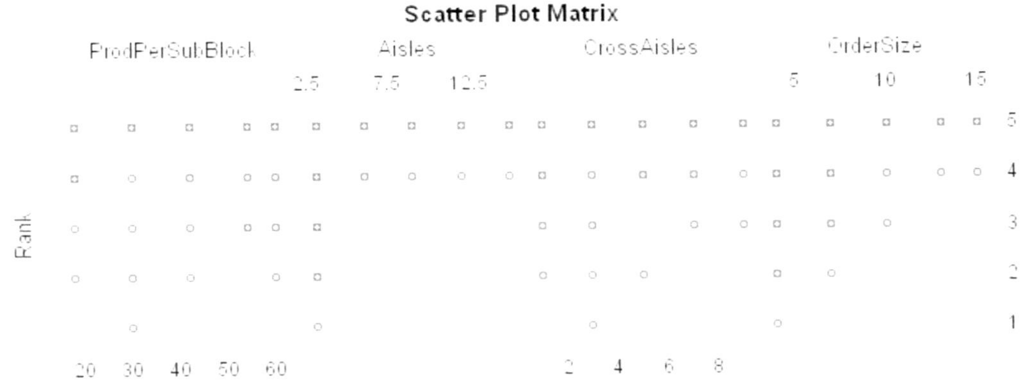

c) Midpoint

Figure 16 contd. ...

d) Return

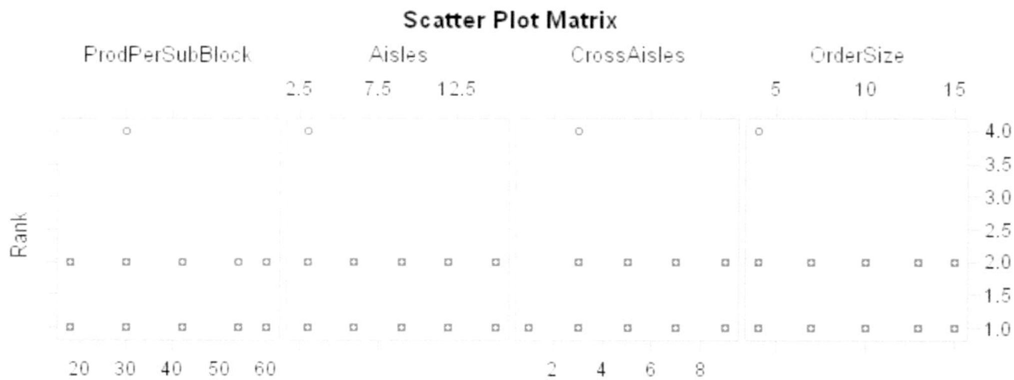

e) Combined

Figure 16: Pearson correlation coefficient Scatter Plot Matrix.

6. Conclusions and Directions for Future Work

In this chapter, we studied and developed five routing policies to apply them in order to multiple scenarios and situations and generate enough information to find trends and understand the behavior of these heuristics.

These experiments provide evidence that the performance of these policies can be highly sensitive to the different characteristics of the warehouse; for example, in the case of the S Shape routing policy, its performance is mainly affected by (1) the number of locations per sub-block, and (2) the size of orders of products to be picked. In the case of Largest Gap, the most marked trend is defined by the number of aisles, affecting its performance. Similarly, although not as marked as in Largest Gap, in Midpoint, the number of aisles tends to reduce its efficiency. The Return policy seems to be highly sensitive to the number of locations per sub-aisle, with performance deteriorating in a highly marked way. On the contrary, Combined improves as the number of locations increases.

These insights can be used in the future to design a heuristic enriched with these key elements about routing policies. In this way, such a heuristic can foresee aspects of the test instances that can affect its performance and mitigate the consequences.

References

[1] Ochoa Ortiz-Zezzatti, A., Rivera, G., Gómez-Santillán, C., Sánchez–Lara., B. Handbook of Research on Metaheuristics for Order Picking Optimization in Warehouses to Smart Cities. Hershey, PA: IGI Global, 2019. doi.org/10.4018/978-1-5225-8131-4.

[2] Tompkins, J.A., White, J.A., Bozer, Y.A. and Tanchoco, J.M.A. 2010. Facilities planning. New York, John Wiley & Sons.

[3] Petersen, C.G. and Aase, G. 2004. A comparison of picking, storage, and routing policies in manual order picking. Int. J. Production Economics, 92: 11–19.

[4] De Koster, R., Le-Duc, T. and Roodbergen, K. 2007. Design and control of warehouse order picking: A literature review. European Journal of Operational Research, 182: 481–501.

[5] Theys, C., Bräysy, O., Dullaert, W. and Raa, B. 2010. Using a TSP heuristic for routing order pickers in warehouses. European Journal of Operational Research, 200(3): 755–763.

[6] Pansart, L., Nicolas, C. and Cambazard, H. 2018. Exact algorithms for the order picking problem. Computers and Operations Research, 100: 117–127.

[7] Scholz, A. 2016. An exacto solution approach to the single-picker routing problem in warehouses with an arbitrary block layout. Working Paper Series, 6.

[8] Henn, S., Scholz, A., Stuhlmann, M. and Wascher, G. 2015. A New Mathematical Programming Formulation for the Single-Picker Routing Problem in a Single-Block Layout, 5: 1–32.

[9] Ratliff, H.D. and Rosenthal, A. 1983. Order-picking in a rectangular warehouse: a solvable case of the traveling salesman problem. Operations Research, 31(3): 207–521.

[10] Gu, J., Goetschalckx, M. and McGinnis, L.F. 2007. Research on warehouse operation: A comprehensive review. European Journal of Operational Research, 177(1): 1–21. doi.org/10.1016/j.ejor.2006.02.025.

[11] Hong, S. and Youngjoo, K. 2017. A route-selecting order batching model with the S-shape routes in a parallel-aisle order picking system. European Journal of Operational Research, 257: 185–196.

[12] Cano, J.A., Correa-Espinal, A.A. and Gomez-Montoya, R.A. 2017. An evaluation of picking routing policies to improve warehouse. International Journal of Industrial Engineering and Management, 8(4): 229–238.

CHAPTER-14

Color, Value and Type Koi Variant in Aquaculture Industry Economic Model with Tank's Measurement Underwater using ANNs

Alberto Ochoa-Zezzatti,[1,*] *Martin Montes-Rivera*[2] and *Roberto Contreras-Masse*[1]

1. Introduction

A fish tank can be installed in various spaces, from the living room of a home, a consulting room, a restaurant, an aquarium or a hotel. There are more than 400 ornamental species that have commercial relevance: zebrafish, angel, Japanese, molly or sword. So, the possibilities of this agro-business, whose demand is growing in the Mexican market, are numerous. Commissioner Mario Aguilar Sánchez, during the closing of the First National Watercolor Expo in the Federal District [1], noted that 60 million organisms are produced each year, worth 4.5 billion MXN, from about 700 productive units. He affirmed that the national production is developed in 23 entities, where 160 multi-species species are cultivated, such as koi carp, guppy, molly, angelfish, platy, danio zebra, tetra, cichlid, betta, gurami, sword, nun, oscar, plecos, catfish, shark, sumatra, dragon and red seal. The national production of ornamental fish is a business with prospects of social and economic growth that is developed in 23 entities, where 160 species and varieties are cultivated by aquarists, said the national commissioner of Aquaculture and Fisheries, Mario Aguilar Sánchez.

However, according to various groups of breeders and authorities of the Federal Government, the great challenge for this segment to take off and generate wealth at the local level, consists of strengthening the breeding, sale, and distribution of fish, since it is now possible to satisfy the demand only by importing animals in large volumes.

Among the existing varieties, one of the most popular is the so-called Goldfish or Japanese fish. The conservation and breeding of cold-water fish are not new concepts, since from ancient times in the Asian continent, particularly in distant China, people began to select beautiful specimens. In this research we focus on colorful Koi carp species, were often introduced in small outdoor ponds or ceramic pots. These animals were not only raised for ornamental purposes but also had a practical purpose since their conservation in captivity facilitated the ability to eat fresh fish at any time without the difficulty of capture in the wild. With the passage of time and the boom aquariofilia

[1] Juarez City University.
[2] Universidad Politécnica de Aguascalientes.
[*] Corresponding author: alberto.ochoa@uacj.mx

acquires, it gives way to the selective breeding of specimens, producing a great variety of fish, both in colors and in certain peculiar characteristics of their phenotype.

The state of Morelos is an ideal setting for the breeding of Japanese ornamental fish. Many of the businesses, most are familiar and small. When you launch into this world of aquiculture do it without any system of decision making that allows them to go a good way to obtain greater profits and in the shortest possible time [2].

But other states could have several problems implementing fish breeding which could benefit the economy of several places, that is why the cultivation of ornamental species had a considerable increase in Mexico. One of the states where this economic activity is emerging is Chihuahua, which, despite not having the most appropriate weather conditions although it has the physiological spaces for it. Thinking in the costs of implementing a tank for different states and the required technologies we proposed with this research to determine which is the ideal model to optimize breeding and development processes of the different species of Koi Fish. Determining the ideal and optimal value of a tank of koi fish is essential to specify the marginal gain of this type of project in aquaculture but the adaptation of the thank depends on several factors like the size of the carps in the tank and its quantity.

Problem Statement

A project is a temporary effort, with a variety of resources that seeks to satisfy several specific objectives in a given time. Innovation is the creation and use of new ideas that give value to the client or business. Proper planning of a Japanese fish breeding project depends on many factors. Detailed planning should be done, foreseeing risks that may arise. Technological Innovation Projects Scheduling Problems (TI-PSP) are a variant of Project Scheduling Problems (PSP). The PSP is a generic name that gives to a whole class of problems in which the best form, time, resources and costs for the programming of projects are necessary. The problem studied in the present investigation corresponds to a PSP theme because it involves variables of resource allocation to tasks and processes (computation).

1.1 Justification and purpose of our investigation

At present, there are no mathematical models to optimize resources of Japanese fish breeding projects, so the present research is a milestone in the subject seeking a mathematical solution. Additionally, although it will be tested on Japanese fish farms projects, research will be the basis for any fish breeding problem that requires cost optimization. In addition, it will serve to centralize the use of economic resources in a more real way. The main social contribution of the research will be to expand the coverage of the budget presented by the breeders and increase the possibility of earning profits in less time. It will be achieved as the budget of each project is determined in real form, to avoid overcharging and underestimating them.

Therefore, we propose to implement a method based on a generated dataset for identifying the size in the carps underwater using Artificial Neural Networks (ANNs) so that only carps over 10 cm size be detected with cameras helping to maintain the correct quantity of liters in the tanks, for the proposed economic model. We selected ANNs because there are good regression models as mentioned and they have been used before for smart farming, monitoring of the water quality environments and aquaculture operations, like shown in Figures [4–6].

2. Cluster Analysis and Artificial Neural Networks (ANNs)

Cluster analysis [3,4] is an unsupervised learning technique that aims to divide a dataset into groups or clusters. It belongs to, like other typologies and that discriminant analysis, to the set of techniques that aims to classify individuals. The fundamental difference between cluster and discriminant analysis is that in cluster analysis the groups are unknown a priori and are precisely what we want to determine. While in discriminant analysis, the groups are known and what we want to know is the extent to which the available variables discriminate against these groups and can help us to classify or assign the individuals to the given groups.

Observations in the same group are similar (in a sense) to each other and are different in the other groups. Clustering methods can be divided into two basic types: hierarchical and partitioned grouping. Hierarchical clustering can be achieved with the agglomerated algorithm. This algorithm starts with disjoint clusters (n objects in a single group) and gradually proceeds to merge objects or clusters of more similar objects into a cluster.

Algorithm 1. Basic Algorithm of Hierarchical Cluster Clusters

```
Get the proximity matrix
Repeat
    Merge the two closest clusters
    Update the proximity matrix that reflects the proximity
    between the new cluster and the new cluster
Until there is a single cluster.
```

For the decision-making, a cluster analysis using hierarchical clustering and the agglomerated algorithm was employed. It was used as the input variables' initial budget and space to mount the Japanese fish farm (m2). The hierarchical cluster generates a dendrogram, which is a tree diagram frequently used to illustrate the arrangement of the clusters produced. A dendrogram is a diagram showing the attribute distances between each pair of merged classes in a sequential fashion. To avoid crossing lines, the diagram is graphically exposed such that the members of each pair of classes that merge are close elements. Dendrograms are often used in computational biology to illustrate the grouping of genes or samples, sometimes on top of heatmaps. After obtaining the dendrogram, a mathematical model will be applied to each element of the dendrogram. This will reveal the optimal values for the implementation of the project according to the values of entry of budget and quantity of square meters.

ANNs are mathematical for representing biological neurons that were proposed in the 1950s , their applications cover several areas including regression models [4]. The activation functions are the hyperbolic tangent sigmoid for hidden layers, and the output layer is the linear activation function, which builds a good approximator for functions with finite discontinuities [7]. In this research, we train the neural network with the Scaled Conjugate Gradient (SCG) backpropagation and the Mean Square Error (MSE) as the expectation function in the equation (3). The SCG Backpropagation is a modified backpropagation proposed by Moller in 1993 that calculates the gradient in specific conjugate directions increasing convergence speed. With as the number of samples, as the target, and as the computed output of the FNNs.

3. Mathematical Model

In this section, the mathematical model that was used to optimize the costs of the initial investments for breeding goldfish is addressed. The model analyzes the main elements necessary to start a business of Japanese fish farming.

Each investment project includes material resources that are divided into infrastructure elements, equipment needed for cultivation, cost of young small fish (approximately 2 months old) and cost of fish feed. Most of the costs were obtained from the website Mercado Libre [5], except for costs for the construction of tanks [6].

The Objective Function corresponds to the budget necessary for the cultivation of Japanese fish and is formulated as follows:

$$Rb = Af * Cf + Cff + I \tag{1}$$

Table 1: Elements of the mathematical model.

Acronym	Concept
Ib	Initial budget
Rb	Real budget
Nml	Number of meters long
Nmw	Number of meters width
Nma	Number of square meters available
Nms	Number of square meters suggested by the model
Af	Amount of fish to buy
Nft	Number of fish per tank 3 m x 2 m x 0.5 m
Lf	Quantity of liters needed by a fish of 10 or more centimeters (constant value)
Cf	Cost of small fish (3-4 cm)
Cff	Cost of food for all fish
I	Infrastructure cost
Tc	Tank cost of 3 m x 2 m x 0.70 m
Ce	Cost per equipment
Nt	Number of tanks
Mt	Square meters of a tank (constant value 3m x 2m)

Where **Af** corresponds to the quantity of fish, **Cff** is the general cost of food for the fish, **I** corresponds to the estimated cost of infrastructure for the crop and **Cf** is the cost per Japanese fish where the average value is 10 MXN per fish.

Model Restrictions:

The budget **Rb** cannot exceed the initial budget that is denoted by **Ib**.

$$Ib \geq Rb \tag{2}$$

Similarly, the restriction is made that the number of square meters used **Nms** cannot exceed the available **Nma**:

$$Nms \geq Nma \tag{3}$$

For this, first determine the amount of m^2 available **Nma**:

$$Nma = Nmw * Nml \tag{4}$$

Where **Nmw** corresponds to the width of the available space and **Nml** to the length

$$Nt = \frac{Nma}{Mt + 2} \tag{5}$$

$$Nms = Nt * (Mt + 2) \tag{6}$$

Nt refers to the number of tanks, **Mt** to the area each tank occupies in m^2 and 2 is the value that is due to the space that must be left by each tank (2 meters wide).

$$Af = \frac{Nt * 3000}{Lf} \tag{7}$$

Where **Af** refers to the quantity of fish to be bought according to the quantity of tanks of 3m x 2m x 0.5m (3000 liters) and **Lf** is the quantity of liters required by Japanese fish of size greater than or equal to 10 cm,

$$Nft = \frac{3000}{Lf} \tag{8}$$

Nft is the number of fish to be placed per tank,

$$Tc = 2\,((l + a) * h) * 208.41 + ((2(l + a) * h) + l * a) * 125.61 \tag{9}$$

In this, **Tc** is the cost per tank, where **l** is the length, **a** is the width and **h** the height of the tank, in this case **l** = 3 m, **a** = 2 m and **h** = 0.7 m. See table # 2.

$$Cff = \frac{Af * 0.408}{1.5} * 170 \tag{10}$$

Table 2: Description of the elements for the construction of the tanks.

Concept	Costs
Annealed wall of 5.5 cm thick, common finish in flat areas	208.41
Polished planar with wood plane in walls, with mortar, cement-sand in proportion 1:6 of 2.0 cm of thickness, includes the repelled	125.61

Where **Cff** is the cost per food according to **Af** that is the quantity of fish, in a period of 4 months, period in which they must have 10 or more centimeters. See table # 3. For this computation [5,6] were used as reference where in the research the best results were obtained feeding the goldfish twice a day, where the amount is calculated by 2% of body mass.

$$Ce = Mph + Wt + Ctf + Wp \tag{11}$$

$$Ctf = Clf * 16000 + Csf * 4800 \tag{12}$$

Table 3: Measuring Elements for Goldfish Food.

Concept	Amount
1.5 kg food package cost	170 MXN
Amount of food per fish in 4 months	0.408 kg

Ce is the cost for equipment. See table # 4.

Table 4: Equipment and costs.

Acronym	Concept	Amount
Mph	PH meter	225 MXN
Wt	Water Thermometer	80 MXN
Ctf	Cost per filters	$Ctf * 16000 + Csf * 48000$
Clf	Number of large filters de 45000 liters	$\frac{Nt * 3000}{45000}$
Csf	Number of small filters of 6000 liters	$\frac{Nt * 3000 - 45000 * C_l}{6000}$
Wp	Water Pump, (1/2) Hp Siemens Centrifuga	1297.60 MXN

The following equation is the one that is responsible for calculating the cost of infrastructure (I).

$$I = Ce + Tc * Nt \tag{13}$$

Model for determining the location of a koi carp to determine its size underwater

Based on the method described in [8], suppose an observer at the edge of a pool perceiving an object immersed in water located at a distance and at a depth of incidence of the light on the water, which are determined using the law of refraction.

The apparent position of an object seen by the observer is in the direction of the refracted ray and the object is located at the origin at a depth of but the observer perceives it at h-associated with incidence of the light on the water. A ray (red color) starts from the object and forms an angle of incidence θi. The refracted ray forms an angle with the normal one (Figure 1). According to the law of refraction, and visualized in Figure 1:

$$n\sin\theta i = \sin\theta r \qquad (14)$$

where n = 1.33 is the water refractive coefficient and the angle of refraction θr is greater than the incident θi.

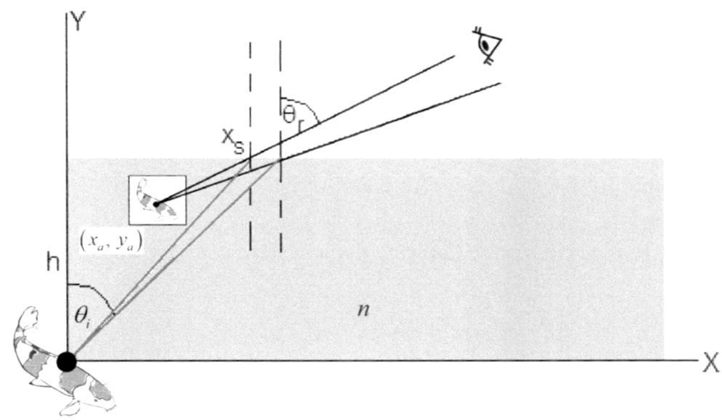

Figure 1: Apparent position of koi carp under water based on refraction incidence.

The direction of the refracted ray and its extension passes through the point (xs,h) and its slope is 1/tanθr. Knowing that xs=htanθi. The equation for this line is

$$y - h = \frac{x - x_s}{\tan\theta_r}$$

From the object, a ray (blue) forms an angle of incidence θ'i. The refracted ray forms an angle θ'r with the normal one.

The direction of the refracted ray and its extension passes through the point (x's,h) and its slope is 1/tanθ'r. Knowing that x's=htanθ'i. The equation for this line is

$$y - h = \frac{x - x'_s}{\tan\theta'_r}$$

The extensions of the refracted rays are cut at the point indicated in blue coordinates

$$x_a = \frac{\tan\theta_i \tan\theta'_r - \tan\theta_r \tan\theta'_i}{\tan\theta'_r - \tan\theta_r} h$$

$$y_a = \left(\frac{\tan\theta'_i - \tan\theta'_i}{\tan\theta'_r - \tan\theta_r}\right) h$$

This is the apparent position (xa, already) of the object as seen by an observer in the direction of the refracted beam. Where θ'i = θi+δ, where δ is a small angle increment.

We represent the apparent position (xa, ya) of an object located at the origin, for various angles of incidence.

The depth of the object is h = 1 m, the angle increment $\delta = 0.01$ degrees. The arrows indicate the direction of the refracted beam, the direction of observation of the object, as is possible see in Figure 2.

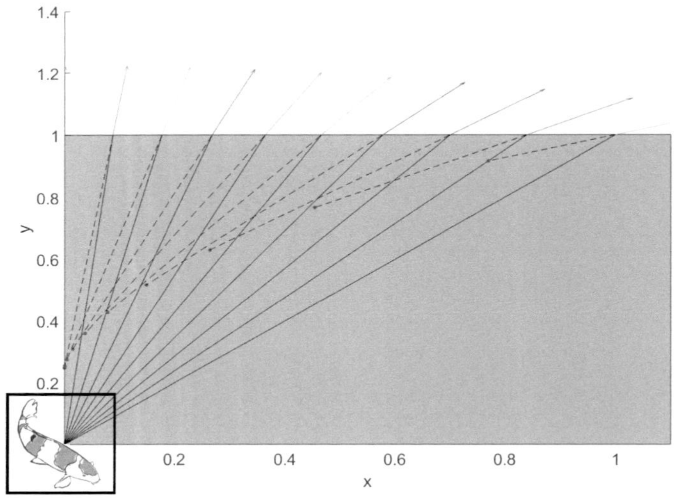

Figure 2: Apparent positions of koi carp under water as observed position varies.

The position of the observer is fixed

In this section we will describe the apparent position of objects immersed in a pool in relation to an observer on the edge (Figure 3).

The Y-axis coincides with the position of the swimmer and the X-axis with the surface of the water. The position of the swimmer's eyes is (0,y0) and the position of the submerged object (red dot) in the water is (xb, yb), the apparent position of the object (blue dot) is (xa, already). An incident beam from the object is refracted at a point on the surface separating the two media at xs, forming an angle θi with the normal. The refracted ray reaches the eyes of the observer forming an angle θr with the normal.

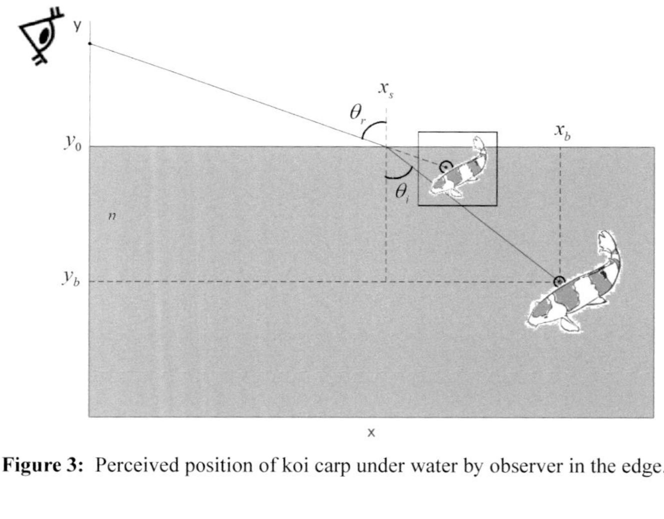

Figure 3: Perceived position of koi carp under water by observer in the edge.

Knowing the position of the object (xb, yb) and that of the swimmer's eyes (0,y0), from the law of refraction, nsinθi = sinθr, we will calculate the xs position, where the ray of light coming from the object is refracted, and the angles of incidence θi and the refracted ray θr. In the figure, we see that

$$\tan\theta_i = \frac{x_b - x_s}{-y_b} \qquad \tan\theta_r = \frac{x_s}{y_0}$$

Eliminating xs and using the law of refraction

$$x_b - y_0 \tan\theta_r + y_b \frac{\sin\theta_i}{\sqrt{1-\sin^2\theta_i}} = 0$$

$$x_b - y_0 \frac{\sin\theta_r}{\sqrt{1-\sin^2\theta_r}} + y_b \frac{\sin\theta_r}{\sqrt{n^2-\sin^2\theta_r}} = 0$$

We solve this transcendent equation, to calculate θr, then xs and θi
The equation of the direction of the refracted beam and its extension is

$$y = -\frac{x - x_s}{\tan\theta_r}$$

$$x_b - x = y\tan\theta_r - y_b\tan\theta_i$$

To determine the apparent position of the object, we need to draw one more ray of refraction angle θ'r= θr+δ, (being δ a very small angle, infinitesimal) and find the intersection of the extension of the two refracted rays, as shown exaggeratedly in the figure. The equations of the red and blue lines are, respectively

$$\begin{cases} x_b - x = y\tan\theta_r - y_b\tan\theta_i \\ x_b - x = y\tan\theta'_r - y_b\tan\theta'_i \end{cases}$$

We clear the apparent position xa as it is the point of intersection of the two lines

$$\begin{cases} x_a = x_b - y_b \dfrac{\tan\theta'_i\tan\theta_r - \tan\theta'_r\tan\theta_i}{\tan\theta'_r - \tan\theta_r} \\ y_b = y_b \dfrac{\tan\theta'_i - \tan\theta_i}{\tan\theta'_r - \tan\theta_r} \end{cases}$$

We calculate the order already

$$\frac{\tan\theta'_i - \tan\theta_i}{\tan\theta'_r - \tan\theta_r} = \frac{\dfrac{\sin\theta'_r}{\sqrt{n^2-\sin^2\theta'_r}} - \dfrac{\sin\theta_r}{\sqrt{n^2-\sin^2\theta_r}}}{\tan\theta'_r - \tan\theta_r} =$$

$$\frac{\dfrac{\sin(\theta_r+\delta)}{\sqrt{n^2-\sin^2(\theta_r+\delta)}} - \dfrac{\sin\theta_r}{\sqrt{n^2-\sin^2\theta_r}}}{\tan(\theta'_r+\delta) - \tan\theta_r}$$

We make the approach

$$f(\theta_r+\delta)\approx f(\theta_r)+\frac{df}{d\theta_r}\delta$$

$$\tan(\theta_r+\delta)\approx\tan\theta_r+\frac{\delta}{\cos^2\theta_r}$$

$$\frac{\sin(\theta_r+\delta)}{\sqrt{n^2-\sin^2(\theta_r+\delta)}}\approx\frac{\sin\theta_i}{\sqrt{n^2-\sin^2\theta_r}}+\frac{n^2\cos\theta_r}{(n^2-\sin^2\theta_r)^{3/2}}\delta$$

The final result is

$$y_a=y_b\frac{\left\{\frac{\sin\theta_r}{\sqrt{n^2-\sin^2\theta_r}}-\frac{n^2\cos\theta_r}{(n^2-\sin^2\theta_r)^{3/2}}\delta\right\}-\frac{\sin\theta_r}{\sqrt{n^2-\sin^2\theta_r}}}{\left\{\tan\theta_r+\frac{\delta}{\cos^2\theta_r}\right\}-\tan\theta_r}$$

$$=y_b\frac{n^2\cos^3\theta_r}{(n^2-\sin^2\theta_r)^{3/2}}$$

We calculate the abscissa xa

$$\frac{\tan\theta'_i\tan\theta_r-\tan\theta'_r\tan\theta_i}{\tan\theta'_r-\tan\theta_r}=\frac{\frac{\sin\theta'_i}{\sqrt{1-\sin^2\theta'_i}}\tan\theta_r-\tan\theta'_r\frac{\sin\theta_i}{\sqrt{1-\sin^2\theta_i}}}{\tan\theta'_r-\tan\theta_r}=$$

$$\frac{\frac{\sin\theta'_r}{\sqrt{n^2-\sin^2\theta'_r}}\tan\theta_r-\tan\theta'_r\frac{\sin\theta_r}{\sqrt{n^2-\sin^2\theta_r}}}{\tan\theta'_r-\tan\theta_r}=$$

$$\frac{\frac{\sin(\theta_r+\delta)}{\sqrt{n^2-\sin^2(\theta_r+\delta)}}\tan\theta_r-\tan(\theta+\delta)\frac{\sin\theta_r}{\sqrt{n^2-\sin^2\theta_r}}}{\tan(\theta+\delta)-\tan\theta_r}\approx$$

$$\frac{\left\{\frac{\sin\theta_r}{\sqrt{n^2-\sin^2\theta_r}}+\frac{n^2\cos\theta_r}{(n^2-\sin^2\theta_r)^{3/2}}\delta\right\}\tan\theta_r-\left\{\tan\theta_r+\frac{\delta}{\cos^2\theta_r}\right\}\frac{\sin\theta_r}{\sqrt{n^2-\sin^2\theta_r}}}{\left\{\tan\theta_r+\frac{\delta}{\cos^2\theta_r}\right\}-\tan\theta_r}$$

The final result is

$$x_a=x_b+y_b\frac{(n^2-1)\sin^3\theta_r}{(n^2-\sin^2\theta_r)^{3/2}}$$

Calculation of the apparent position of a submerged object as different Koi carps issues inner diverse distances form the water tank in Figure 4.
Let y0=1.5 be the height of the swimmer's eyes, the position of the object (xb, yb) is (5,-2)m

We solve the transcendent equation to calculate the angle of the refracted beam θr and the xs position on the water surface where the incident beam is refracted

$$x_b - y_0 \frac{\sin\theta_r}{\sqrt{1-\sin^2\theta_r}} + y_b \frac{\sin\theta_r}{\sqrt{n^2-\sin^2\theta_r}} = 0$$

and then the apparent position (xa, ya) of an object in the position (xb, yb)

$$\begin{cases} x_a = x_b + y_b \dfrac{(n^2-1)\sin^3\theta_r}{(n^2-\sin^2\theta_r)^{3/2}} \\ y_a = y_b \dfrac{n^2\cos^3\theta_r}{(n^2-\sin^2\theta_r)^{3/2}} \end{cases}$$

We trace the incident beam, the refracted beam and its extension to the apparent position of the object (Koi carp) in Figure 6. After obtaining last equation, we get a mathematical model for determining the apparent position of the koi carps, but we perceive images of the apparent position through the camera, therefore, we could fix the equations for obtaining the real position of the carp with the apparent position but in the real life identifying is a difficult task.

Alternatively, we propose to generate a dataset using a simulation with the equation (28) and varying its parameters, then train an ANN with the structure in section 2 for determining the real position with the input of the apparent position of carps, then that information will be used for determining the size of a carp.

4. Results and Discussion

Design of experiment

We built dataset with 100,701 points for training the ANN and generating the model, it was generated by varying in 0.01 m the real position of the carps (x_b, y_b) in ranges $0 \le x_b \le 5$ and $-2 \le y_b \le 0$, considering a 2 m deep thank and 5 meters as maximum distance for perceiving the carp. All those positions are show in Figure 4.

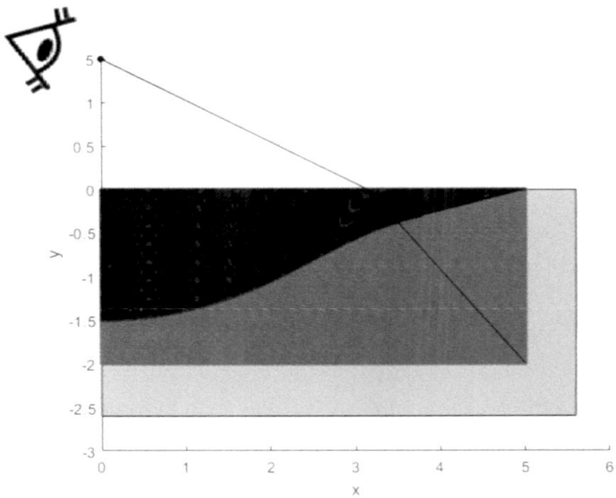

Figure 4: Variations of position of koi carp underwater for generating the dataset.

We made a script to draw the appearance of a circular object of radius 0.5.

After training the ANNs with the different architectures we used 10-fold cross-validation obtaining the cross-validated errors in Table 5, that allow us to define the best architecture for the neural network as 2 hidden layers and 8 neurons per layer, as is shown in Figure 5.

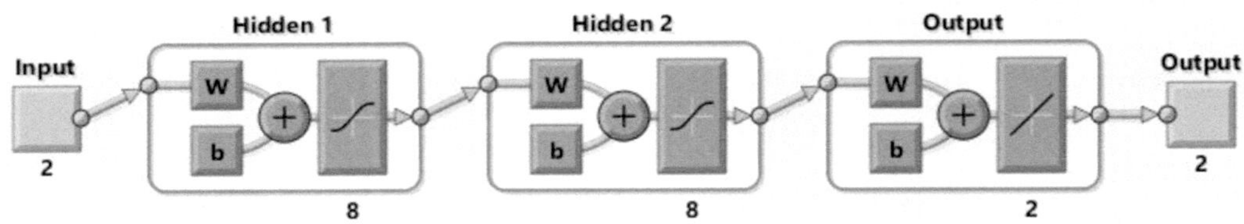

Figure 5: 10-fold cross-validated architecture for the ANN model.

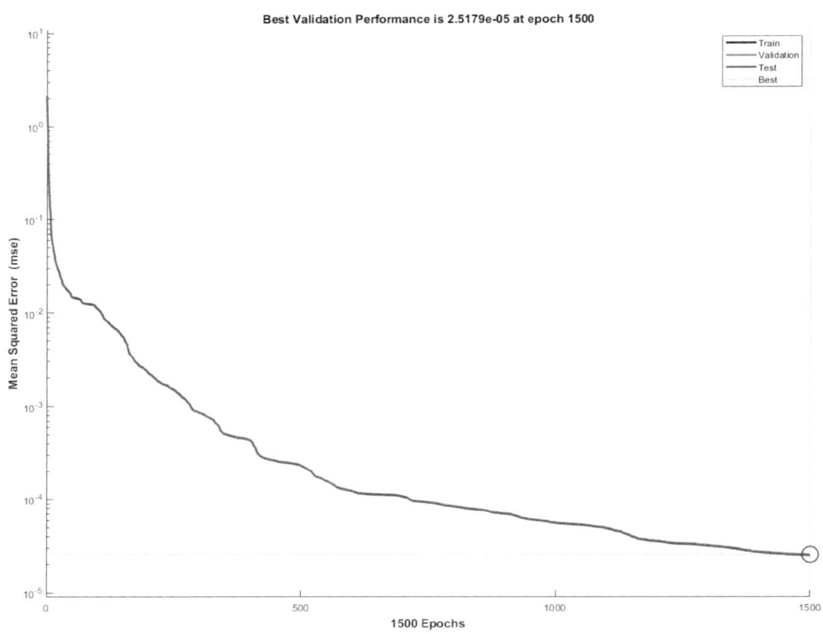

Figure 6: Training Performance for ANN model.

Finally, we drew the apparent shape of the bottom of a pool as seen by the swimmer on the edge. The real shape is described by the function

$$y_b = \begin{cases} -0.9 - \dfrac{0.5x_b}{15.5} & 0 \le x_b < 15.5 \\ \dfrac{-1.6x_b + 21.02}{2.7} & 15.5 \le x_b < 18.2 \\ -3 & 18.2 \le x_b < 25 \end{cases}$$

The histogram comparison the performance in the training which also support the suppression of overfitting is shown in Figure 7.

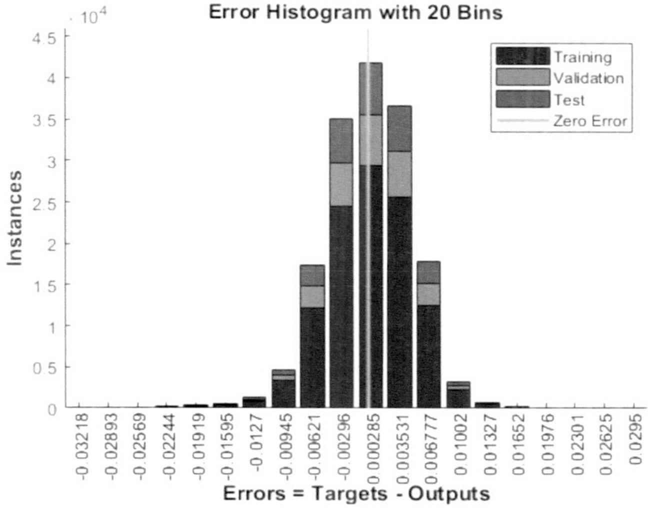

Figure 7: Histogram Performance for ANN model.

Features of this tank water with Koi carps.

The regression responses for the outputs x_b and y_b showing a correct response based on the values specified during the training are shown in Figure 8.

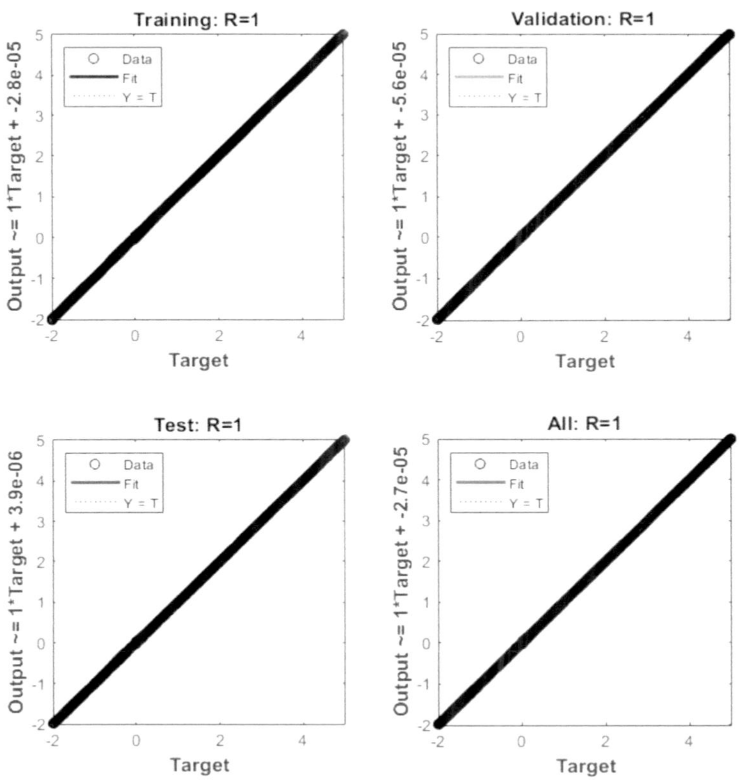

Figure 8: Regression Performance for ANN model.

The Aquaculturist does not perceive the bottom of the fresh water tank from a certain distance xb of about 6 m, for which it is already almost zero (see in Figure 9).

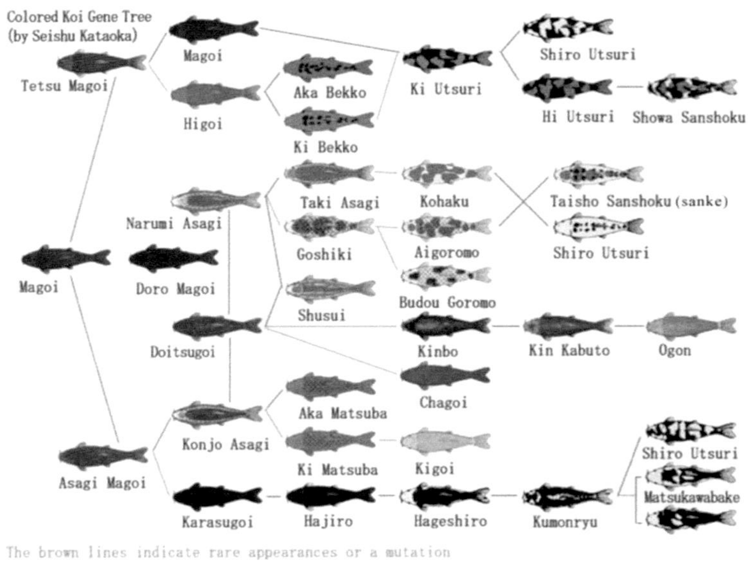

Figure 9: Evolution of Koi fish and its possible genetic combinations.

Determining the value of each issue is a complicated task, mainly because many of the issues are subspecies of other species and the valuation models are different for each species, as shown in Figure 10.

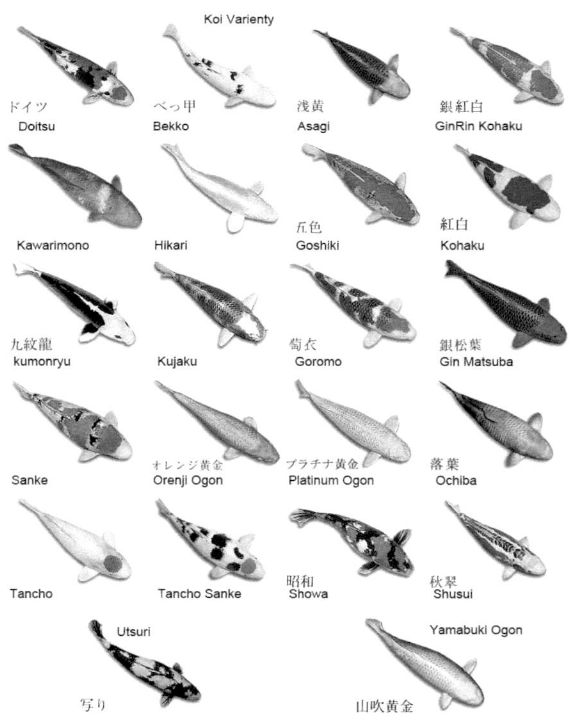

Figure 10: Different species analyzed in our study.

5. Experimentation

In order to be able to simulate the most efficient arrangement of individuals in a social network, we developed an atmosphere able to store the data of each one of them representing individuals of each society, this with the purpose of distributing in an optimal form to each one of the evaluated societies. One of the most interesting characteristics observed in this experiment was the diversity of the cultural patterns established by each community. After identifying the best architecture, we trained the neural with the 80% for training dividing it in 70% for training, 15% for cross-validation in the training and 15% for testing, finally with the trained model we test the performance in the 20% reserved at the beginning as test set. The training results comparing the training performance, cross-validated performance, and test performance are shown in Figure 6, showing that there is not overfitting in the ratio between train, validation, and test responses. The generated configurations can be metaphorically related to the knowledge of the behavior of the community with respect to an optimization problem (to select Aquaculture societies, without being of the same quadrant [3]). The main experiment consisted of detailing each one of the 21 koi carp's variant. This allowed us to generate the best selection of each Quadrant and their possible location in a Koi Fish Pond, which was obtained after comparing the different cultural and social similarities from each community, and to evaluate each one of them with Multiple Matching Model. Using ANNs we determine the correct species, size and relatively the possible weight of an issue as is possible to see correctly in Figure 11.

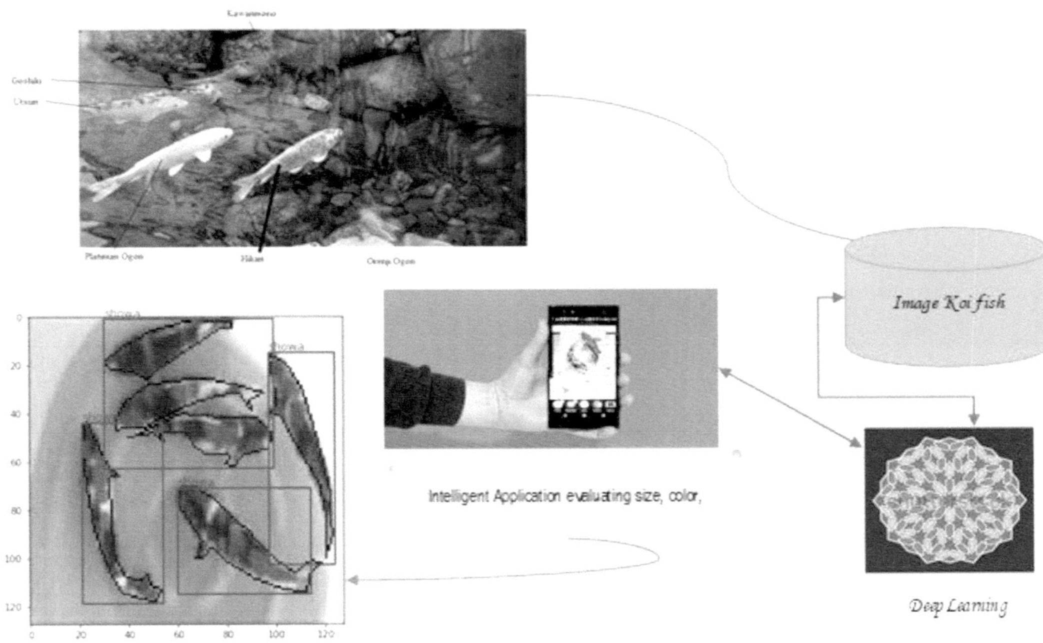

Figure 11: Intelligent application to determine the correct parameters associated with the final price of an issue of Koi Fish species determined.

The developed tool classified each one of the societies pertaining to each quadrant, with the proposed result of obtaining real position of carps based on the apparent position the future research for this work must be to extract the apparent positions from images captured with a digital camera or a cellphone picture, this should be done by using Deep Learning, specifically Convolutional Neural Networks (CNNs), because they are good classifiers in object recognition, which could identify the koi carps in images and its position, after that the obtained location could be send to our ANN trained of obtaining the real position of the coordinates in the boxes and the calculate the sizes of the carps based on the corners of its box transformed into real positions..

The design of the experiment consists of an orthogonal array test, with the interactions between the variables: socialization, the temperature required, adult size, cost of food, maintenance in a freshwater tank, growing time, fertility rate, valuation in a sale. These variables are studied in a range of colors (1 to 64). The orthogonal array is L-N (2**8), in other words, 8 factors in N executions, N

Table A: Orthogonal array associated with our research.

No.	Factors							Weather measurements	
	A	B	AB	C	D	BC	E		
	1	2	3	4	5	6	7	1	2
1	1	1	1	1	1	1	1	26	38
2	1	1	1	2	2	2	2	16	6
3	1	2	2	1	1	2	2	3	17
4	1	2	2	2	2	1	1	18	16
5	2	1	2	1	2	1	2	0	5
6	2	1	2	2	1	2	1	0	1
7	2	2	1	1	2	2	1	4	5
8	2	2	1	2	1	1	2	5	3

Table B: Types of fish of kind Koi.

Instance	Crappie	Socialization	Temperature	Size(cm)	Cost of food	Maintenance	Growing time	Fertility rate	Valuation
1	Doitsu	4	21 °C a 31°C	13	0.72	0.23	0.14	0.72	10
2	Bekko	3	24°C a 27°C	6.5	0.91	0.92	0.92	0.77	10
3	Asagi	5	25.5 °C	10 a 13	0.43	0.94	0.33	0.98	15
4	GinRin Kohaku	6	26.5°C	5	0.18	0.67	0.79	0.74	15
5	Kawarimono	5	27°C a 31°C	7.5	0.85	0.52	0.74	0.27	20
6	Hikari	3	25°C a 31 °C	10 a 15	0.32	0.47	0.71	0.96	20
7	Goshiki	5	22°C a 30°C	10 a 13	0.66	0.82	0.36	0.17	15
8	Kohaku	6	15°C a 32°C	4 a 7	0.33	0.47	0.54	0.24	10
9	Kumonryu	7	21°C a 27°C	5 a 7.5	0.55	0.89	0.43	0.48	10
10	Kujaku	5	13°C a 27°C	10	0.44	0.87	0.47	0.26	25
11	Goromo	6	20°C a 30°C	10 a 13	0.88	0.27	0.22	0.42	20
12	Gin Matsuba	6	24°C a 28°C	25 a 60	0.72	0.23	0.19	0.44	20
13	Sanke	7	22°C a 28°C	6	0.91	0.92	0.47	0.71	20
14	Orenji Ogon	2	22°C a 28°C	5	0.43	0.94	0.23	0.68	20
15	Platinum Ogon	7	24°C a 26.5°C	4	0.18	0.67	0.58	0.27	20
16	Ochiba	6	26.5°C	5	0.85	0.52	0.38	29	20
17	Tancho	5	20°C a 30°C	27	0.32	0.47	0.51	12	20
18	Tancho Sanke	3	20°C a 25°C	15 a 20	0.66	0.82	0.18	34	30
19	Showa	5	18°C a 25°C	7	0.33	0.47	0.84	14	50
20	Shisui	5	20°C a 30°C	40 a 50	0.55	0.89	0.18	79	50
21	Utzuri	4	24°C a 28°C	25 a 30	0.44	0.87	0.86	60	35
22	Yamabuki Ogon	3	22°C a 28°C	14	0.88	0.27	0.64	64	40

is defined by the combination of possible values of the 8 variables and the possible range of color (see Table A and the importance of each issue in Table B, socialization attribute is a Lickert model with better socialization 7 and poor socialization 1). Considering features of porosity in a Fresh Water tank with many koi fishes and when weather conditions affect these.

Conclusions and Future Research

Using ANNs, we improved the understanding substantially to obtain the change of "best paradigm", because we appropriately classified the agent communities basing to us on an approach to the relationships that keep their attributes, this allowed us to understand that the concept of "negotiation" exists with base in the determination of the function of acceptance on the part of the rest from the communities to the propose location for the rest of the same ones. ANNs offers a powerful alternative to optimization problems and redistribution of the clustering technique. For that reason, this technique provides a quite comprehensible panorama with a model which implies maintaining the size of the carps upper 10 cm, making that if is used a digital camera outside of the water detecting positions of the carps will be affected by the refraction of water and phenomenon represented [7]. This technique allows including the possibility of generating experimental knowledge created by the ANNs for a novel dominion of application. The analysis of the level and degree of cognitive knowledge of each community is an aspect that is desired to evaluate for future work. The answer can reside between the similarity that exists in the communication between two different cultures and as these are perceived [9]. On the other hand, to understand the true similarities that have different societies with base in the characteristics that make them contributor of a cluster and it as well allows him to keep his own identity, demonstrates that the small variations go beyond phenotypes characteristics and are mainly associate to tastes and similar characteristics developed through the time [6] to diverse variant of Koi carps.

Future Research

With the proposed result of obtaining real position of carps based on the apparent position the future research for this work must be to extract the apparent positions from images captured with a digital camera or a cellphone picture, this should be done by using Deep Learning, specifically Convolutional Neural Networks (CNNs), because they are good classifiers in object recognition, which could identify the koi carps in images and its position, after that the obtained location could be send to our ANN trained of obtaining the real position of the coordinates in the boxes and the calculate the sizes of the carps based on the corners of its box transformed into real positions. Deep Learning offers a powerful alternative for object recognition. On the other hand, with the sizes identified in the tank it is possible to send koi carps to different tanks went its size is not over 10 cm. The general description of future research is shown in Figure 11.

References

[1] Conapesca. Nuestros mares, sinónimo de abundancia y diversidad de alimentos. Rev. Divulg. Acuícola, vol. 4, no. 38, p. 8, 2017, [Online]. Available: http://divulgacionacuicola.com.mx/revistas/36-Revista Divulgación Acuícola Julio2017.pdf.

[2] Hernández-Pérez, E., Gónzalez-Espinosa, M., Trejo, I. and Bonfil, C. 2011. Distribución del género Bursera en el estado de Morelos, México y su relación con el clima. Rev. Mex. Biodivers., 82(3). [Online]. Available:

[3] Salam, H.J., Hamindon, W. and Badaruzzaman, W. 2011. Cost Optimization of Water Tanks Designed according to the ACI and EURO Codes. doi: 10.13140/RG.2.1.2102.8329.

[4] Goodfellow, I., Bengio, Y. and Courville, A. 2016. Deep Learning. The MIT Press.

[5] Hsu, W.C., Chao, P.Y., Wang, C.S., Hsieh, J.C. and Huang, W. 2020. Application of regression analysis to achieve a smart monitoring system for aquaculture. Inf., doi: 10.3390/INFO11080387.

[6] Yang, X., Ramezani, R., Utne, I.B., Mosleh, A. and Lader, P.F. 2020. Operational limits for aquaculture operations from a risk and safety perspective. Reliab. Eng. Syst. Saf., doi: 10.1016/j.ress.2020.107208.
[7] Hagan, M.T., Demuth, H.B., Beale, M.H. and De Jesús, O. 1996. Neural Network Design, 2nd ed. 1996.
[8] Suresh, S., Westman, E. and Kaess, M. 2019. Through-water stereo slam with refraction correction for AUV Localization. IEEE Robot. Autom. Lett., doi: 10.1109/LRA.2019.2891486.
[9] Berrar, D. 2018. Cross-validation. In Encyclopedia of Bioinformatics and Computational Biology: ABC of Bioinformatics.

CHAPTER-15

Evaluation of a Theoretical Model for the Measurement of Technological Competencies in the Industry 4.0

Norma Candolfi-Arballo,[1,4] *Bernabé Rodríguez-Tapia,*[1,4,*] *Patricia Avitia-Carlos,*[1,4] *Yuridia Vega*[1,3] and *Alfredo Hualde-Alfaro*[2]

This chapter presents the design and validation of a measuring instrument using the digital questionnaire evaluation technique, oriented to the self-perception of business leaders, to diagnose the current state of company work dynamics regarding the use, incorporation, learning, and technological appropriation. From the study carried out, a theoretical model capable of measuring the technological competencies of business leaders is obtained to diagnose the current state of companies' work dynamics regarding use, incorporation, learning, and technological appropriation.

1. Introduction

Industry 4.0 was defined due to the growing trends in the use of ICT for industrial production, based on three main components: the internet of things (IoT), cyber-physical systems (CPS) and smart factories [1]. Industry 4.0 undoubtedly generates numerous new opportunities for companies, but several automation and digitalization challenges arise simultaneously [2]. Therefore, management, as well as employees, must not only acquire specific technical skills but appropriate them [3].

Multiple studies have been developed around the growth of industries generated by the technological factor [4,5,6,7,8,9,10,11,12,13,14,15], pointing to the challenges faced by underdeveloped countries to achieve high levels of competitiveness, industrial scaling and similar scopes to those registered by developed countries, and these refer to a vision regarding the proposals of technological appropriation in the productive processes on which the context is prioritized [16,17]. The research highlights the importance of clear top-down governance to succeed in the appropriate use of technologies since an "uncoordinated bottom-up series" would block the path to Industry 4.0.

The following chapter shows the design and validation of a measuring instrument, using the digital questionnaire evaluation technique, oriented to the self-perception of business leaders to diagnose the current state of the companies work dynamics regarding the use, incorporation, learning, and technological appropriation.

[1] Autonomous University of Baja California, Blvd. University, Valle de las Palmas, 1000 Tijuana, Baja California, México.
[2] Department of Social Studies at COLEF. México.
[3] Industrial Processes Research Group.
[4] Distance Continuing Education Research Group.
* Corresponding author: rodriguez.bernabe@uabc.edu.mx

The measuring instrument is composed of five dimensions called: technological competences, environment and internal communication, environment and external communication, training and updating, and innovation factors in the company; these were constructed from a theoretical review of the literature on technological competencies, considerations of the technological competences concept in the industry, international considerations, such as the European Framework of Competencies, Emerging Markets and national considerations regarding current technical and technological knowledge in the industry.

The instrument was validated through expert judgment, by experts on social studies topics in the industry, and with experience in anthropological studies, management leadership, market momentum, digital marketing, economics, innovation management, and human capital in organizations.

The reliability of the instrument is also performed by calculating Cronbach's alpha, which is an internal consistency indicator that measures the degree to which the items are correlated; that is, the items' homogeneity in the measurement instrument.

2. Industry Technological Adoption

The formation of human capital is one of the focuses of attention priorities, specifically regarding to the updating of knowledge about technological equipment, development policies and everything implied from the planning of projects in the field of information technology, communication and collaboration to the evaluation of results, which should be related to the increase in the level of competitiveness, industrial scaling and improvement opportunities for companies [18,19], achieving active participation in the global field [17]. In that sense, analyzing the human capital of the industry allows the development of proposals aimed to strengthen their labor competencies and the evaluation of the advantage taken from the technological equipment used, considering it a continuous improvement activity [12,13,20,21,22].

The author Hobday (1995) in [16] describes the concept of Technology as a resource that combines physical capital and human skills, representing the dynamic capacity to create and increase the skills and knowledge in the industry, so it will allow a company to improve its skills, permitting a specific production process or development of new products to be integrated and adapted [16]. An essential part of Technology, the Information and Communication Technologies (ICT) or, in a more extended concept, Information, Communication and Collaboration Technologies (ICCT), as described in [23], refers to the possibilities to develop collaborative experiences by breaking time and space barriers by modifying industrial processes, forcing the need for changes in organizational structures and allowing new mechanisms of interaction and communication between company members and even between companies, promoting national and international cooperation.

ICTs are currently a relevant issue in multiple areas of impact from the educational, productive and governmental sectors. Emphasizing the productive sector reveals that industry plays a key role in the development and implementation of ICT and vice versa since they are directly linked to production processes, innovation, solutions, and transformation of the goods and services that the market requires. Nowadays, the ICT industry is developing faster, driven by Asian countries such as China and Japan. The growth in the ICT industry has not only resulted in the development and production of equipment, but also in services that transform the global distribution of software production [11]. In [9], some advantages of incorporating ICT are highlighted:

- Increase the efficiency of industrial and business processes, through updated information, historical information, indicators comparative, collaboration among employees, generation, and dissemination of knowledge, as well as the monitoring of profits and investments.
- Communication with suppliers, minimizing delivery times and accelerating operations for the acquisition of required in time inputs.

- Digital integration of a client portfolio, a historical selection of products, reports generation, administration, and selection of media.

In this sense, it is considered necessary to analyze the adoption of technology in the industry to establish objectives that lead to an increase in the incorporation and appropriation of ICT, considering studies that are aware of the context in which companies are developing within their country. Thus, it is also necessary to analyze human capital concerning technological competences to face the changes in the already defined processes.

To establish and promote a proposal for the inclusion of the ICT within a certain productive sector, it is considered necessary to explore and characterize the various profiles, obtaining indicators about competencies and ICT vision. That is, describe that business or leader profile and its relationship with ICT.

There is a need for a structured methodology that describes the steps to build an ideal profile in terms of technology for an industrial leader that allows him to favor the inside of his company with the incursion of technology.

3. Technological Competencies Studies on Industry from Global to Regional Perspective

The evaluation of technological competencies in the industry has a history of study and implementation. In Europe, programs are developed under the European Framework of e-competencies/*European e-Competence Framework* e-CF [24], a program composed of forty competencies in information and communication technologies for the industry. In the e-CF initiative, competencies are listed in five levels, aimed to cover technological needs. The actions are oriented to raise awareness, certification, and training of human capital, participation in innovative teaching-learning programs, mobility, and practices to attract young people to join technological careers at universities and to increase awareness about the importance of technological skills. Continuing with the international scenario, another interesting setup is the structure known as *Information Technology Infrastructure Library* (ITIL), which promotes the information technologies services within companies while considering quality standards. To assure this quality, ITIL proposes a series of standards for the development of infrastructure, appropriation, and operations based on information technologies. To date, there are several versions of ITIL, in the latest version of ITIL, the topics are classified into service strategies, service design, service transition, services operation and continuous improvement of services [25].

In Latin America, other case studies have been developed with quantitative, qualitative, or mixed evaluations. The proposed dimensions can be directed to perception, social activity, interactivity, use of the content, updating of practices, among others [26,27,28,29,30,31,32,33,34].

In the case of Mexico, [35] describes the assessment of the need for competencies to cover profiles in administrative positions where the attributes of advanced technology skills and computational knowledge are relevant indicators. In [36], an analysis of the learning of technological competencies in the maquiladora industry is shown. The analysis characterizes the export maquiladora industry and its industrial, technological and productive scaling. The results indicate that technological and productive scaling present weaknesses. The author proposes growth strategies, such as supplier development programs, public investment in innovation and development links between universities and companies, as well as the need for clarity in the objectives and nature of the industry in the electronics sector Mexico's northern border.

In Baja California, studies analyzing various sectors of the industry have been conducted in the productive sector, mainly oriented to the Software, Electronics and Aerospace industries, studying their components, development and future vision under a socio-economic and cultural analysis approach. In [11], human capital skills and labor abilities related to a particular sector are identified; in [12] diagnosis of the Aerospace industry is made and in [10] the policies for business development in the state are reviewed. On the other hand, at the Autonomous University of Baja California, while [5] have worked on models of competitiveness based on the information and communication technologies knowledge; [6], applied the matrix of technological capabilities into the industry of Baja California; and [37] described a systematic review of the literature on the concept of technological competence in the industry, rethinking the meaning of the term in knowledge areas seldom explored.

4. Evaluation Delimitations

The study is aimed at the Renewable Energy Sector industry in the state of Baja California, to which belongs a group of companies identified as a new and promising national investment strategy. The population of interest is professionals within the Renewable Energy Sector, named for this project business leader, who is currently in a managing responsibility position in Small and Medium Enterprises registered in the state.

The purpose of the evaluation focuses on the description of the behavior of an industrial sector in the state, regarding the technological competencies that their leaders demonstrate, without emphasizing the particularities of each company, that is to say, it is not intended to evaluate each leader individually and point differences in performance and levels of knowledge among the companies analyzed, on the contrary, the proposal corresponds to a comprehensive evaluation of the sector, demonstrating as final evidence a situational analysis and behavior in a diagnostic manner.

4.1 Methodological perspective

The evaluation is structured under a methodological perspective of the ethnography of digital culture, where literacy and digital awareness processes are explored, as well as the ethnography of innovation, dedicated to social participation in technological innovation processes. An analysis is determined from complexity, inquiring about business leaders and its relationship with higher education institutions, government agencies, and business clusters. Likewise, the social impact and the conditions of the industry in a border area are reviewed, relating the variables to innovation factors and/or company development. The focus of the research is oriented to a quantitative study under a methodological analysis of the organizations [38]. The analysis in the industrial sector considers indicators attached to administrative, economic, and engineering sciences that provide concepts for intervention, attention, and follow up of studies in organizations.

4.2 Dimensions definition and construction of evaluation variables

The dimensions and evaluation variables are structured based on a theoretical literature review of the evaluation of technological competencies [26,28,31,32,33,34,39,40,41]; the conceptualization of the term technological competencies in the industry [37]; international considerations, such as the European Framework of e-Competences (e-CF); the SFIA Reference Model (Skills Framework for the Information Age); emerging markets; Information Technology Infrastructure Library—ITIL [24,25,28,42,43,44,45,46,47,48,49,50]; and national considerations regarding technical and technological knowledge in the current Renewable Energy Sector industry [4,5,6,8,9,13,35,36,51, 52,53,54,55,56,57,58,59,60,61,62].

From the theoretical review, five evaluation dimensions are defined. The dimensions are the technological knowledge dimension, oriented to measure the use and mastery of electronic devices, the manipulation of specialized software and Web 2.0 applications; the dimension of environment and internal communication, where the internal structure of the company is evaluated, human relations and learning in terms of communication, knowledge acquisition, department analysis, training and promotion of human capital, professional degrees, and the impulse and technological vision of the company leaders are analyzed; the environment dimension and external communication, linkage, means and devices that make communication effective are assessed, collaboration between companies or sectors, working in networks for the company growth and recognition; the training and updating dimension, based on the follow-up of the leaders on issues to update technological knowledge, the preferred training modality, the training institutions, the periodicity of the training and the immediate application of the knowledge acquired in the training; and the company innovation factors' dimension, analyzed in terms of company innovation, from the products construction proposals, as their impact in the global market, patent registration, certifications acquisition, the administrative flexibility of the company's structure that allows the incorporation of an innovation and development group or department for the creation of new products, or, if such a group already exists, analyze conditions and context, as well as its impact on the company objectives. In Figure 1 the theoretical model of the evaluation is represented graphically.

Regarding the measurement variables defined in the dimension of technological knowledge, in the *software and hardware update*, the version control in the programs and equipment is reviewed, as well as the constant revision of the market proposals; on *interoperability and security*, integrity and transfer of information; in *collaboration and mobile applications*, the technological tools that are used. In the *environment and internal communication* dimension, in the internal structure organization variable, it is reviewed what is related to the strategic planning of the company and its administrative conditions based on technological elements; the *technological culture* refers to the behavior of the leader within his work community, his relationship with technology and the diffusion in the use of it; the *digital resilience* considers the possibilities that the company shows to face computer problems, solve and quickly reorganize them, without affecting the processes and projects in execution. In the *environment and external communication* dimension, client tracking analyzes the structure defined for the acquisition and growth of the client portfolio; in the *distribution strategies*, the communication and collaboration methods with distributors or possible distributors are reviewed; in *digital marketing*, the promotion of the company is analyzed through social media and the market strategies used; in *group participation*, the collaboration and contributions of the leader in governmental, academic and industrial groups are investigated. In the *training and updating* dimension, the questions in training strategies are oriented to know how much the company leader promotes and receives updates; in *innovative training practices*, the modality in which the update courses are promoted and if mixed programs are considered is reviewed. Finally, in the *company innovation factors* dimension, the *patents and new products/services* variable is defined to review the results of the company on the design and registration of new products, models and/or services; in *innovation and development*, the organizational conditions for the creation of spaces for development; in *certification and regulation*, the attention of human capital to the validation of knowledge by means of certifications and their knowledge about pre-established mechanisms in the sector for the regulation of processes.

The quantitative measurement approach is proposed, and an evaluation instrument oriented towards self-perception of leaders is developed, using the digital questionnaire evaluation technique, with a nominal response scale. The evaluation instrument is constructed to diagnose the work dynamics current state on the state Small and Medium Enterprises companies of the Renewable Energy Sector, regarding the use, incorporation, learning, and technological appropriation. Table 1 shows the structure of the measuring instrument.

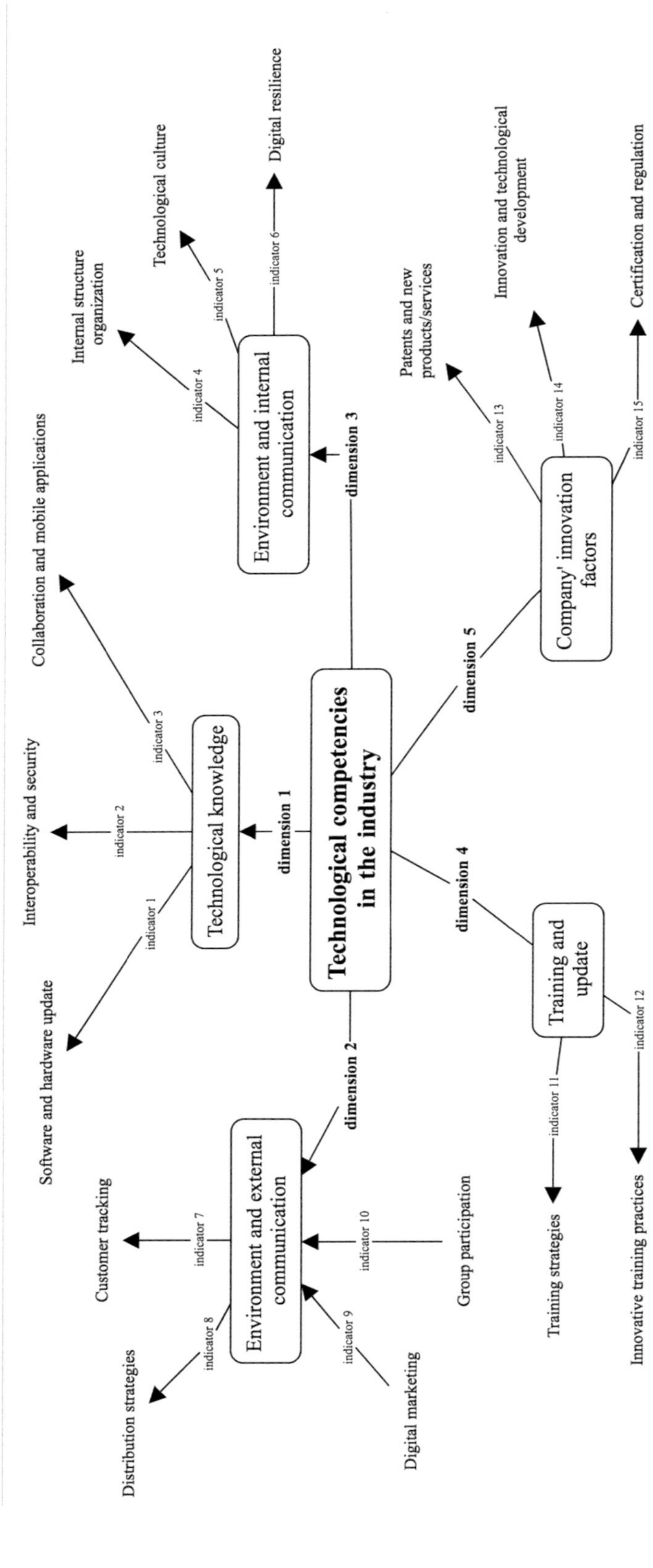

Figure 1: Theoretical evaluation model.

4.3 Variable operationalization

The first version of the evaluation instrument was composed of 15 variables and 80 indicators divided into five dimensions, which were distributed as follows:

Table 1: Evaluation instrument structure.

Evaluation dimension	Operational Variables	Indicators
Technological Knowledge	Software and hardware update	1,2,4
	Interoperability and security	5,6,7
	Collaboration and mobile applications	3,8,9,10,11
Environment and internal communication	Internal structure organization	12,13,14,15,16,25,26,27
	Technological culture	17,32,33,34,35,36,37
	Digital resilience	18,19,20,21,22,23,24,28,29,30,31
Environment and external communication	Customer tracking	38,39,40,41
	Distribution strategies	42,43,44,45,46,47
	Digital marketing	48,49,50,51,52
	Group participation	53,54,55,56,57,58,49
Training and updating	Training strategies	60,61,62,63,64,70,71,72
	Innovative training practices	65,66,67,68,69
Company' innovation factors	Patents and new products/services	73,77
	Innovation and technological development	74,75,76,80
	Certification and regulation	78,79

- Dimension 1 – Technological competencies, 4 variables, 11 indicators.
- Dimension 2 – Environment and internal communication, 5 variables, 26 indicators.
- Dimension 3 – Environment and external communication, 4 variables, 22 indicators.
- Dimension 4 – Training and updating, 2 variables, 13 indicators.
- Dimension 5 – Company' innovation factors, 2 variables, 8 indicators.

In addition to the mentioned items, the company identification data section is integrated into the instrument and the personal data of the business leader, with seven and four items, respectively. In Figure 1, the theoretical model of the evaluation is presented graphically, and Table 1 shows the indicators associated with each variable and evaluation dimension.

5. Validation of the Evaluation

5.1 Variable operationalization

The Content Validity by Expert Judgment is applied [44,63,64], through a group of six judges with expertise in social studies topics in the industry and with anthropological studies experience, management leadership, market impulse, digital marketing, economics, innovation management and human capital in organizations. A table is used for the review, integrated by the dimension of the evaluation, the item number, the item, the item relevance (essential, useful and useless), clarity of the item redaction (legible or illegible) and general observations.

Once validated, the Content Validity Reason (CVR) and the Content Validity Index (CVI) are calculated using the Microsoft Office Excel program. The CVI of the instrument is obtained by validating each item to determine if it is acceptable or unacceptable, indexes higher than 0.5823 are expected, otherwise, the item must be removed from the instrument [64]. The operation variables are shown in Equations 1 and 2.

$$CVR = \frac{ne}{N} \tag{1}$$

Equation (1): ne = Number of expert judges who agreed – essential, useful and useless; and N = Total number of judges.

$$\text{CVI} = \frac{\sum_{i-1}^{M} CVRi}{M} \tag{2}$$

Equation (2): CVRi = Content validity ratio of acceptable items; and M = Total acceptable items on the instrument.

5.2 Item reliability

Once the content validity test was made, a second reliability analysis of the instrument was carried out, through a pilot test, by non-probabilistic convenience sampling to 29% of the managers and/or executives from the 46 companies detected as potential for the analysis.

The reliability of the instrument was carried out in the statistical software "Statistical Package for the Social Sciences (SPSS)", version 24. Through Cronbach's Alpha calculation, which is an indicator of internal consistency that measures the degree to which the items are correlated, that is the homogeneity of the items in the measuring instrument [65]. Its value ranges from 0 to 1, where the closer to zero the higher the percentage of the error in the measurement, while the instrument reliability is greater when closer to one [66]. An alpha greater than 0.7 is considered acceptable, greater than 0.8 is good and greater than 0.9 is excellent [67].

6. Results

6.1 Validation of the measuring instrument results

The CVI global value was calculated at 0.93, based on Tristán's proposal [64], the result is catalogued as acceptable. Results show that 48 indicators from the total 80 obtained a CVR value of 1 the maximum scale score. Table 2 shows the CVR average per dimension on the evaluation instrument validation by the six expert judges; likewise, the items that were suggested to modify due to a lack of legibility are indicated.

Table 2: Evaluation instrument validation.

Dimension	Average validation calculation (CVR)	Item illegibility
Technological knowledge	0.89	1,2,4,5,7,8,9
Environment and internal communication	0.928	12,17,25,29,33
Environment and external communication	0.916	38,47,48,51,57
Training and updating	0.910	71
Company' innovation factors	0.937	73,79
Global CVI =	**0.93**	

6.2 Instrument reliability results

As can be seen in Table 3, a global Cronbach's Alpha of 0.972 was obtained in the instrument, which is excellent according to the acceptance values, and the individual values for each dimension are presented in it, showing an excellent result in the Environment and internal communication dimension as well as in the External communication dimension (0.968 and 0.913), good result for

Table 3: Internal consistency analysis by dimension.

Dimension	**Cronbach's Alpha**
Technological competencies	0.896
Environment and internal communication	0.968
Environment and external communication	0.913
Training and updating	0.868
Company' innovation factors	0.745
Instrument Total	**0.973**

Technological competences and Training and updating dimensions (0.896 and 0.868) and acceptable result in the company's innovation factors dimension (0.745). For this last case, the authors point out that results below 0.8 require reviewing the items writing [68], since it may not be understandable to the respondent.

6.3 Items reliability

The review of the item's reliability is another relevant issue for the design of the instrument since, through this, it allows us to analyze if the items are consistent with the measurement of the instrument. For this test, the total item correlation indicator was used, which ranges from –1 to 1, and it measures the correlation of the items with all the others. Three criteria are considered if the correlation is close to zero, the question does not contribute to the scale if the value is negative, it is a question that is wrongly formulated or ambiguous, and if it is positive it is well related to the instrument; the closer to one, the greater the strength [69].

In Table 4, each item's correlations are shown, as well as the Cronbach's alpha corrected for the case of eliminated items. Therefore, question 40 is highlighted to be eliminated and, thus, increase Cronbach's alpha to 0.975. Besides, questions 69 and 70 were reviewed for being close to zero.

7. Discussion and Conclusions

From the study carried out, a theoretical model capable of measuring technological competencies of business leaders is obtained, to diagnose the current state of the company's work dynamics regarding use, incorporation, learning and technological appropriation. The main dimensions identified and validated are the *technological knowledge dimension,* oriented to the measurement of the use and mastery of electronic devices, the manipulation of specialized software and Web 2.0 applications; the *environment and internal communication dimension*, where the internal company structure is evaluated, human relations and learning in terms of communication, acquisition of knowledge, analysis of departments, training and promotion of human capital, academic degrees, and the impulse and technological vision of the leaders in the company are analyzed; the *environment and external communication dimension*, the linkage is assessed, the means and devices that make communication effective, collaboration between companies or sectors, work in networks for the growth and recognition of the company; the *training and updating dimension*, based on the follow-up of the leaders on topics of technological knowledge updating, the preferred training modality, the training institutions, the periodicity of the training and the immediate application of the knowledge acquired in the training; and the *factors of innovation in the company dimension*, are analyzed in terms of company innovation, from the construction of proposals and products, as the impact of the same in the global market, the registration of patents, the acquisition of certifications, the administrative flexibility of the structure the company that allows the incorporation of a group or department innovation and development for the creation of new products, or, if that group already exists, analyze the conditions and the context, as well as the impact of the same on the objectives

Table 4: Total items correlation and Cronbach's alpha corrected.

Items	Total items correlation	Cronbach's alpha if the element is removed	Items	Total items correlation	Cronbach's alpha if the element is removed	Items	Total items correlation	Cronbach's alpha if the element is removed
P1	0.193	0.974	**P27**	0.822	0.973	**P53**	0.527	0.973
P2	0.589	0.973	**P28**	0.649	0.973	**P54**	0.886	0.973
P3	0.518	0.973	**P29**	0.873	0.972	**P55**	0.195	0.974
P4	0.734	0.973	**P30**	0.756	0.973	**P56**	0.662	0.973
P5	0.801	0.973	**P31**	0.799	0.973	**P57**	0.652	0.973
P6	0.821	0.973	**P32**	0.831	0.973	**P59**	0.429	0.973
P7	0.679	0.973	**P33**	0.648	0.973	**P60**	0.726	0.973
P8	0.532	0.973	**P34**	0.863	0.973	**P61**	0.341	0.973
P9	0.758	0.973	**P35**	0.794	0.973	**P62**	0.687	0.973
P10	0.324	0.973	**P36**	0.639	0.973	**P63**	0.469	0.973
P11	0.869	0.973	**P37**	0.455	0.973	**P64**	0.752	0.973
P12	0.506	0.973	**P38**	0.231	0.973	**P65**	0.589	0.973
P13	0.676	0.973	**P39**	0.601	0.973	**P66**	0.282	0.974
P14	0.767	0.973	**P40**	-0.310	0.975	**P67**	0.252	0.974
P15	0.575	0.973	**P41**	0.837	0.973	**P68**	0.282	0.974
P16	0.808	0.973	**P42**	0.719	0.973	**P69**	0.062	0.974
P17	0.805	0.973	**P43**	0.689	0.973	**P70**	0.062	0.974
P18	0.863	0.973	**P44**	0.397	0.973	**P71**	0.632	0.973
P19	0.677	0.973	**P45**	0.724	0.973	**P72**	0.291	0.973
P20	0.666	0.973	**P46**	0.680	0.973	**P73**	0.530	0.973
P21	0.481	0.973	**P47**	0.694	0.973	**P74**	0.802	0.973
P22	0.627	0.973	**P48**	0.553	0.973	**P75**	0.206	0.974
P23	0.732	0.973	**P49**	0.464	0.973	**P76**	0.487	0.973
P24	0.600	0.973	**P50**	0.755	0.973	**P77**	0.485	0.973
P25	0.436	0.973	**P51**	0.294	0.973	**P78**	0.427	0.974
P26	0.579	0.973	**P52**	0.554	0.973	**P79**	0.312	0.973

of the company. These dimensions have operational variables, such as *software and hardware update, interoperability and security, collaboration and mobile applications, internal structure organization, technological culture, digital resilience, customer tracking, distribution strategies, digital marketing, group participation, strategies for training, innovative training practices, patents and new products/services, innovation and technological development and certification and regulation*. The final product has a measuring instrument, using the 79-item digital questionnaire evaluation technique.

The theoretical conceptual model shown in Figure 1 was duly validated by the expert judgment employing the Content Validity Reason (CVR) and the Content Validity Index (CVI) obtaining the value of 1 for CVR and 0.93 for CVI. The measuring instrument was corrected in the reagents that had a lack of readability and then validated by Cronbach's alpha. It is important to note that Cronbach's global calculation of alpha was 0.972, which according to the acceptable values is considered excellent.

As future work, it is necessary to validate the model by applying the measurement instrument to a considerable sample and thus be able to define a useful organizational diagnostic methodology to define a technological profile of a business leader concerning the incorporation, learning, and technological appropriation.

References

[1] Ghadimi, P., Wang, C., Lim, M.K. and Heavey, C. 2019. Intelligent sustainable supplier selection using multi-agent technology: Theory and application for Industry 4.0 supply chains. Computers & Industrial Engineering, (127): 588–600. https://doi.org/10.1016/j.cie.2018.10.050.

[2] Hecklau, F., Galeitzke, M., Flachs, S. and Kohl, H. 2016. Holistic Approach for Human Resource Management in Industry 4.0. Procedia CIRP, (54): 1–6. https://doi.org/10.1016/j.procir.2016.05.102.

[3] Schneider, P. 2018. Managerial challenges of Industry 4.0: An empirically backed research agenda for a nascent field. Review of Managerial Science, 12(3): 803–848. https://doi.org/10.1007/s11846-018-0283-2.

[4] Asociación Mexicana de la Industria de Tecnologías de la Información. Alianza Mundial de Servicios de Tecnologías de la Información, 2011.

[5] Ahumada, E., Zárate, R., Plascencia, I. and Perusquia, J. 2012. Modelo de competitividad basado en el conocimiento: El caso de las PyMEs del Sector de Tecnologías de la Información en Baja California. Revista Internacional Administración & Finanzas, pp. 13–27.

[6] Brito, J., Garambullo, A. and Ferreiro, V. 2014. Aprendizaje y acumulación de capacidades tecnológicas en la industria electrónica den Tijuana. Revista Global de Negocios, 2(2): 57–68.

[7] Buenrostro, M.E. 2013. Experiencias y desafíos en la apropiación de las TICs por las PyME Mexicanas - Colección de Memorias de Seminarios, INFOTEC.

[8] Carrillo, J. and Hualde, A. 2000. El desarrollo regional y la maquiladora fronteriza: Las peculiaridades de un Cluster Electrónico en Tijuana. Mercado de valores, (10): 45–56.

[9] Carrillo, J. and Gomis, R. 2003. Los retos de las maquilladoras ante la pérdida de competitividad. Comercio Exterior, (53): 318–327.

[10] Fuentes, N. 2008. Elementos de la política de desarrollo empresarial: El caso de Baja California, México. Reglas, Industria y Competitividad, pp. 152–172.

[11] Hualde, A. and Díaz, P.C. 2010. La Industria de software en Baja California y Jalisco: dos experiencias contrastantes. Estrategias empresariales en la economía basada en el conocimiento, ISBN 978-607-95030-7-9.

[12] Hualde, A., Carrillo, J. and Dominguez, R. 2008. Diagnóstico de la industria Aeroespacial en Baja California. Características productivas y requerimientos actuales y potenciales de capital humano. Tijuana: Colegio de la Frontera Norte.

[13] Instituto Mexicano para la Competitividad. Visión México 2020. Políticas Públicas en materia de Tecnologías de la Información y Comunicaciones para impulsar la Competitividad de México. México: Concepto Total S.A de C.V., 2006.

[14] Marzo, N.M., Pedreja, I.M. and Rivera, T.P. 2006. Las competencias profesionales demandadas por las empresas: el caso de los ingenieros, pp. 643–661.

[15] Núñez-Torrón, S.A. Las 7 competencias imprescindibles para la transformación digital, 2016.

[16] Sampedro, José Luis. Contribucción de capacidades de innovación en la industria de Software a través de la creación de interfases: estudio de caso de empresas mexicanas. Economía y Sociedad (11)(17) (2006).

[17] Pérez-Jácome, D. and Aspe, M. Agenda Digital.mx México: Secretaría de Comunicaciones y Transportes (1)(2012).

[18] Porter, M. 2002. Ventaja Competitiva: creación y sostenimiento de un desempeño superior. Compañía Editorial Continental, pp. 556.

[19] Porter, Michael E. Competing to Change the World: Creating Shared Value. Rotterdam School of Management, Erasmus University, Rotterdam, The Netherlands, 2016.

[20] PYME. Fondo de apoyo para la Micro, Pequeña y Mediana Empresa, 2014. URL http://www.fondopyme.gob.mx/.

[21] INEA. Instituto Nacional de Educación para Adultos, 2010. URL http://www.inea.gob.mx/.

[22] Instituto PYME. Acceso a la Tecnología, 2015. URL: http://www.institutopyme.org.

[23] Lloréns, B.L., Espinosa, D.Y. and Castro, M.M. 2013. Criterios de un modelo de diseño instruccional y competencia docente para la educación superior escolarizada a distancia apoyada en TICC. Sinéctica Revista Electrónica de Educación.

[24] eCompetence. European e-Competence Framework, 2016. URL: www.ecompetence.eu.

[25] ITIL. Training Academy. The Knowledge Academy, 2017. URL: https://www.itil.org.uk/.

[26] Patel, P. and Pavitt, K. 1997. The technological competencies of the world's largest firms: complex and path-dependent, but not much variety. Research Policy, 26(2): 141–156.

[27] Renaud, A. 1990. Comprender la imagen hoy. Nuevas imágenes, nuevo régimen de lo visible, nuevo imaginario, en A.A.V.V. Video culturas de Fin de Siglo, Madrid, Cátedra.

[28] Urraca, R.A. 2013. Especialización tecnológica, captura y formación de competencias bajo integración de mercados; comparación entre Asia y América Latina. Economía y Sociedades, (22)(3): 641–673.

[29] Chomsky, N. 1965. Aspects of the Theory of Syntax, Cambridge, MA: MIT Press.

[30] Hymes, D. 1974. Pidginization and creolization of languages: Proceedings of a conference held at the University of the West Indies Mona, Jamaica. Cambridge University Press.

[31] González, J.A. 1999. Tecnología y percepción social evaluar la competencia tecnológica. Estudios sobre las culturas contemporáneas, (9): 155–165.

[32] Cabello, R. 2004. Aproximación al estudio de competencias tecnológicas. San Salvador de Jujuy, 2004.

[33] Motta, J.J., Zavaleta, L., Llinás, I. and Luque, L. 2013. Innovation processes and competences of human resources in the software industry of Argentina. Revista CTS, (24): 147–175.

[34] Romijn, H. and Albadalejo, M. 2002. Determinants of Innovation capability in small electronics and software firms in southeast England, Research Policy (31)(7): 1053–1067.

[35] García Alcaraz, J.L. and Romero González, J. 2011. Valoración subjetiva de los atributos que los ingenieros consideran requerir para ocupar puestos administrativos: Un estudio en empresas maquiladoras de Ciudad Juárez. Revista mexicana de investigación educativa, (16)(48): 195–219.

[36] Hernández, J. Sampedro and Vera-Cruz, A. 2003. Aprendizaje y acumulación de capacidades tecnológicas en la industria maquiladora de exportación: El caso de Thomson-Multimedia de México, Espacios. Espacios (24).

[37] Candolfi Arballo, N., Chan Núñez, M. and Rodríguez Tapia, B. 2019. Technological Competences: A Systematic Review of the Literature in 22 Years of Study. International Journal Of Emerging Technologies In Learning (14)(04): pp. 4–30. http://dx.doi.org/10.3991/ijet.v14i04.9118.

[38] Colobrans, J. 2011. Tecno-Antropología, Etnografies de la Cultura Digital i Etnografies de la Innovación. Revista d'Etnologia de Catalunya.

[39] Villanueva, G. and Casas, M.D. 2010. e-Competencias: nuevas habilidades del estudiante en la era de la educación, la globalidad y la generación de conocimiento. Signo y pensamiento, pp. 124–138.

[40] Gil Gómez H. 2003. Aprendizaje Interorganizativo en el entorno de un Centro de Investigación Tecnológico. Aplicación al sector textil de la Comunidad Valenciana. Universidad Politécnica de Valencia.

[41] Ordoñez, J.E., Gil-Gómez, H., Oltra, B.R. and González-Usach, R. 2015. Importancia de las competencias en tecnologías de la información (e-skills) en sectores productivos. Propuesta de investigación en el sector transporte de la comunidad Valenciana. 3Ciencias TIC, (4)(12): 87–99.

[42] Burillo, V., Dueñas, J. and Cuadrado, F. 2012. Competencias profesionales ETIC en mercados emergentes. Fundación Tecnologías de la Información. Madrid: FTI-AMETIC.

[43] Crue-TIC Y Rebiun. Competencias informáticas e informacionales en los estudios de grado, 2009, España. URL:http://www.rebiun.org/doc/documento_competencias_informaticas.pdf.

[44] Díaz, Y.E. and Báez, L.L. 2015. Exploración de la capacidad de liderazgo para la incorporación de TICC en educación: validación de un instrumento/Exploring the leadership to incorporate TICC in education: validation of an instrument. Revista Latinoamericana de Tecnología Educativa-RELATEC, (14)(3): 35–47.

[45] European Commission. e-Skills: The international dimension and the impact of globalisation. European Commission DG Enterprise and Industry, 2014.

[46] ITE (2011). Competencia Digital. Instituto de Tecnologías Educativas. Departamento de Proyectos Europeos, 2011. URL: http://recursostic.educacion.es/blogs/europa/.

[47] OCDE. Digital Economy Outlook 2017. Organización para la Cooperación y el Desarrollo Económicos, 2017.

[48] Ukces. Information and Communication Technologies: Sector Skills Assessment, 2012. UK Commission for Employment and Skills.

[49] Urraca, R.A. 2007. Patrones de inserción de las empresas multinacionales en la formación de competencias tecnológicas de países seguidores. Revista Brasileira de Innovación.

[50] Cabello, R. and Moyano, R. 2012. Tecnologías interactivas en la educación. Competencias tecnológicas y capacitación para la apropiación de las tecnologías. Buenos Aires, Argentina: Universidad Nacional de General Sarmiento.

[51] Barajas, M., Carrillo, J., Casalet, M., Corona, J., Dutrénit, G. and Hernández, C. 2000. Protocolo de Investigación Aprendizaje Tecnológico y Escalamiento Industrial: Generación de Capacidades de Innovación en la Industria Maquiladora de México.
[52] Barroso, R.S. and Morales, Z.D. 2012. Trayectoria de acumulación de competencias tecnológicas y procesos de aprendizaje. Propuesta de un modelo analítico para agencia de viajes y operadoras turísticas. Estudios y perspectivas en turismo, (21): 515–532.
[53] CEMIE. Centros Mexicanos de Innovación en Energía, 2015. URL: https://www.gob.mx/sener/articulos/centros-mexicanos-de-innovacion-en-energia.
[54] CFE. Comisión Federal de Electricidad, 2017. URL: http://www.cfe.gob.mx/.
[55] Chan, M.E. 2016. Virtualization of Higher Education in Latin America: Between Trends and Paradigms, (48): 1–32.
[56] CONUEE. 2008. Comisión Nacional para el Uso Eficiente de la Energía, 2008. URL https://www.gob.mx/conuee.
[57] Dutrénit, G. 2000. Learning and knowledge management in the firm: from knowledge accumulation to strategic capability. Edward Elgar.
[58] Dutrénit, G. 2004. Building technological capabilities in latecomer firms: a review essay. Science Technology Society, 9: 209–241.
[59] Gasca, L.K. 2015. Reforma Energética en México. México: SENER.
[60] Gobierno de Baja California. Programa Especial de Energía 2015–2019. Mexicali, BC, México, 2015.
[61] Instituto Federal de Telecomunicaciones. Adopción de las TIC y uso de internet en México, 2018.
[62] Secretaría de Energía. Comisiones Estatales de Energía. Secretaría de Energía, 2016. URL http://www.conuee.gob.mx/wb/Conuee/comisiones_estatales_de_energia.
[63] Lawshe, C.H. 1975. A quantitative approach to content validity. (D. 10.1111/j.1744-6570.1975.tb01393.x, Ed.) Personnel Psychology, pp. 563–575.
[64] Tristán, A. 2008. Modificación al modelo de Lawshe para el dictamen cuantitativo de la validez de contenido de un instrumento objetivo, pp. 37–48.
[65] Valdivieso, C. 2013. Efecto de los métodos de estimación en las modelaciones de estructuras de covarianzas sobre un modelo estructural de evaluación del servicio de clases. Comunicaciones en Estadística, (6)(1): 21–44.
[66] Hernández Sampieri, R., Fernández Collado, C. and Baptista Lucio, P. 2006. Capítulo 1, Similitudes y diferencias entre los enfoques cuantitativo y cualitativo. En McGraw-Hill (Ed.), 2006. https://doi.org/10.6018/turismo.36.231041.
[67] Castillo-Sierra, D.M., González-Consuegra, R.V. and Olaya-Sánchez, A. 2018. Validity and reliability of the Spanish version of the Florida Patient Acceptance Survey. Revista Colombiana de Cardiologia (25)(2): 131–137. https://doi.org/10.1016/j.rccar.2017.12.018.
[68] Rositas, J., Badii, M.H. and Castillo, J. 2006. La confiabilidad de las evaluaciones del aprendizaje conceptual: Indice Spearman-Brown del metodo split-halves (Reliability of the evaluation of conceptual learning: index of Spearman-Brown and the split-halves method). Innovaciones de Negocios, (3)(2): 317–329.
[69] Morales, Pedro. 2012. Análisis de ítems en las pruebas objetivas. Madrid: Universidad Pontificia Comillas.

CHAPTER-16

Myoelectric Systems in the Era of Artificial Intelligence and Big Data

Bernabé Rodríguez-Tapia,[1,2,*] *Angel Israel Soto Marrufo*,[1] *Juan Miguel Colores-Vargas*[2] and *Alberto Ochoa-Zezzatti*[1]

The technological progress, particularly in the implementation of biosignal acquisition systems, big data, and artificial intelligence algorithms, has enabled the gradual increase in the use of myoelectric signals. Its applications range from monitoring and diagnosing neuromuscular diseases to myoelectric control to assist the disabled. This chapter describes the proper treatment of EMG signals such as detection, processing, characteristics extraction techniques and classification algorithms.

1. Introduction

The technological progress has made it possible for intelligent devices such as smartphones, tablets, and phablets to use sensors (like accelerometer, triaxial, gyroscope, magnetometer, and altimeter) to give the consumer a very intuitive sense of the virtual environment [1], but beyond the implementation of sensors in different devices, it has started the digitization of health. The Health and Healthcare in the Fourth Industrial Revolution article, published recently by the World Economic Forum, highlights that social networks, internet of things (IoT), wearables, sensors, big data, artificial intelligence (AI), augmented reality (AR), nanotechnology, and 3D printing are about to drastically transform society and health systems.

Prominent leaders in health sciences and informatics have stated that AI could have an important role in solving many of the challenges in the medical sector. [2] mentions that almost all clinicians, from specialized physicians to paramedics, will use artificial intelligence technology in the future, especially for in-depth learning. A significant niche of this technological advance is related to the development of portable systems that allow the monitoring of biosignals and devices that can assist disabled people.

Biosignals have been used in healthcare and medical domains for more than 100 years, among the most studied ones are electroencephalography (EEG) and electrocardiography (ECG), however, due to the development of commercial technologies for myoelectrical (EMG) signal acquisition, data storage, and management, monitoring and control based on EMG signals has increased [3]. Real-time evaluation of these signals may be essential for musculoskeletal rehabilitation or for preventing muscle injury. On the other hand, muscle activation monitoring is useful for the diagnosis of neuromuscular disorders.

[1] Universidad Autónoma de Ciudad Juárez, Av. Hermanos Escobar, Omega, 32410 Cd Juárez, Chihuahua, México.
[2] Universidad Autónoma de Baja California, Blvd. Universitario #100. Unidad Valle de las Palmas, 21500 Tijuana México.
* Corresponding author: rodriguez.bernabe@uabc.edu.mx

Recently, the HMI (Human-Machin-Interface) and IT communities have started using these signals for a wide variety of applications, such as muscle-computer interfaces. Sensors on human body extremities enable the use of exoskeletons, electric chair control, prosthesis control, myoelectric armbands, handwriting identification, and silent voice interpretation.

Characteristics of EMG signals

The EMG signal is known as the electrical manifestation of the neuromuscular activation associated with a contracting muscle. [4] defines it as "the current produced by the ionic flow through the membrane of muscle fibers that spread across intermediate tissues reaching the surface for the detection of an electrode", therefore, it is a signal which is affected by the anatomical and physiological properties of muscles, the control scheme of the nervous system, as well as the characteristics of the instrumentation that is used to detect and register it.

The EMG signal consists of the action potentials of muscle fiber groups organized into functional units called *motor units* (MUs), this signal can be connected with sensors placed on the surface of the skin or through needle or wire sensors into muscle tissue. A graph of surface EMG signal decomposition at its motor unit action potentials is displayed in Figure 1. It is often desirable to review the data gathered at the time of individual motor unit discharges, in order to assess the degree of dysfunction in diseases such as cerebral palsy, Parkinson's disease, amyotrophic lateral sclerosis (ALS), stroke, and other diseases. Nonetheless, from a practical perspective, it is desirable to obtain such data from the signal detected from a single sensor that is as subtle as possible and that detects EMG signals high in MU rather than multiple sensors that detect EMG signals low in MU [5].

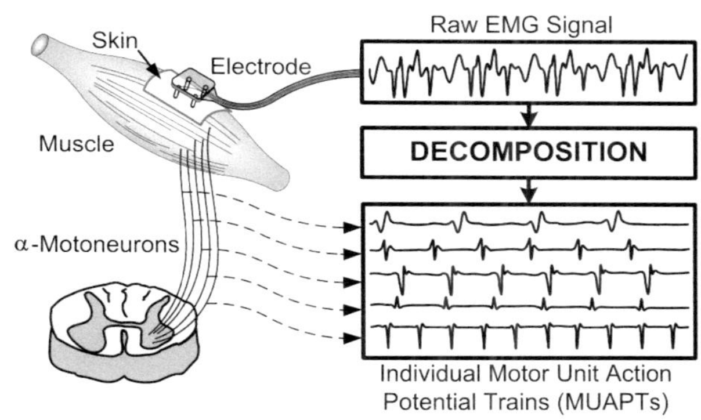

Figure 1: EMG signal and MUAP decomposition [5].

Electrical characteristic of EMG signals

According to several authors' research, EMG signals can vary in amplitude from 0 to 10 mV and their energy density can be scaled from 0 to 500 HZ.

An experiment performed by [11] identifies that the highest quantity of energy of an EMG signal ranges in a frequency scale from 20 to 150 Hz, making it very vulnerable to noise and interference.

EMG signal contaminants

[7] identifies two major issues of concern that influence signal accuracy. The first is the signal-to-noise ratio. In other words, the ratio of the energy in the EMG signals to the energy in the noise signal. The second concern is signal distortion, which means that the relative contribution of any frequency component in the EMG signal must not be altered.

Table 1: Electrical characteristics of EMG signals.

Amplitude	Frequency	Author
0–6 mV	0 to 500 Hz	[6]
0–10 mV	-	[7]
-	20 to 400 Hz	[8]
10 uv–10 mV	10–500 Hz	[9]
1–10 mV	20–500 Hz	[10]

Source: *Author's own compilation*

On the other hand, [12] points out that the identity of a real EMG signal originated in the muscle is lost due to two main effects: the attributes of the EMG signal that rely on the individual's internal structure, including individual skin formation, bloodstream speed, skin temperature, tissue structure (muscle, fat, etc.), and external contaminants in EMG recordings, including inherent electrode noise, device motion, electric line interference, analog-to-digital conversion cutoff, quantization error, amplifier saturation, and electrocardiographic (ECG) interference. The major external contaminants are displayed in Table 2.

Table 2: External contaminants of EMG signal.

Contaminants	Authors
Device motion	
Line interference	
Amplifier saturation	[13–15]
Physiological interference (e.g. EGC)	
Noise (additive white gaussian noise saturation)	

Source: *Author's own compilation*

Due to the inherent characteristics of EMG signals, proper processing is necessary for its correct interpretation.

An overall system based on pattern recognition consists of three stages:

(1) Processing stage: The signal is collected with electrodes and preprocessed with amplifiers and filters and then converted into digital data. A raw signal is output as segments.

(2) Extraction and characteristics reduction stage. It involves transforming the raw signal into a characteristics vector in order to highlight important data. At its output, there is a reduced vector of characteristics.

(3) Classification stage. Classification algorithms are used to distinguish different categories between the reduced vector of characteristics. The categories obtained will be used for stages such as control commands or diagnostics.

The following sections describe the main considerations for signal processing at each stage.

2. EMG Signal Processing: Signal Acquisition and Segmentation

EMG signal processing consists of a series of stages that enable the information generated by muscle contractions to be processed and interpreted properly. The block diagram in Figure 2 clearly illustrates this required transformation.

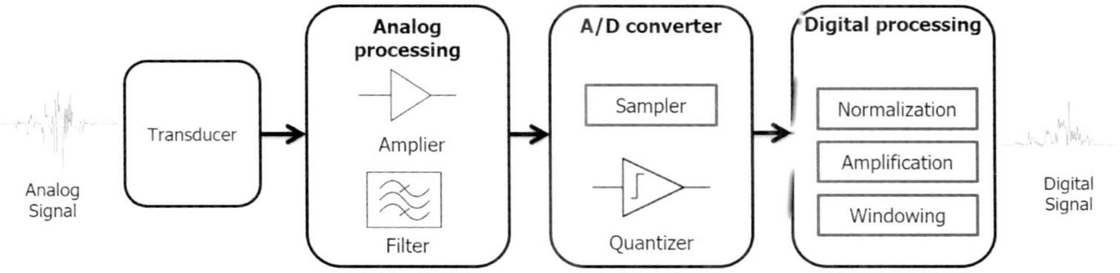

Figure 2: Block diagram of signal acquisition and processing.

2.1 Signal acquisition

2.1.1 Detection stage

Two main techniques are used for EMG signal detection: Non-invasive, using surface electrodes (on the skin), and invasive, by inserting electrodes directly into the muscle (wire or needle type). Electrodes are normally used individually or in pairs, these configurations are called monopolar and bipolar, respectively [4]. The electrodes chosen for muscle and nerve registration will vary depending on the purpose of the research or the numbers of fibers to be analyzed.

Superficial technique

There are two categories of surface electrodes: passive and active. The passive electrode consists of a conductive surface (usually metal) that detects the current in the skin through its skin-electrode interface. The active electrodes contain a high input impedance electronic amplifier. This arrangement makes it less sensitive to impedance. The current tendency is towards active electrodes.

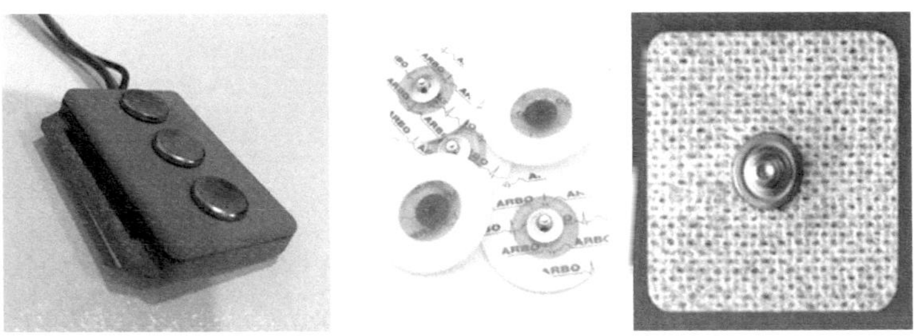

Figure 3: Different types of surface electrodes.

The disadvantages of surface electrodes lie in their restriction to surface muscles and that they cannot be used to selectively detect signals from small muscles or adjacent muscles. However, they are useful in myoelectric control for the physically disabled population, studies of motor behavior, when the activation time and magnitude of the signal contains the required information, or in studies with children or other people opposed to the insertion of needles [4].

Intramuscular technique

The most common electrode is the needle type, like the "concentric" electrode used by clinicians. This monopolar configuration contains an isolated wire and a bare tip to detect the signal. The bipolar configuration contains a second wire and provides a second surface detection.

[4] mentions that the needle electrode has two distinct advantages. One is that it allows the electrode to detect individual MUAPs during relatively low force contractions. The other is that the electrodes can be conveniently re-positioned within the muscle.

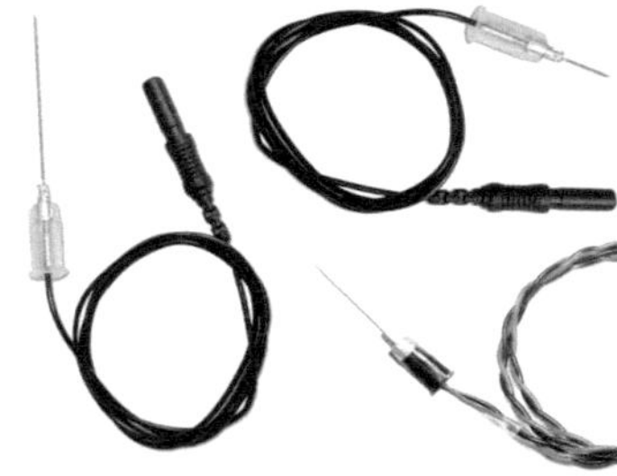

Figure 4: Unipolar and Concentric Bipolar Needle Electrodes (BIOPAC®).

2.1.2 Sensor characteristics

The experimental protocol for EMG signal detection plays an important role in giving greater reliability to the signal taken by the electrode, that is why it is necessary to take care of the properties of the sensor, skin preparation technique, sensor placement in the muscles, and electrodes fixing.

Properties of the sensor

In 1996, the Surface Electromyography for Noninvasive Assessment of Muscles (SENIAM) association was created with the objective of developing recommendations on key elements to allow a more useful exchange of data obtained by sEMG. After analyzing 144 studies, [16] points out the most used criteria regarding the configuration used, material, shape, size and distance between electrodes. Table 3 summarizes the desirable characteristics of the analyzed authors.

Table 3: Users' desirable characteristics in sensor properties.

Configuration	**Bipolar**
Material	Ag/AgCl
Shape and size	Round from 8 to 10 mm
Electrode distance	20 mm

Source: *Author's own compilation*

Placement procedure

The most commonly used skin preparation techniques include: shaving, cleansing the skin with alcohol, ethanol or acetone, and gel application [16].

Sensor placement

Three strategies can be identified for placement of a pair of electrodes [17].

- In the center or most prominent lump of the belly muscle
- Someplace between the innervation zone and the distal tendon
- At the motor point

The reference electrode is placed over inactive tissue (tendons or osseous areas), often at a certain distance from active muscles. The "popular" locations for placing the reference electrode have been the wrist, waist, tibia, sternum, and spinal process [16].

Fixing of electrodes

The way the sensor is connected to the body is known as "fixation", this facilitates good and steady contact between the electrode and the skin, a limited risk of the sensor moving over the skin and a minimal risk of pulling the wires. Some methods may include adhesive tape (double-sided) or collar, elastic bands, and keeping the sensor in the desired placement by hand [16].

2.1.3 Amplifier stage

The quality of an EMG signal taken from an electrode will rely on the properties of the amplifiers, due to its nature, the amplitude of the EMG signals is weak and the amplifier gain must be in the range of 1000 to 10000. The consideration of incorporating amplifiers makes it necessary to have a high common-mode rejection ratio (CMRR), a high input impedance, a short distance to the source signal, and a strong direct current signal suppression [18].

2.1.4 Filtering stage

Analog filtering, usually bandpass, is applied to the raw signal before it is digitized. Bandpass filtering eliminates low and high frequencies from the signal. The low-frequency cut-off of the bandpass filter eliminates interference associated with motion, transpiration, etc., and any direct current (DC) compensation. Typical values for low-frequency cut are 5 to 20 Hz. The high-frequency cut-off of the bandpass filter eliminates high-frequency noise and prevents an alias from occurring in the sampled signal. The high-frequency cut-off must be high enough for the rapid on and off bursts of the EMG to remain easily identifiable. Typical values are 200 Hz–1 KHz [19]. The recommendations made by SENIAM for surface EMG are high pass with 10–20 Hz cut and low pass near 500 Hz [16]. The recommendation given by the International Society of Electrophysiology and Kinesiology (ISEK) for surface EMG are high pass 5 Hz and low pass with 500 Hz cut [20].

2.1.5 A/D converter stage

In the computer processing of EMG signals, the unprocessed EMG (after amplification and bandpass filtering) should be stored in the computer for digital processing. The minimum acceptable sampling is at least twice the highest frequency cut of the bandpass filter. For instance, if a 10–500 Hz bandpass filter was used, the minimum rate used to store the signal in the computer should be at least 1000 Hz, as indicated by the Nyquist sampling theorem, and preferably higher to improve accuracy and resolution, besides, the number of bits, model, and manufacturer of the A/D card used to display data in the computer should be provided [20].

It is desirable that as much information as possible be available to facilitate the interpretation of muscle contraction, however, the higher the sampling frequency, the more data will be collected in units of time, and this translates into more strict requirements for hardware equipment. Consequently, the cost can increase significantly, hence, appropriate reduction of sampling frequency is a highly desirable option [11]. Due to physical, processing, data transmission, and power consumption limitations, portable acquisition systems often sample EMG signals at a lower frequency than clinically performed (e.g., 200 Hz for the MYO armband or 250 Hz for the OpenBCI Cyton). In this sense, [21] developed a research to test the effects of frequency in the classification of basic movements of the hand and fingers of healthy subjects. Specifically, the study compared the effects of precision classification between 1000 Hz, frequency used in clinical acquisition systems, and 200 Hz, used in portable systems, finding that if the sampling frequency is lower than the one specified by the Nyquist theorem, there is an effect on the precision classification, however, it can be considered to work on the segmentation of data by analyzing small windows to analyze the data.

2.1.6 Amplitude analysis stage

The EMG signal has a variation of amplitude in time, if a correct analysis in time is desired, the average of this signal will not provide useful information, since it presents variations above and below the zero value, that is why different methods are used for the correct analysis of amplitude.

Rectification: The rectification process is carried out before any relevant analysis method is performed. It entails the concept of rendering only positive deviations of the signal, it is achieved by eliminating negative values (half-wave rectification) or reversing negative values (full-wave rectification), the latter is the preferable procedure as it preserves all the energy of the signal [4].

Root mean square average (rms): An alternative to capture the envelope is calculating the value of the root mean square (rms) within a window that "slides" through the signal [19]. This approach is mathematically different from the rectification and filtering approach. [4] points out that, due to the parameters of the mathematical operation, the rms value provides the most rigorous measure of the information content of the signal, because it measures the energy of the signal.

2.2 Signal segmentation

A segment is a subset of samples from a signal in which characteristics are extracted, and these characteristics are provided to a pattern classifier. The analysis window should have the following two considerations: the window time, considering the processing times of a classifier in real time, and the segmentation techniques, which can be adjacent or overlapping.

Windows size. Due to real-time limitations, an adjacent segment length and the processing time to generate classified control commands must be equal to or less than 300 ms. In addition, the length of a segment must be appropriately large, as the bias and variance of characteristics increase as the length of the segment decreases, and consequently degrade the performance of the classification [22]. The same author notes that, due to real-time computing and high-speed microprocessors, processing time is usually less than 50 ms, and segment length can vary between 32 and 250 ms.

Adjacent window. Disjointed adjacent segments with a predefined length are used for characteristic extraction; a classified movement emerges after some delay in processing. It is considered the easiest approach (used in the original description of the continuous classifier).

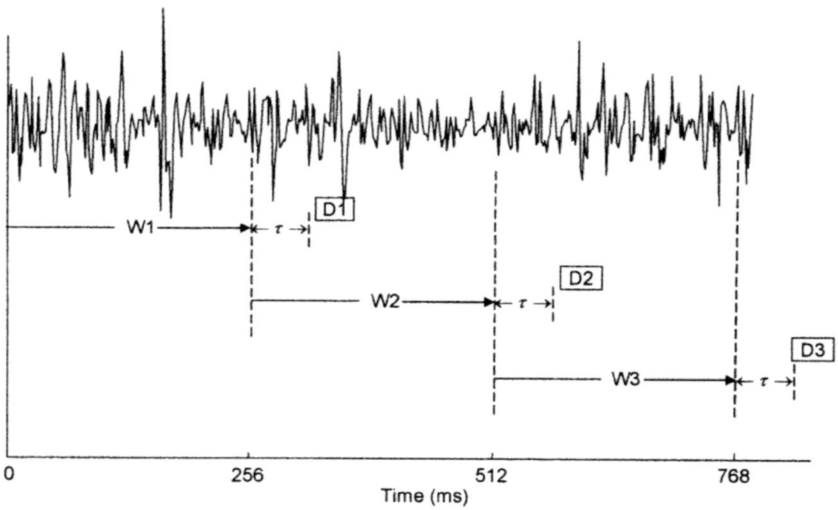

Figure 5: Adjacent window technique for an EMG signal channel. The data windows (W1, 12 and W3) are adjacent and disjointed. For each data window a classification decision is made (D1, D2 and D3) in time t, the processing time required of a classifier [23].

Overlapping windows. In this technique, the new segment slides over the existing segment, with a shorter increase time than the segment length.

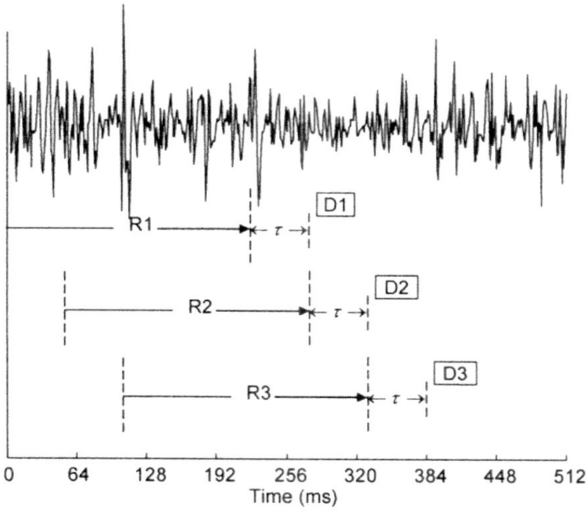

Figure 6: Overlapping window technique for an EMG channel. Window diagram that maximizes computing performance and produces the most possible dense decision flow [23].

According to research performed by [23] and [24] regarding the effect of both techniques, they conclude that: overlapping segmentation increases processing time and produces no significant improvement, the segmentation of adjacent windows seems to achieve an increase in sorting performance. In this technique a smaller segment increase produces a denser but semi-redundant class decision flow that could improve response time and accuracy. [24] observed that a window of less than 125 ms produces high variation in frequency domain characteristics.

3. Extraction Methods for EMG Characteristics

In the interpretation of EMG signals, characteristic extraction methods aim to transform the recorded and preprocessed signal, better known as "raw signal", and transform it into a relevant data structure, known as "characteristics vector"; in addition to reducing data dimensionality, these methods eliminate redundant data [3]. In this sense, the selection or extraction of highly effective characteristics is one of the most critical stages to improve sorting efficiency by [22].

According to [25], there are three sets of characteristics: (a) time domain, (b) frequency domain, and (c) time-frequency domain. Time domain characteristics are often calculated rapidly because they do not need a transformation. Frequency domain characteristics are based on the estimated power spectral density (PSD) of the signal, calculated by periodograms or parametric methods; these require more calculations, and time to be calculated. The characteristics in the time-frequency domain can locate the signal energy both in time and frequency, allowing a precise description of the physical phenomenon, generally requiring a transformation that could be computationally heavy.

3.1 Time domain

This group of characteristics is widely used for pattern recognition in the detection of muscle contraction and muscle activity [26]. Due to their computational simplicity, time domain characteristics, also known as line techniques, are the most popular ones for EMG signal pattern recognition. They can all be done in real time and electronically, and their implementation is simple [27]. Characteristics in time domain are displayed in Table 4.

Table 4: Characteristics in time domain.

Integrated EMG (IEMG)	$IEMG_k = \sum_{i=1}^{N} \lvert x_i \rvert$
Mean absolute value (MAV)	$MAV_k = \frac{1}{N}\sum_{i=1}^{N} \lvert x_i \rvert$
Modified mean absolute value 1 (MMAV1)	$MMAV1_k = \frac{1}{N}\sum_{i=1}^{N} w_i \lvert x_i \rvert$ $w(i) = \begin{cases} 1, & 0.25N \leq i \leq 0.75N \\ 0.5, & otherwise \end{cases}$
Modified mean absolute value 2 (MMAV2)	$MMAV_k = \frac{1}{N}\sum_{i=1}^{N} w_i \lvert x_i \rvert$ $\mathrm{w(i)} = \begin{cases} 1, & 0.25N \leq i \leq 0.75N \\ 4i/N, & 0.25N > i \\ 4(i-N)/N, & 0.75 < i \end{cases}$
Mean absolute value slope (MAVS)	$MAVS_k = MAV_{k+1} - MAV_k$
Root mean square (RMS)	$RMS_k = \sqrt{\frac{1}{N}\sum_{i=1}^{N} x_i^2}$
EMG variance (VAR)	$VAR_k = \frac{1}{N}\sum_{i=1}^{N} x_i^2$
Wavelength (WL)	$WL_k = \sum_{i=1}^{N-1} \lvert x_{i+1} - x_i \rvert$
Zero crossing (ZC)	$\{x_i > x_{i-1}\ and\ x_i > x_{i+1}\}\ or\ \{x_i < x_{i-1}\ and\ x_i < x_{i+1}\}$ AND $\lvert x_i - x_{i+1} \rvert \geq \epsilon\ or\ \lvert x_i - x_{i-1} \rvert \geq \epsilon$
Wilson amplitude (WAMP)	$WAMP_i = \sum_{i}^{N} f(\lvert x_i - x_{i-1} \rvert)$ $f(x) = \begin{cases} 1 & x > \epsilon \\ 0 & otherwise \end{cases}$
Simple square integral (SSI)	$SSI_k = \sum_{i=1}^{N} (\lvert x_i^2 \rvert)$
EMG histogram (HEMG)	HEMG divides the elements in EMG signal into b equally spaced segments and returns the number of elements in each segment

Source: [3]

3.2 Frequency domain

Frequency domain characteristics are mostly used for detection of muscle fatigue and neuronal anomalies [26]. They are based on power spectral density (PSD) and are calculated by periodogram or parametric methods [27]. These characteristics require more calculations and time to be calculated in comparison to the characteristics in time domain. The main methods are described in Table 5.

Table 5: Characteristics in frequency domain.

Coefficients autoregressive (AR)	$x_k = -\sum_{i=1}^{N} a_i x_{k-i} + e_k$
Frequency median (FMD)	$F_{MD} = \frac{1}{2}\sum_{i=1}^{M} PSD_i$
Frequency mean (FMN)	$F_{MN} = \frac{\sum_{i=1}^{M} f_i PSD_i}{\sum_{i=1}^{M} PSD_i}$
Modified frequency median (MFMD)	$MFMD = \frac{1}{2}\sum_{i=1}^{M} A_j$
Modified frequency mean (MFMN)	$MFMN = \frac{\sum_{j=1}^{M} f_j A_j}{\sum_{j=1}^{M} A_j}$
Frequency ratio (FR)	$FRj = \frac{\lvert F(\cdot)\rvert_{jlowfreq}}{\lvert F(\cdot)\rvert_{jhighfreq}}$

Source: *Author's own compilation*

3.3 Time-frequency domain

The characteristics in the time-frequency domain can locate the signal energy both in time and frequency, allowing a precise description of the physical phenomenon, generally requiring a transformation that could be computationally heavy. The primary methods are shown in Table 6.

Tablet 6: Characteristics in the time-frequency domain.

Short-term Fourier transform (STFT)	$STFT_x(t, \omega) = \int W^*(\tau - t)\, x(\tau)\, e^{-j\omega\tau}\, d\tau$
Wavelet transform (WT)	$W_x(a, b) = x(t)\left(\frac{1}{\sqrt{a}}\right)\Psi^*\left(\frac{t-b}{a}\right)dt$
Wavelet packet transform (WPT)	WPT is a generalized version of the continuous wavelet transform and the discrete wavelet transform. The basis for the WPT is chosen using an entropy-based cost function.

Source: *Author's own compilation*

The main difference between STFT, WT and WPT is how each one divides the time-frequency plane. The STFT has a static pattern, each cell has an identical aspect ratio; while the WT has a variable pattern and the cell aspect ratio varies in a way that the frequency resolution is proportional to the center frequency. Lastly, the WPT has an adaptive pattern, which offers several tilt alternatives [28].

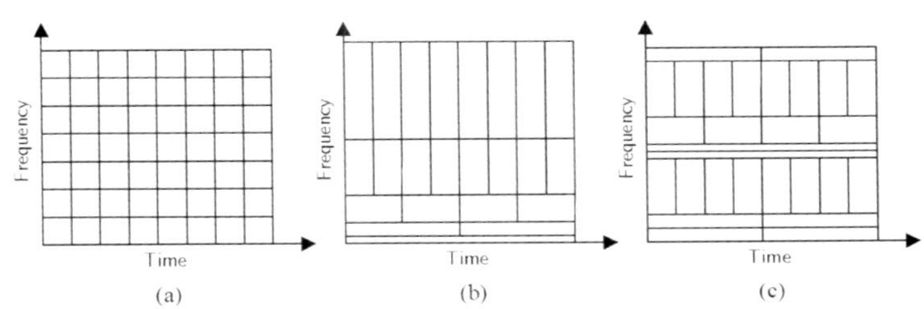

Figure 7: Time-frequency pattern of (a) STFT, b) WT and (c) WPT [28].

3.4 Dimensionality reduction

Dimensionality reduction is fundamental to increase the performance in the classification stage. In this process, the characteristics that best describe the behavior of the signal are preserved while the number of dimensions is reduced. There are two main strategies for dimensionality reduction [29].

Characteristics projection: this strategy consists of identifying the better combination of the original characteristics to form the new set of characteristics, usually smaller than the original, the principal component analysis (PCA) can be used as a characteristic's projection technique [25]. PCA produces a set of uncorrelated characteristics by projecting the data into the vectors of the covariance matrix [30].

Characteristics selection: This strategy selects the better subset of the original characteristics vector according to certain criteria to assess whether one subset is better than another. The ideal criteria for classification should be to minimize the probability of misclassification, although simpler criteria based on class separability are generally selected [25].

4. Classification Algorithms

Once the characteristics of a recorded EMG signal have been retrieved and the dimensionality has been reduced, some classification algorithms must be implemented. [22] advises that, due to the nature of the myoelectric signal, it is reasonable to expect a wide variation in the value of a particular characteristic. In addition, there are external factors such as changes in electrode position, fatigue, or sweating that cause changes in a signal pattern over time. However, a classifier should be able to cope optimally with such variable patterns; it must be fast enough to comply with restrictions in real-time. There are several classifier approaches such as neural network, Bayes classifier, fuzzy logic, linear discriminant analysis, support vector machine, hidden Markov model and k-nearest neighbors [3]. The summary of the main classification algorithms is displayed in Figure 8. Examples of the uses of the different classifiers are shown in Table 7.

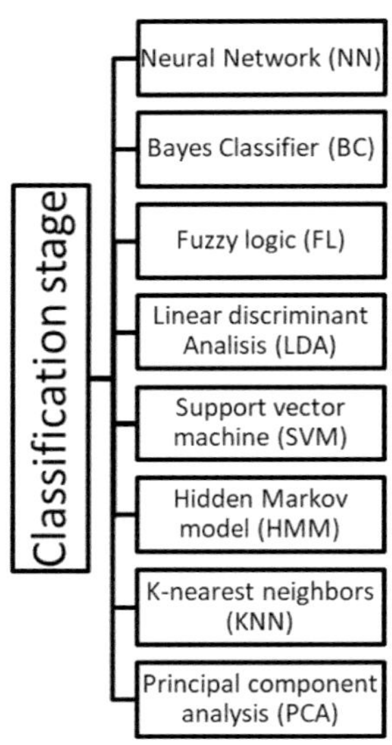

Figure 8: Summary of the classification stage.

Table 7: Classifier Usage.

Classifier	Application
SVM, LDA and MLP	Evaluating upper limb motions using EMG
NN	EMG-based computer interface
FL	Control of a robotic arm for rehabilitation
SVM	Post-stroke robot-aided rehabilitation
LDA, and SVM	Classification of muscle activity for robotic device control
NN	Hand motion detection from EMG
BN, and a hybrid of BN and NN	EMG-based human–robot interface
NN, BN and HMM	HCI system
FL	Classification of arm movements for rehabilitation

Source: [31]

5. Conclusion

The development of technology and portable system applications to monitor and control through myoelectric signals is possible thanks to acquisition systems, real-time processing, and classification algorithms, associated with the analysis of large amounts of data. This has made it possible to detect, process, analyze and control signals as small and complex as those generated by any muscle contraction.

Knowing each stage in the processing of these signals allows us to identify criteria for the design of new human-computer interfaces, more efficient and useful for the user.

No doubt there is a need for proper detection and ergonomic systems, despite the efforts of communities such as SENIAM and ISEK, the mapping for the location of sensors are still being studied. On the other hand, portable acquisition systems must be developed with adequate characteristics in the sampling frequency, in order to decrease computational costs in processing and time, but without losing vital frequency spectra for the correct monitoring and interpretation of patterns. The development of statistical algorithms, data analysis, and artificial intelligence are making possible the optimization of relevant characteristics in the interpretation of patterns, allowing the reduction of the raw data dimensionality of the sampled signals to facilitate their interpretation by the different classification algorithms. The difficulty of systems that can interpret patterns through myoelectric signals lies in the diversity of the anatomical set of users, the placement of sensors, and the relevant characteristics, which is why the algorithms of machine learning or deep learning can allow greater progress in dealing with each of these variables.

References

[1] Athavale, Y. and Krishnan, S. 2017. Biosignal monitoring using wearables: observations and opportunities. Biomedical Signal Processing and Control, 38: 22–33. https://doi.org/10.1016/j.bspc.2017.03.011.

[2] Topol, E.J. 2019. High-performance medicine: the convergence of human and artificial intelligence. Nat Med, 25(1): 44–56. https://doi.org/10.1038/s41591-018-0300-7.

[3] Rechy-Ramirez, E.J. and Hu, H. 2015. Bio-signal based control in assistive robots: a survey. Digital Communications and Networks, 1(2): 85–101. https://doi.org/10.1016/j.dcan.2015.02.004.

[4] De Luca, C.J. 2006. Electromyography. Encyclopedia of Medical Devices and Instrumentation. John Wiley Publisher, 98–109.

[5] De Luca, C.J., Adam, A., Wotiz, R., Gilmore, L.D. and Nawab, S.H. 2006. Decomposition of surface EMG signals. Journal of Neurophysiology, 96(3): 1646–1657. https://doi.org/10.1152/jn.00009.2006.

[6] Betancourt, O., Gustavo, A., Suárez, G., Franco, B. and Fredy, J. 2004. Disponible En: Http://Www.Redalyc.Org/Articulo.Oa?Id=84911640010.

[7] Raez, M.B.I., Hussain, M.S., Mohd-Yasin, F., Reaz, M., Hussain, M.S. and Mohd-Yasin, F. 2006. Techniques of EMG signal analysis: detection, processing, classification and applications. Biological Procedures online, 8(1): 11–35. https://doi.org/10.1251/bpo115.
[8] Supuk, T., Skelin, A. and Cic, M. Design 2014. Development and testing of a low-cost SEMG system and its use in recording muscle activity in human gait. Sensors, 14(5): 8235–8258. https://doi.org/10.3390/s140508235.
[9] Fuketa, H., Yoshioka, K., Shinozuka, Y. and Ishida, K. 2014. Measurement sheet with 2 V organic transistors for prosthetic hand control. IEEE Transactions on Biomedical Engineering, 8(6): 824–833. https://doi.org/10.1109/TBCAS.2014.2314135.
[10] Prince, N., Nadar, S., Thakare, S., Thale, V. and Desai, J. Design of Front End Circuitry for Detection of Surface EMG Using Bipolar Recording Technique. 2016 International Conference on Control Instrumentation Communication and Computational Technologies, ICCICCT 2016, 2017, 594–599. https://doi.org/10.1109/ICCICCT.2016.7988019.
[11] Chen, H., Zhang, Y., Zhang, Z., Fang, Y., Liu, H. and Yao, C. Exploring the Relation between EMG Sampling Frequency and Hand Motion Recognition Accuracy. In 2017 IEEE International Conference on Systems, Man, and Cybernetics (SMC); IEEE: Banff, AB, 2017; pp 1139–1144. https://doi.org/10.1109/SMC.2017.8122765.
[12] Chowdhury, R., Reaz, M., Ali, M., Bakar, A., Chellappan, K. and Chang, T. 2013. Surface electromyography signal processing and classification techniques. Sensors, 13(9): 12431–12466. https://doi.org/10.3390/s130912431.
[13] Chan, A. and MacIsaac, D. CleanEMG: Assessing the Quality of EMG Signals. 34th Conference of the Canadian Medical & …, 2011, No. November, 17–20.
[14] McCool, P., Fraser, G.D., Chan, A.D.C., Petropoulakis, L. and Soraghan, J.J. 2014. Identification of Contaminant Type in Surface Electromyography (EMG) Signals. IEEE Transactions on Neural Systems and Rehabilitation Engineering, 22(4): 774–783. https://doi.org/10.1109/TNSRE.2014.2299573.
[15] Rosli, N.A.I.M., Rahman, M.A.A., Mazlan, S.A. and Zamzuri, H. 2014. Electrocardiographic (ECG) and Electromyographic (EMG) Signals Fusion for Physiological Device in Rehab Application. 2014 IEEE Student Conference on Research and Development, SCOReD 2014. https://doi.org/10.1109/SCORED.2014.7072965.
[16] Hermens, H.J. 2000. Development of Recommendations for SEMG Sensors and Sensor Placement Procedures, 10: 361–374.
[17] Mesin, L., Merletti, R. and Rainoldi, A. 2009. Surface EMG: The issue of electrode location. Journal of Electromyography and Kinesiology, 19(5): 719–726. https://doi.org/10.1016/j.jelekin.2008.07.006.
[18] Wang, J., Tang, L. and Bronlund, J.E. 2013. Surface EMG signal amplification and filtering. International Journal of Computer Applications, 82(1): 15–22. https://doi.org/10.5120/14079-2073.
[19] Rose, W. 2016. Electromyogram Analysis. Online course material. University of Delaware. Retrieved July, 2011, 5.
[20] Merletti, R. and Di Torino, P. 1999. Standards for Reporting EMG Data. J Electromyogr Kinesiol, 9(1): 3–4.
[21] Phinyomark, A., Khushaba, R.N. and Scheme, E. 2018. Feature extraction and selection for myoelectric control based on wearable EMG sensors. Sensors, 18(5). https://doi.org/10.3390/s18051615.
[22] Asghari Oskoei, M. and Hu, H. 2007. Myoelectric control systems-a survey. Biomedical Signal Processing and Control, 2(4): 275–294. https://doi.org/10.1016/j.bspc.2007.07.009.
[23] Englehart, K. and Hudgins, B. 2003. A robust, real-time control scheme for multifunction myoelectric control. IEEE Trans Biomed Eng, 50(7): 848–854. https://doi.org/10.1109/TBME.2003.813539.
[24] Rainoldi, A., Nazzaro, M., Merletti, R., Farina, D., Caruso, I. and Gaudenti, S. 2000. Geometrical factors in surface EMG of the vastus medialis and lateralis muscles. Journal of Electromyography and Kinesiology, 10(5): 327–336. https://doi.org/10.1016/S1050-6411(00)00024-9.
[25] Zecca, M., Micera, S., Carrozza, M.C. and Dario, P. 2002. Control of multifunctional prosthetic hands by processing the electromyographic signal. Critical Reviews in Biomedical Engineering, 30(4–6).
[26] Veer, K. and Sharma, T. 2016. A Novel feature extraction for robust EMG pattern recognition. Journal of Medical Engineering & Technology, 40(4): 149–154. https://doi.org/10.3109/03091902.2016.1153739.
[27] Oskoei, M.A. and Hu, H. 2006. GA-Based feature subset selection for myoelectric classification. In 2006 IEEE International Conference on Robotics and Biomimetics; IEEE: Kunming, China, pp. 1465–1470. https://doi.org/10.1109/ROBIO.2006.340145.
[28] Englehart, K., Hudgins, B., Parker, P.A. and Stevenson, M. 1999. Classification of the myoelectric signal using time-frequency based representations. Medical Engineering & Physics, 21(6-7): 431–438. https://doi.org/10.1016/S1350-4533(99)00066-1.
[29] Englehart, K. 1998. Signal Representation for Classification of the Transient Myoelectric Signal. Ph. D. thesis, University of New Brunswick.
[30] Bishop, C.M. et al. 1995. Neural Networks for Pattern Recognition; Oxford university press.
[31] Spiewak, C. 2018. A Comprehensive Study on EMG Feature Extraction and Classifiers. OAJBEB, 1(1). https://doi.org/10.32474/OAJBEB.2018.01.000104.

CHAPTER-17

Implementation of an Intelligent Model based on Big Data and Decision Making using Fuzzy Logic Type-2 for the Car Assembly Industry in an Industrial Estate in Northern Mexico

José Luis Peinado Portillo,[1,*] *Alberto Ochoa-Zezzatti*,[1] *Sara Paiva*[2] and *Darwing Young*[3]

These days, we are living in the epitome of Industry 4.0, where each component is intelligent and suitable for Smart Manufacturing users, which is why the specific use of Big Data is proposed to determine the continuous improvement of the competitiveness of a car assembling industry. The Boston Consulting Group [1] has identified nine pillars of I4.0, which are: (i) Big Data and Analytics, (ii) Autonomous Robots, (iii) Simulation, (iv) Vertical and Horizontal Integration of Systems, (v) Industrial Internet of Things (IoT for its acronym in English), (vi) Cybersecurity, (vii) Cloud or Cloud, (viii) Additive Manufacturing including 3D printing, and (ix) Augmented Reality. These pillars are components of the Industry 4.0 that can be implemented as models of continuous competitiveness. In Industry 4.0, the Industrial IoT is a fundamental component and its penetration in the market is growing. Car manufacturers, such as General Motors or Ford, expect that by 2020 there will be 50 billion (trillion in English) connected devices, Ericsson Inc. estimates 18 billion. These estimated quantities of connected devices will be due to the increase in technological development, development in telecommunications and adoption of digital devices, and this will invariably lead to the increase in the generation of data and digital transactions, which leads to the mandatory increase in regulations, for security, privacy and informed consent in the integration of these diverse entities that will be connected and interacting among themselves and with the users. Finally, the use of Fuzzy Logic type 2 is proposed to adopt the correct decision making and achieve the reduction of uncertainty in the car assembly industry in the Northeast of Mexico.

1. Introduction

Today, technology is an important part of everyday life, from the way we communicate to the different types of technologies that allow us to carry out many types of processes in different industries.

[1] Universidad Autónoma de Ciudad Juárez, Av. Hermanos Escobar, Omega, 32410 Cd Juárez, Chihuahua, México.
[2] Universidad de Portugal.
[3] Centro CONACYT
* Corresponding author: jose_peinado@utcj.edu.mx

The Mexican industry, particularly the automotive industry, is not exempt from these technological advances, which are part of industry 4.0 (I4.0), and has an endless number of technologies that make it competitive in the market. However, these technologies are not effective enough to meet the demands of today's world, therefore, this chapter will show a literature review of the concepts that will be the basis for the proposal of a new intelligent model that is able to combine cutting-edge technologies and optimize processes and resources within the automotive industry in northern Mexico.

2. Literature Review

This section shows the main concepts of this article and how they have been generating and evolving throughout history. This section gives us an idea of what exists with respect to the technologies mentioned as Industry 4.0, Big Data, Fuzzy Logic Type-2.

2.1 Industry 4.0

Industry 4.0 (I4.0) is the latest standard for data and computation-oriented advanced manufacturing [2]. The term "Industry 4.0" originated from a project initiated by High-tech strategy of the German government to promote the computerization of manufacturing. Industry 4.0 is considered as the next phase in the digitization of the manufacturing sector, and it is driven by four characteristics: the amount of data produced, the increasing requirements of computational power, the usage of artificial intelligence techniques, and connectivity to high-speed networks [3]. The I4.0 was named thusly because it is the fourth industrial revolution, the first one (I1.0) refers to the first revolution which occurred in the 1800s, where the most important change was mechanical manufacturing, then in the 1900s the second revolution heralded the arrival of the assembly line, leading to an increase in mass production, and the third revolution occurred around 1970 with the introduction of robots that improved production efficiency. All this information is presented in the next table.

Table 1: Technology evolution from Industry 1.0 to Industry 4.0 [2].

Time	Evolution Transition	Defining technology
1800s	Industry 1.0	Mechanical Manufacturing
1900s	Industry 2.0	Assembly Line (mass production)
1970	Industry 3.0	Robotic Manufacturing (Flexible Manufacturing)
2010	Industry 3.5	Cyber Physical Systems
2012 onward	Industry 4.0	Virtual Manufacturing

As mentioned before, the I4.0 is based on nine pillars; this was written by [1] and the pillars are:

1. Big Data and Analytics
2. Autonomous Robots
3. Simulation
4. Horizontal and Vertical System Integration
5. The Industrial Internet of Things
6. Cybersecurity
7. The Cloud
8. Additive Manufacturing
9. Augmented Reality

2.2 Big data

One of the most important parts of I4.0 is Big Data and Analytics, which normally is associated with the result of the use of internet, sensors, management systems, but big data isn't about a big group of data, it is a model named "Model of 3 v's", i.e., Volume, Velocity, Variety [4]. Then this model was increased with a new "v", variability [5] for the "Model 4 v's", the next suggested for the "Model 5v's" was value, and over time this model has been increasing to the last model named "3v^2 Model" and is mentioned by Wu et al. [6], and they show us the next Venn Diagram:

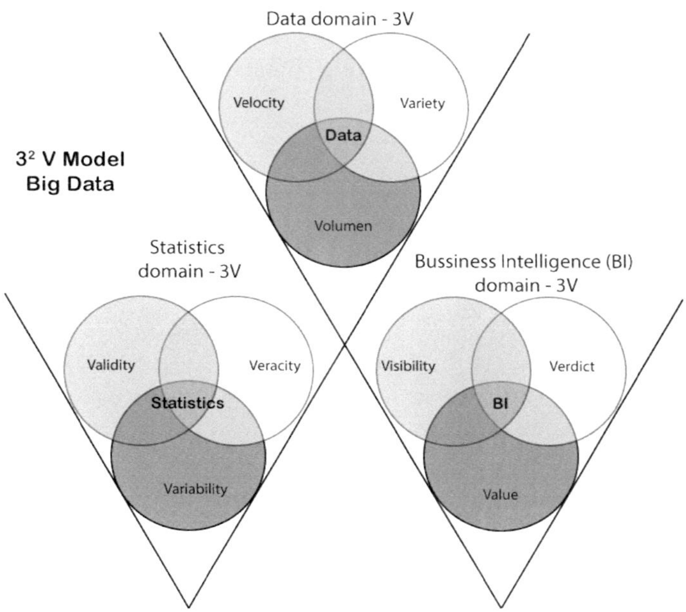

Figure 1: 3^2 v's Model for Big Data.

Some of the authors like Zhang et al. [7] talk about the use of Big Data in the automobile industry. They propose that the use of big data helps determine the characteristics that a user searches for in a car, in addition to predicting how sales will be in the coming months.

Otherwise, Kambatla et al. [8] talk about the future to big data. They give us an idea of what the use of big data implies, from the type of hardware that is needed to apply this technology, be it the use of memory or the hierarchy of memory that this implies, to the types of network and systems distributed that enable the application of big data for companies.

Furthermore, Philip Chen and Zhang [9] mention that in order to be competent, the use of big data is a big part of innovation, competition, and production for any company and that the use of big data should include the use of cloud computing, quantum computation and biological computation, besides that, the development of tools is an important part of the use of these technologies.

2.3 Fuzzy logic type-2

Fuzzy logic has attracted the attention of researchers for the last couple of decades. It has opened new horizons in both the academia and industry sectors, although, conventional fuzzy systems

(FSs), or so-called type-1 FSs, are capable of handling input uncertainties, they are not adequate to handle all types of uncertainties associated with knowledge-based systems [10]. The type-2 provide additional design degrees of freedom fuzzy logic systems, which can be very useful when such systems are used in situations where lots of uncertainties are present. The resulting type-2 fuzzy logic systems (T2 FLS) have the potential to provide better performance than a type-1 (T1) FLS [11]. A type-2 fuzzy set is characterized by a fuzzy membership function, i.e., the membership value (or membership grade) for each element of this set is a fuzzy set in [0,1], unlike a type-1 fuzzy set where the membership grade is a crisp number in [0,1] [12].

Membership functions of type-1 fuzzy sets are two-dimensional, whereas membership functions of type-2 fuzzy sets are three-dimensional. It is the new third-dimension of type-2 fuzzy sets that provides additional degrees of freedom that make it possible to directly model uncertainties [11].

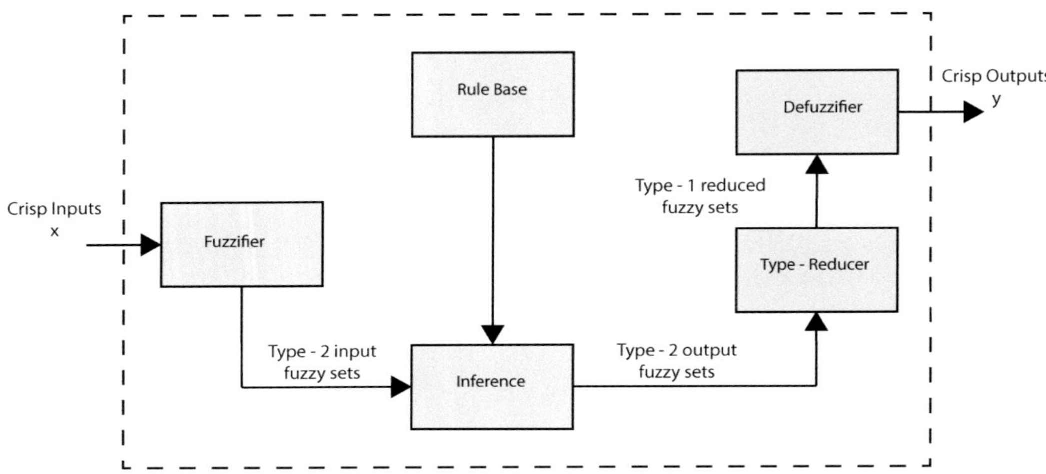

Figure 2: Diagram of a fuzzy logic controller.

3. Discussion

The automobile assembly industry today has multiple options for the assembly, from different models of cars, different types between these models, even the color of these is an important factor for decisions within companies.

On the other hand, currently, companies use different mathematical models as a solution for decision making, which, although useful and functional, only present between 60% and 65% of success in them, showing a little less than half of the failure within the decisions for the company.

Consider, a car is assembled in 7 stages and this passes through 4 work stations, only the assembly of this car has as result 28 critical points, now if 3 different models are made at the same time, and what happens if 4 cars are made of each model, the number of variables and critical points of the process grow significantly (Figure 3), so the mathematical and stochastic models are not being practical enough for this type of companies, representing 40% of losses or inefficiencies in the production of final products.

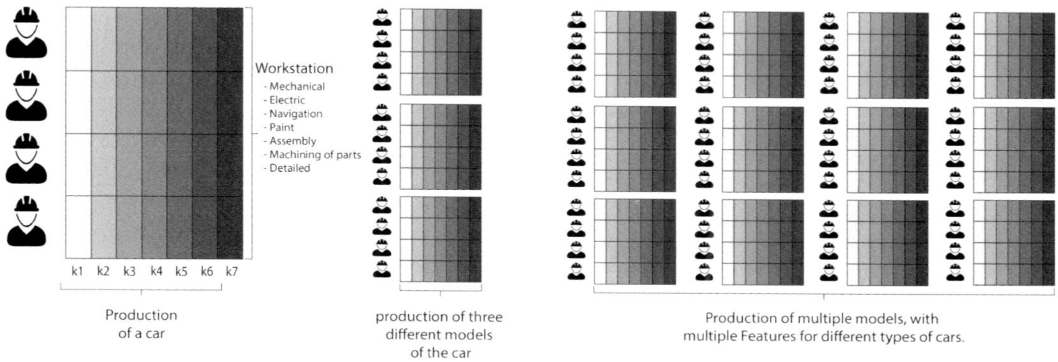

Figure 3: A multiple production of cars with multiple variables produce multiple critical points within the company.

4. Proposed Methodology

The proposal to help the way to optimize resources in the supply chain of a company is the realization of an intelligent model based on Big Data, which will be the technology responsible for generating the best options to optimize the use of materials in the warehouse of a car assembly industry in north-eastern Mexico (Figure 4), as well as a great help in making decisions for the company. Once the analysis through Big Data and the best options generated are available, Fuzzy Logic Type 2 technology will be integrated to determine the best way to use the company's resources or the best decision for the company.

The combination of these cutting-edge technologies would represent an improvement for many of the warehouses within the assembly industry within Mexico; this model can even be adaptable to other industries and government agencies or any business that has a warehouse and involves decision making in it since the goal of this intelligent model is to increase the optimization of resources and the effectiveness of decisions made by the company by up to 85%.

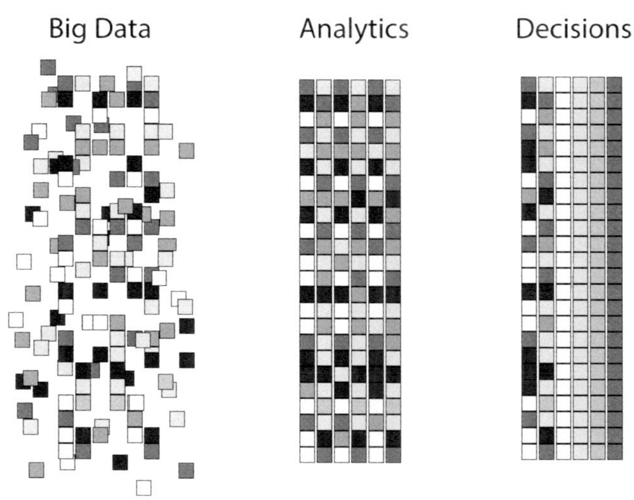

Figure 4: Use of big data for sorting and generation of options.

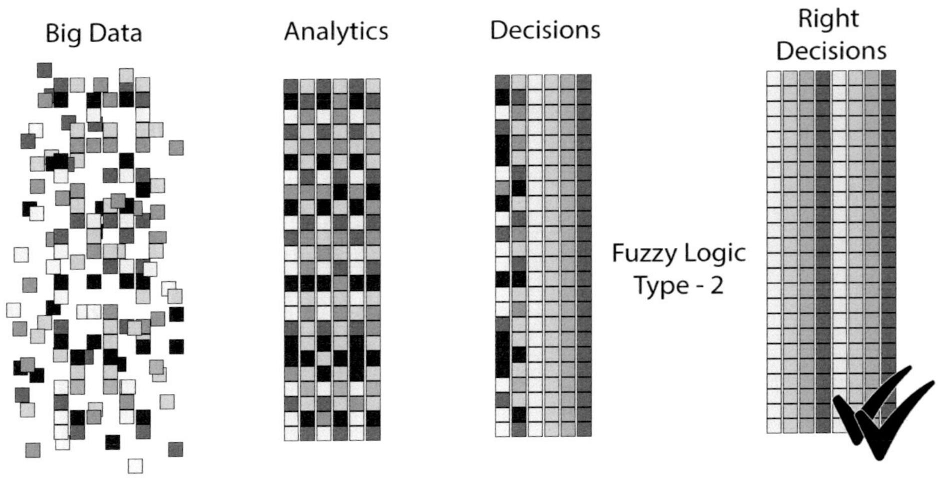

Figure 5: Integration of Fuzzy Logic Type-2 for the choice of the best option.

5. Conclusion and Future Research

There are many scientific articles that enable the research and development of the intelligent model to continue; it is worth mentioning that, although there are articles related to Big Data, other Fuzzy Logic Type-2, there is not much about the combination of both technologies, so it is thought that the development of a hybrid intelligent model could be a great revolution in the management of decisions and warehouses within the industry.

References

[1] Rüßmann, M. et al. 2015. Industry 4.0: Future of productivity and growth in manufacturing. Bost. Consult. Gr., no. April, p. 20.
[2] Govindarajan, U.H., Trappey, A.J.C. and Trappey, C.V. 2018. Immersive Technology for Human-Centric Cyberphysical Systems in Complex Manufacturing Processes : A Comprehensive Overview of the Global Patent Profile Using Collective Intelligence, vol. 2018.
[3] Sung, T.K. 2018. Industry 4.0: A Korea perspective. Technol. Forecast. Soc. Change, 132, no. October 2017, pp. 40–45.
[4] Khan, M., Jan, B. and Farman, H. 2019. Deep Learning: Convergence to Big Data Analytics. Springer Singapore.
[5] Kaur, N. and Sood, S.K. 2017. Efficient resource management system based on 4Vs of big data streams. Big Data Res., 9, no. February, pp. 98–106.
[6] Wu, C., Buyya, R. and Ramamohanarao, K. 2016. Big Data Analytics = Machine Learning + Cloud Computing, no. Ml.
[7] Zhang, Q., Zhan, H. and Yu, J. 2017. Car sales analysis based on the application of big data. Procedia Comput. Sci., 107, no. Icict, pp. 436–441.
[8] Kambatla, K., Kollias, G., Kumar, V. and Grama, A. 2014. Trends in big data analytics. J. Parallel Distrib. Comput.
[9] Philip Chen, C.L. and Zhang, C.Y. 2014. Data-intensive applications, challenges, techniques and technologies: A survey on Big Data. Inf. Sci. (Ny).
[10] Zamani, M., Nejati, H., Jahromi, A.T., Partovi, A., Nobari, S.H. and Shirazi, G.N. 2008. Toolbox for Interval Type-2 Fuzzy Logic Systems.
[11] Mendel, J.M., John, R.I. and Liu, F. 2006. Interval type-2 fuzzy logic systems made simple. IEEE Trans. Fuzzy Syst.
[12] Hagras, H.A. 2004. A hierarchical type-2 fuzzy logic control architecture for autonomous mobile robots. IEEE Trans. Fuzzy Syst., 2004.

CHAPTER-18

Weibull Reliability Method for Several Fields Based Only on the Modeled Quadratic Form

Manuel R. Piña-Monarrez[1,*] and *Paulo Sampaio*[2]

In this chapter, a practical and dynamic method to determine the reliability of a process (or product) is presented. The novelty of the proposed method is that it let us to use the Weibull distribution to determine the reliability index, by using only the quadratic form of the analyzed process (or product) as an input. So, since this polynomial can be fitted by using, e.g., simulation, mathematical and/or physical modeling, empirical experimentation and/or any optimization algorithm, the proposed method can easily be implemented in several fields of the smart manufacturing environment. For example, in the industry 4.0 framework, the proposed method can be used to determine, in dynamic form, the reliability of the analyzed product, and to give instantaneous feedback to the process. Therefore, to show the efficiency of the proposed method to determine the reliability in several fields, it is applied to the design, the quality and the monitoring product phases as well as to the fatigue (wearout and aging) phase. In order to let readers adapt the given theory to their fields and/or research projects, a detailed step by step method to determine the Weibull parameters directly from the addressed quadratic form is given for each one of the presented fields.

1. Introduction

Nowadays smart manufacturing (*SM*) is empowering businesses and achieving significant value by leveraging the industrial internet of things. Therefore, because process and products are now more complex and multifunctional, more accurate, flexible and dynamic analysis' tools are needed in the *SM* environment. For example, these technical tools are now being implemented into the industry 4.0 framework to evaluate and to make instantaneous feedback in the *SM* environment. Therefore, in this chapter a method to determine and/or to design a product or process with high reliability (*R(t)*) is presented. More importantly, since the proposed method is based on the Weibull distribution [1], then based on its Weibull shape parameter (β), the proposed method allows us to evaluate the reliability of the process or product in either of their principal phases; to know the design phase, which occurs for $\beta < 1$, the production phase which occurs for $\beta = 1$, and the wearout and aging phase which occurs for $\beta > 1$ [2]. Hence, due to the flexibility given by the β parameter, the proposed method can be used in the *SM* environment to evaluate in dynamic form the reliability of any *SM* process for which we know the optimal function.

[1] Universidad Autónoma de Ciudad Juárez, Av. Hermanos Escobar, Omega, 32410 Cd Juárez, Chihuahua, México.
[2] Salvador University (UNIFACS), Brazil..
* Corresponding author: manuel.pina@uacj.mx

The novelty of the proposed method is that it lets us to determine the Weibull parameters directly from the quadratic form elements of the optimal polynomial function used to represent the analyzed process (or product). Thus, since in the proposed reliability method, its input are only the elements of the quadratic form, its integration into the industry 4.0 paradigm is direct, and it will leave it to the decision maker managers to continuously determine the reliability that their processes present.

On the other hand, it is important to highlight that, because the proposed method can be applied based only on the quadratic form of *any optimization function*, then, since this optimization function can be determined by several mathematical, physical, statistical, empirical, and simulations tools, as they can be a genetic algorithm, mathematical and physical modeling, empirical experimentation [3] and [4], finite element analysis, and so on, readers easily will be able to adapt the given method to determine the reliability in their field and/or their projects. Therefore, with the objective that everyone can adapt the given method, in Sections 2 and 3 the theorical bases on which the proposed method was formulated, the references, where a detailed explanation of the technical formulations can be found, as well as the formula to determine the Weibull scale value which let us to determine the mean and the standard deviation of the input data, are all given. And to show how the proposed method works in several different fields, its application is presented in Section 4 to the mechanical stress design field [5]. In Section 5, it is applied to the quality field analysis [6]. In Section 6, it is applied to the multivariate statistical process control field [7]. In Section 7, it is applied to the physical field by designing and performing a random vibration test analysis for both the normal and the accelerated conditions. Finally, in Section 8, it is applied to the Fatigue (wear and aging) field.

Additionally, to facilitate its application to the fields or projects of the readers, in each one of the above mentioned field applications, a detailed step by step formulation to fit the Weibull parameters which represent (1) the random behavior of the applied stress, and (2) the Weibull-q parameters from which We can validate that the estimated Weibull stress distribution accurately represents the random behavior of the applied stress, are both derived. And its validation is made by demonstrating that by using the expected stress values given by the Weibull-q parameters, we can accurately derive both the mean and the standard deviation values of the Q elements from which the Weibull parameters were determined.

2. Weibull Generalities

This section has the objective of presenting the characteristics of the Weibull distribution that we can use to determine its parameters directly from an observed set of lifetime data or the known log-mean and log-standard deviation of the analyzed process. The main motivations to do this are (1) the Weibull distribution is very flexible to model all life phases of a products and processes, (2) in either phase of a process (or product), as can be among others, design, analysis, improvement, forecasting or optimization, both the region which contains the optimum (minimum or maximum) and the variable levels (values) at which the process represents the optimum, *must be considered*, and *3)* because it is always possible to model the optimal region by using a homogeneous second order polynomial model of the form

$$\hat{Y} = b_0 + b_1X_1 + b_2X_2 + b_{12}X_1X_2 + b_{11}X_1^2 + b_{22}X_2^2 \tag{2.1}$$

Therefore, because from Equation (2.1), the optimum of the analyzed process is determined from the quadratic form of the fitted optimal polynomial, we can use its quadratic form Q to determine the Weibull parameters. The quadratic form Q in terms of the interaction (b_{ij}) and quadratic effects (b_{jj}) of the fitted polynomial [8] is given as

$$Q = \begin{bmatrix} b_{11} & \frac{1}{2}b_{12} \\ \frac{1}{2}b_{21} & b_{22} \end{bmatrix} \tag{2.2}$$

Here, it is important to notice that, when the interaction effects of Q are zero ($b_{ij} = 0$), the optimum occurs in the normal plane (see Figure 1), and when they are not zero ($b_{ij2} = 0$), the optimum occurs in a rotated plane (see Figure 2).

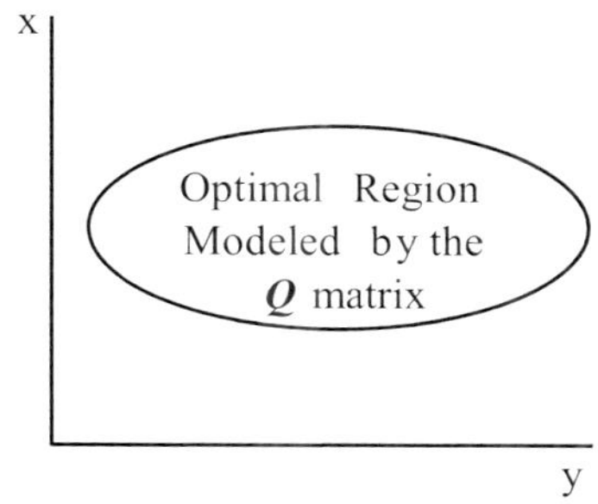

Figure 1: Normal plane.

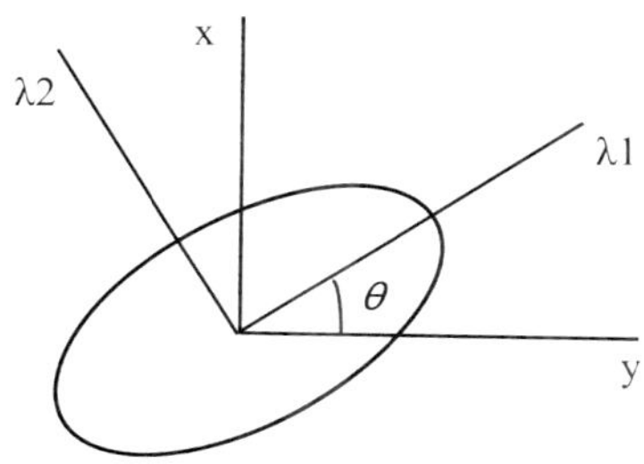

Figure 2: Rotated plane.

Thus, because the rotated plane is represented by the eigenvalues (λ_1 and λ_2) of the Q matrix, in any optimization process analysis when $b_{ij2} \neq 0$ both λ_1 and λ_2 and the rotation angle θ (see Figure 2) *must be estimated*, and then they are used to determine the optimal of the process. Even more, because from the eigenvalues of Q, the corresponding angle θ is unique, then the corresponding eigenvalues λ_1 and λ_2 are both unique also. Consequently, in this chapter, both λ_1 and λ_2 and θ are used to determine the corresponding Weibull shape β and scale η parameters. Therefore, because λ_1, λ_2 and θ are unique, β and η are unique also.

On the other hand, notice because λ_1 and λ_2 are the axes of the rotated plane (see Figure 2), then the forms of the analyzed system before and after the rotation are different. Thus, since *the normal distribution does not have a shape parameter*, then the normal distribution should not be used to model the Q form when its interaction elements are not zero ($b_{ij2} \neq 0$). In contrast, also notice that because θ completely determines λ_1 and λ_2 (see Equation (3.4)), and they can also be determined by the logarithm of the collected data as in [9], the probabilistic behavior of Q easily can be modeled by using the Weibull distribution [1] given by

$$f(t) = \frac{\beta}{\eta}\left(\frac{t}{\eta}\right)^{\beta-1} \exp\left\{-\left(\frac{t}{\eta}\right)^{\beta}\right\} \tag{2.3}$$

Moreover, since for different β values the Weibull distribution can be used to model the whole life of any product or process [2], the use of the Weibull distribution to model the quadratic form Q, fitted from data of several fields, is direct. So, since β and η are both time and stress dependent parameters, then the Weibull distribution is efficient to predict through the time the random behavior of the λ_1 and λ_2 values of Q. The analysis to estimate β and η directly from the λ_1 and λ_2 values of Q is as follows.

2.1 Weibull parameter estimation

In this section, the Weibull β and η parameters are determined from a set of collected lifetimes data by using the linear form of the Weibull reliability function given by

$$R(t) = \exp\left\{-\left(\frac{t}{\eta}\right)^{\beta}\right\} \tag{2.4}$$

Since the linear form of Equation (2.4) is of the form

$$Y = b_0 + \beta x \tag{2.5}$$

then the estimation of the unknown b_0 and β parameters is performed by using the well-known *least square method* given by.

$$\hat{\beta} = (X^t X)^{-1} X^t Y \tag{2.6}$$

And, since in Equations (2.5) and (2.6), the elements of the vector Y are unknown, then in the estimation process the median rank approach [10] is used to estimate them. The steps are as follows.

2.2 Steps to estimate β and η from a set of collected lifetime data

Step 1. If you are going to collect the data, then determine the desired *R(n)* index for the analysis. Then, based on the *R(n)* index, determine the corresponding sample size n to be collected [11] as

$$n = \frac{-1}{\ln(R(t))} \tag{2.7}$$

In contrast, if you are analyzing a set of n collected data, then the *R(n)* index which the set of the used n data represents, is determined from Equation (2.7) by solving it to *R(n)*.

Note 1. Here, notice that in Equation (2.7) n is not being used to determine if data ***whether follows or not*** a Weibull distribution. Instead, it is being used only to collect ***the exact amount of data*** which let us accurately fit the Weibull parameters [11].

Step 2. By using the n value estimated in step 1, determine the cumulated failure percentile by using the median rank approach [10] as

$$F(t_i) = (i - 0.3)/(n + 0.5) \tag{2.8}$$

Where $F(t_i) = 1 - R(t_i)$ is the cumulated failure time percentile.

Step 3. By using the $F(t_i)$ elements from step 2, determine the corresponding Y_i elements as

$$Y_i = \ln(-\ln(1 - F(t_i))) = b_0 + B\ln(t_i) \tag{2.9}$$

Note 2. Equation (2.9) is the linear form of Equation (2.4), that was defined in Equation (2.5).

Step 4. From a regression between the Y_i elements of step 3, and the logarithm of the collected lifetimes $X_i = ln(t_i)$ elements, determine the *Weibull-q time* β and η_{tq} values. From Equation (2.9), β is directly given by the slope, and the *Weibull-q* scale value is given as

$$\eta_{tq} = \exp\{-b_0/\beta\} \tag{2.10}$$

The addressed β and η_{tq} parameters are the corresponding ***Weibull-q*** family W(β, η_{tq}) that represents the collected data.

Step 5. From the X_i elements of step 4, determine its corresponding log-mean μ_x and log-standard deviation σ_x values, and determine the *Weibull* scale parameter that represents *Q(x)*

$$\eta_t = exp\{\mu_x\} \tag{2.11}$$

Thus, the addressed β and η_t parameters are the ***Weibull*** family W(β, η_t) that represents the related quadratic form $Q(x)$, as shown in Section 3. At this point, only notice that $\eta_t \neq \eta_{tq}$ because while η_t is directly given by the μ_x value, η_{tq} is given by the collected data.

From this section the general conclusion is that by using Equations (2.9) and (2.10) the ***Weibull-q time*** distribution which represents the collected lifetime data is determined, and by using Equations (2.9) and (2.11) the ***Weibull time*** distribution that represents $Q(x)$ is determined. Now let us present the numerical application.

2.2.1 Practical example

Here, let use data given in Table 1, that was published in [9]. The step by step analysis is as follows:

Step 1. Because in Table 1 $n = 21$, from Equation (2.7), *the reliability of the analysis* is $R(n)$=0.9535. Here observe $R(n)$=0.9535 *is not the reliability of the analyzed product*, instead it can be seen only as the reliability confidence level used in the statistical analysis

Steps 2 and *3*. The $F(t_i)$, Y_i and X_i elements are all given in Table 1.

Table 1: Weibull analysis for collected lifetime data.

Equations											
(2.7) N	(2.8) $F(t_i)$	(2.9) Y_i	(2.12) t_{qi}	(2.9) X_i	(2.13) t_i	(2.7) N	(2.8) $F(t_i)$	(2.9) Y_i	(2.12) t_{qi}	(2.9) X_i	(2.13) t_i
1	0.0327	-3.4034	17.4114	2.8571	13.2298	13	0.5934	-0.1052	91.7387	4.5189	69.6882
2	0.0794	-2.4916	27.5652	3.3165	20.9435	14	0.6401	0.0219	97.8115	4.5830	74.3006
3	0.1261	-2.0034	35.2525	3.5625	26.7831	15	0.6869	0.1495	104.3064	4.6473	79.2335
4	0.1728	-1.6616	41.8781	3.7347	31.8160	16	0.7336	0.2798	111.3853	4.7129	84.6099
5	0.2196	-1.3943	47.9144	3.8694	36.4013	17	0.7803	0.4159	119.2925	4.7815	90.6154
6	0.2663	-1.1720	53.5945	3.9814	40.7158	18	0.8271	0.5625	128.4338	4.8554	97.5581
7	0.3130	-0.9793	59.0583	4.0785	44.8660	19	0.8738	0.7276	139.5757	4.9386	106.0201
8	0.3598	-0.8074	64.4027	4.1651	48.9254	20	0.9205	0.9293	154.5059	5.0402	117.3591
9	0.4065	-0.6504	69.7027	4.2442	52.9511	21	0.9672	1.2296	179.7497	5.1915	136.5305
10	0.4532	-0.5045	75.0229	4.3177	56.9920		μ_y=-0.545624		μ=85.000	μ_x=4.297077	
11	0.5000	-0.3665	80.4249	4.3873	61.0950		σ_y=1.175117		σ=43.0950	σ_x=0.592090	
12	0.5467	-0.2341	85.9727	4.4540	65.3088					Exp(μ_x)=73.4846	

Step 4. By using the Minitab routine, the regression equation is $Y_i = -9.074 + 1.985X_i$. Hence, $\beta = 1.985$ and from Equation (2.10), η_{tq} =exp{–(–9.074/1.98469)} = 96.7372 hrs. Consequently, the *Weibull-q* distribution that represents the life time data is W(β = 1.985, η_{tq} = 96.7372 hrs).

Step 5. Since from Equation (2.11) η_t = 73.4846 hrs, the *Weibull* distribution that represents the related $Q(x)$ form is W(β =1.985, η_t= 73.4846 hrs).

Finally observe, from Equations (2.4) or (2.9) the lifetime which corresponds to the expected $R(t)$ index is given as

$$t_{qi} = \left(\sqrt[\beta]{-\ln(R(t))}\right) * \eta_{tq} = exp\{Y_i/\beta + \ln(\eta_{tq})\} \tag{2.12}$$

For $R(t)$ = 0.9535, t = 20.86 hrs. And the time that corresponds to the expected $R(t)$ index of the related $Q(x)$ form is given by

$$t_i = \left(\sqrt[\beta]{-\ln(R(t))}\right) * \eta_t = exp\{Y_i/\beta + \ln(\eta_t)\} \tag{2.13}$$

For $R(t)$ = 0.9535, it is t_q =15.85 hrs.

From Table 1, we observe that because the mean of the lifetime data of μ = 85 hrs, and the standard deviation of σ = 43.095 hrs, were both generated by the Weibull-q family W(β = 1.985, η_{tq} = 96.7372 hrs), then its corresponding log-mean μ_x = 4.297077 was also generated by the Weibull-q family. Furthermore, using μ_x in Equation (2.11) gives the Weibull scale parameter of η_t =73.4846 hrs of the related Weibull time distribution that represents $Q(x)$, then the Weibull-q family can always be used to validate the η_t parameter.

In Section 3, we will show the elements of the quadratic from which generate the η_t =73.4846 hrs value (λ_1 = 127.72 and λ_2 = 42.28), which corresponds to an angle of Θ = 29.914. However, let us first present how to estimate the *Weibull* time and the Weibull-q families when no experimental lifetime data is available.

2.3 Estimation of β and η without experimental data

When lifetime data is not available, the Weibull parameters β and η are estimated based on the mean μ_y and standard deviation σ_y values of the median rank approached defined in Equation (2.9), and on ***the known*** log-mean μ_x and log-standard deviation σ_x values of the logarithm of the expected lifetimes. Also notice for n = 21, both μ_y = –0.545624 and σ_y = 1.175117 are constant. Based on them, the steps to estimate β and η are as follows.

Step 1. By following steps 1 to 3 of Section 2.1, determine the Y_i elements, and from these elements determine its mean μ_y and its standard deviation σ_y. (from data of Section 2.1.1 the μ_y and σ_y values are given in Table 1).

Step 2. By using the μ_y and σ_y values of step 1 and *the known* log-mean μ_x and log-standard deviation σ_x values, the corresponding expected *Weibull-qβ* and η_{tq} parameters are

$$\beta = \sigma_y / \sigma_x \tag{2.14}$$

$$\eta_{tq} = exp\{\mu_x - Y_i/\beta\} = \ln(\eta_t) - \mu_y/\beta \tag{2.15}$$

Therefore, Equations (2.14) and (2.15) enable us to determine without data the ***Weibull-q*** W(β, η_{tq}) family that represents the expected failure times. And Equations (2.14) and (2.11) enable us to determine without data the *Weibull time* W(β, η_t) family that represents the related *Q(x)* form. Now let us present the numerical application.

2.3.1 Practical example

By using the μ_y and σ_y, and the μ_x and σ_x values from Table 1, in Equations (2.14) and (2.15), the *Weibull-q time* parameters, as in Section 2.2.1, are W(β = 1.984692, η_{tq} = 96.736747 hrs). Similarly from Equation (2.14) and Equation (2.11), the corresponding *Weibull stress* parameters of the expected quadratic form are W(β = 1.984692, η_t = 73.4846 hrs). Additionally, notice from Equation (2.13) that, because σ_x determines the β value, and since the higher the σ_x value the lower the β value, then, as in [12], σ_x must be set as the upper control limit in the control chart used to monitor β. Similarly, because from Equations (2.11) and (2.15), μ_x determines the η_{tq} and η_t the values, and because the lower the μ_x value, the lower the η_{tq} and η_t values, then as in [12], the μ_x value must be set as the minimum allowed value in the control chart used to monitor η_{tq} and η_t. On the other hand, if there is no available experimental ***lifetime*** data, and μ_x and σ_x are unknown, then, based on ***the applied stresses values*** of the analyzed process, the ***Weibull-q and Weibull stress*** parameters are estimated as follows.

3. Weibull Quadratic Stress Form Analysis

The objective of this section is to estimate the *Weibull stress* W(β, η_s), and the *Weibull-q stress* W(β, η_{sq}) parameters directly from the quadratic form elements of the optimal polynomial used to optimize the process. Therefore, from Equation (2.1) the quadratic form *Q(s)* is given by its quadratic and interaction effects as

$$Q(s) = \sum_{i,j=1}^{k} b_{ij} X_i X_j \tag{3.1}$$

And as a consequence of Equation (2.1) being the optimal response surface polynomial model widely used in the experiment design analysis, then based on the B canonical form of Equation (2.1) (see [4] Chapter 10), given by

$$\hat{Y} = Y_s + \lambda_1 \dot{X}_1^2 + \lambda_1 \dot{X}_2^2 \tag{3.2}$$

the *Q(s)* matrix defined in Equation (3.1), in terms of the λ_1 and λ_2 values of Equation (3.2) is given as

$$Q(s) = \sum_{i,j=1}^{k} \lambda_j X_j^2 \tag{3.3}$$

Here, it is important to notice that (1) in this section *Q(s)* instead of representing time represents stress, and (2) that in the case that *Q(s)* has several eigenvalues, then in the analysis we have to use only the maximum $\lambda_1 = \lambda_{max}$ and minimum ($\lambda_2 = \lambda_{min}$) eigenvalues. Therefore, based on the λ_1 and λ_2 values of stress *Q(s)* form, the corresponding Weibull stress β, and η_s parameters that represents *Q(s)* and the η_{sq} that represents the expected stresses values are determined as follows.

3.1 Estimation of β and η from the Q(s) matrix elements

The steps to determine the Weibull-q stress and Weibull stress parameters are:

Step 1. From the *Q(s)* matrix elements of Equation (2.2) or Equation (3.1), determine the eigenvalues λ_1 and λ_2 as

$$\lambda_1, \lambda_2 = \mu \pm \sqrt{\mu^2 - \eta_s^2} \tag{3.4}$$

Where μ is the arithmetic mean, which because the trace of a matrix is invariant, then from the *Q(s)* elements, it is given as

$$\mu = (b_{11} + b_{22})/2 \; \mu = (b_{11} + b_{22})/2 \tag{3.5}$$

Then, η_s is the scale parameter of the *Weibull stress* family, which from the determinant of *Q(s)* is given as

$$\eta_s = \sqrt{b_{11}b_{22} - b_{12}^2} \tag{3.6}$$

Step 2. By using the μ_y value of step 1 of Section 2.2 and the eigenvalues λ_1 and λ_2 of step 1, determine the corresponding β value as

$$\beta = \frac{-4\mu_Y}{0.9947 * \ln(\lambda_1 / \lambda_2)} \tag{3.7}$$

Note 3: From Equations (3.6) and (3.7), the estimated β and η_s values are the *Weibull stress* distribution W(β, η_s) which represents the random expected stresses values.

Step 3. From Equation (3.6) (determinant of *Q(s)*), the expected log-mean μ_x value is given as

$$\mu_x = \ln(\eta_s) = \ln\left(\sqrt{\lambda_1 \lambda_2}\right) \tag{3.8}$$

Step 4. By using the β value of step 2, the value of step 1 of Section 2.2 and the μ_x value of step 3 in Equation (2.15), determine the *Weibull-q stress* η_{sq} parameter which can be used to validate the addressed Weibull stress family.

Note 4: The estimated β and η_{sq} values are the *Weibull-q stress* distribution W(β, η_{sq}). Here remember that for $n = 21$, $\mu_y = -0.545624$ and $\mu_y = 1.175117$ are both constant.

Step 5. By using the β value of step 2, the σ_x value of step 1 of Section 2.2, determine the expected log-standard deviation σ_x as

$$\sigma_x = \sigma_y / \beta \tag{3.9}$$

Finally, note that because both μ_x and σ_x let us to determine the Weibull β and η parameters, and since they are given by the quadratic and interaction effects of the quadratic form *Q(s)*, then in order to control μ_x and σ_x as in [12], the quadratic and interaction effects of *Q(s)*, must be monitored. Or equivalently μ_x and σ_x can be used as the signal parameters in the corresponding dynamic Taguchi analysis [13] to determine the sensibility of μ_x and σ_x to the variation of the *Q(s)* elements.

3.2 Validation that the estimated β and η parameters represents the used Q(s) matrix data

The validation is made in the sense that, by using the expected stress data given by the W(β, η_s) distribution, the eigenvalues λ_1 and λ_2 defined in Equation (3.4), the mean tress μ defined in Equation (3.5), and the η_s stress value defined in Equation (3.6) are all completely determined. This fact can also be seen from Equations (3.7) and (3.8) by noticing that the Weibull-q parameters are determined by using the λ_1 and λ_2 eigenvalues, and by noticing from Equation (3.4), that the λ_1 and λ_2 eigenvalues are determined by using the mean stress μ and η_s the stress value.

Therefore, in order to validate in each application that the addressed Weibull family W(β, η_{sq}) represents the stress data from which it was determined, in the Table of each presented analysis, the expected data which corresponds to the W(β, η_{sq}) family is also given. From this data, observe that the average of the given data is the mean stress μ value defined in Equation (3.5), and that the exponent of the average of the logarithm of these data, is the η_s stress value.

Hence, it is clear that in using these μ and η_s values in Equation (3.4), the corresponding λ_1 and λ_2 eigenvalues are completely determined also. Thus, the conclusion is that in using the expected data of the W(β, η_{sq}) family, the original μ, η_s, λ_1 and λ_2 parameters are all completely determined, then the W(β, η_{sq}) family can be used to validate the W(β, η_s) parameters which determine the random behavior of the applied stresses values given in this section as λ_1 and λ_2. In the next Sections, λ_1 and λ_2 are known as principal stresses σ_1 and σ_2 values ($\lambda_1 = \sigma_1$, $\lambda_2 = \sigma_2$).

4. Mechanical Field Analysis

This section is focused on the design phase of a product or process. In the numerical application, the design of a mechanical element [14] is presented. The analysis is performed based on the quadratic form given by the normal and the shear stress values that are acting on the analyzed mechanical element. Therefore, the *Weibull stressβ* and η_s parameters are both determined based on the steps of Section 3.1 and on the stress *Qs* matrix given by the normal σ_x and σ_y and the shear τ_{xy} stresses values as

$$Qs = \begin{bmatrix} \sigma_1 & \tau_{xy} \\ \tau_{yx} & \sigma_2 \end{bmatrix} \tag{4.1}$$

The steps to determine the *Weibull stress* W(β, η_s) and the Weibull-q W(β, η_{sq}) parameters are as follows.

4.1 Steps to the mechanical design field

Step 1. From the stress analysis of the analyzed component, determine the normal σ_x and σ_y and the shear τ_{xy} stresses values that are acting on the element, and then form the corresponding *Qs* matrix as in Equation (4.1).

Step 2. By using σ_x and σ_y of step 1 in Equation (3.5), determine the arithmetic mean μ.

Step 3. By using the σ_x, σ_y and τ_{xy} values of step 1 in Equation (3.6), determine the *Weibull stress* parameter.

Step 4. By using the μ value of step 2 and the η_s value of step 3 in Equation (3.4), determine the principal stresses $\sigma_1 = \lambda_1$ and $\sigma_2 = \lambda_2$ values.

Step 5. By using the σ_x, σ_y and τ_{xy} values of step 1, determine the principal angle Θ as

$$\theta = 0.5 * \tan^{-1} (2\tau_{xy}/(\sigma_x - \sigma_y)) \tag{4.2}$$

Step 6. By using the principal stresses σ_1 and σ_2 values of step 4, the yield strength S_y value of the used material and the desired safety factor *SF*, in the maximum distortion-energy theory (*DE*) (Von Mises) criterion [15] Section 5.4, given by

$$DE\ theory = \sqrt{\sigma_1^2 - \sigma_1\sigma_2 + \sigma_2^2} < S_y/SF \tag{4.3}$$

And the maximum-shear-stress (MSS) (Tresca theory) [15] Section 5.5, criterion given by

$$\text{MSS}\ theory = \sigma_{max} < S_y/SF \tag{4.4}$$

determine whether the designed element is safe or not.

Step 7. Determine the desired *R(t)* index and, by using it in Equation (2.7), determine the corresponding *n* value.

Step 8. By following steps 1 to 3 of Section 2.1, determine the Y_i elements and from them determine its mean μ_y and its standard deviation σ_y. (Remember that for $n = 21$, $\mu_y = -0.545624$ and $\mu_y = 1.175117$ are both constant).

Step 9. By using the σ_1 and σ_2 values of step 4, and the μ_y value from step 8 in Equation (3.7), determine the Weibull β_s parameter.

Note 5: The η_s parameter of step 3 and the β_s parameter of this step are the *Weibull stress* family $W(\beta_s, \eta_s)$ which determines the random behavior of the applied stress.

Step 10. By using the η_s parameter of step 3, the μ_y value of step 8 and the β_s parameter of step 9 in Equation (2.15), determine the *Weibull-q stress* scale η_{sq} parameter.

Note 6: The η_{sq} parameter of this step and the β_s parameter of step 9 are the *Weibull-q* stress family $W(\beta_s, \eta_{sq})$ which can be used to validate that the addressed $W(\beta_s, \eta_s)$ family completely represents the applied stress values.

Step 11. Determine the *R(t/s,S)* index which corresponds to the yield strength value of used material mentined in step 6, as

$$R(t/s,S) = \frac{S_y^{\beta s}}{S_y^{\beta s} + \eta_s^{\beta s}} \tag{4.5}$$

Note 7: Equation (4.5) is the Weibull/Weibull stress/strength reliability function (see [9] Chapter 6), which is used to estimate the reliability of the analyzed component only when the Weibull shape parameter is the same for both the stress and strength distributions. Here, the *Weibull stress* distribution $W(\beta, \eta_s)$ is given by Equation (3.6) and Equation (3.7), and the *Weibull strength* distribution is given by using S_y as the *Weibull strength* scale parameter. Thus, the *Weibull strength* distribution is $W(\beta_s, S_y = \eta_y)$.

Here, remember that the $R(t) = 0.9535$ index used to estimate *n* in Equation (2.7) is the *R(t)* index of the analysis, and that the *R(t/s,S)* of Equation (4.5) is the reliability of the product. On the other hand, due to ***in any Weibull analysis the σ_{1i} values given by the $W(\beta, \eta_s)$ family, can be used as the S_y value***, then the steps to determine the σ_{1i} values that corresponds to a desired *R(t/s,S)* index are also given.

4.1.1 Additional steps

Step 12. By using the Y_i elements of step 8 and the β_s value of step 9, determine their corresponding Weibull basic elements as

$$\tan(\theta_i) = \exp\{Y_i/\beta_s\} \tag{4.6}$$

Step 13. By using the $\tan(\theta_i)$ values of step 11, and the *Weibull stress* η_s value of step 3, determine the expected pair of principal stresses values σ_{1i} and σ_{2i} for each one of the Y_i elements as

$$\sigma_{1i} = \eta_s/\tan(\theta_i) \text{ and } \sigma_{2i} = \eta_s * \tan(\theta_i) \tag{4.7}$$

Step 14. Determine the reliability $R(t_i/s)$ index for each one of the Y_i elements as

$$R(t_i/s) = \exp\{\tan(\theta_i)\} \tag{4.8}$$

Therefore, the σ_{1i} element of the desired $R(t_i/s)$ index, can be used as the minimum Weibull strength value S_y, at which the mechanical element should be designed. Now let us present the numerical application.

4.2 Mechanical application

Step 1. Let use the normal σ_x and σ_y and shear τ_{xy} stress values given in [5] pg.37. They are $\sigma_x = 90$ mpa, $\sigma_y = 190$ mpa and shear $\tau_{xy} = 80$ mpa. With this data, the *Qs* matrix is [90 80; 80 190].

Step 2. From Equation (3.5), the mean stress is $\mu = (90 + 190)/2 = 140$ mpa.

Step 3. From Equation (3.6), the *Weibull stress* parameter is $\eta_s = (90*190 - 80\text{^}2) = 103.44$ mpa.

Step 4. From Equation (3.4), the principal stresses are $\sigma_1 = 234.34$ mpa and $\sigma_2 = 45.66$ mpa (140 ± 94.34). See Figure 3.

Step 5. From Equation (4.2) $\Theta = 28.99730$.

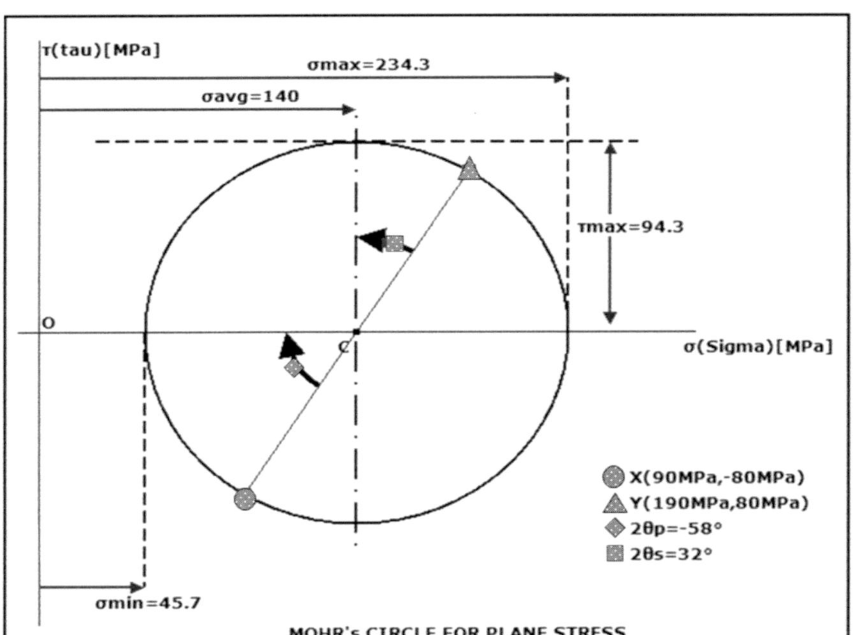

Figure 3: Principal and Shear stress analysis.

Step 6. Suppose after applying the modifier factors, the material' strength is $S_y = 800$ mpa and the safety factor is $SF = 3$. Hence, due to Equation (4.3) $215.2 < 266.7$ and Equation (4.4) $234.3 < 266.7$, the designed element is considered to be safe. See Figure 4.

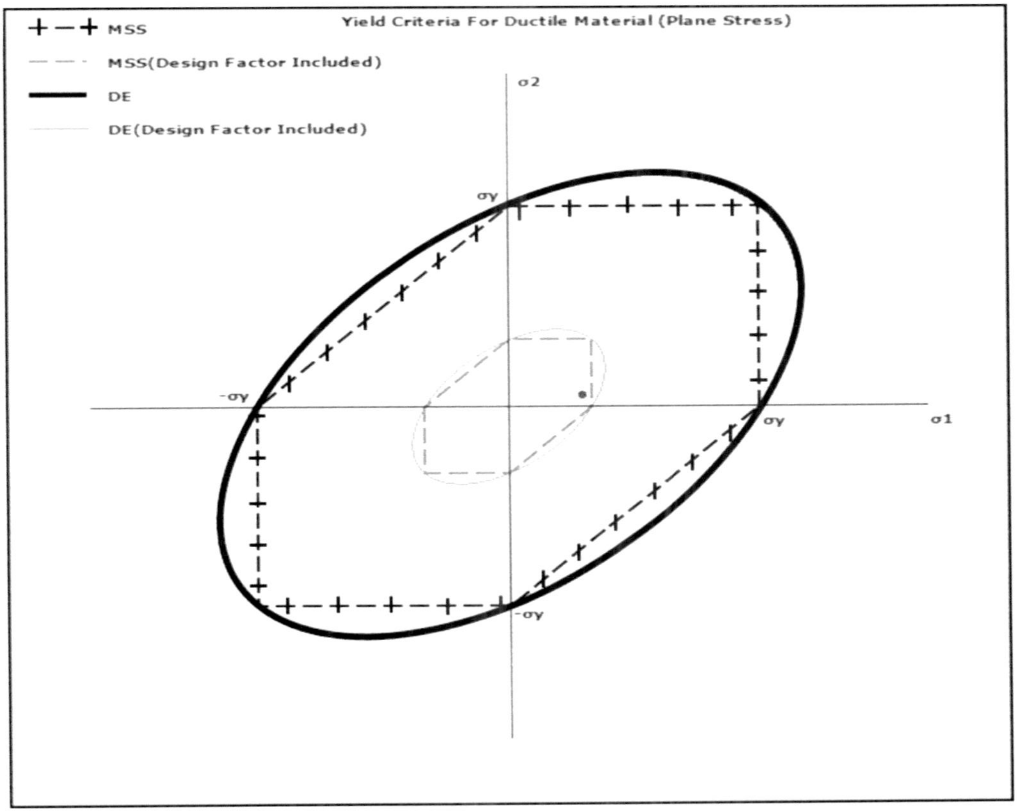

Figure 4: DE and MSS theory analysis.

Step 7. Suppose a reliability analysis with $R(t)$ = 0.9535 is desired, thus, from Equation (2.7), $n = 21$. (Remember that for $n = 21$, $\mu_y = -0.545624$ and $\mu_y = 1.175117$ are both constant).

Step 8. The elements and its corresponding μ_y and σ_y values are given in Table 1.

Step 9. From Equation (3.7), the Weibull shape parameter is $\beta_s = 1.336161$. Therefore, the *Weibull stress* family is W($\beta_s = 1.336161$, $\eta_s = 103.44$ mpa).

Step 10. From Equation (2.15) the *Weibull-q stress* scale parameter is $\eta_{sq} = 155.8244$ mpa. Therefore, the *Weibull-q stress* family is W($\beta_s = 1.336161$, $\eta_{sq} = 155.8244$ mpa).

Step 11. From Equation (4.5) the designed reliability of the mechanical element is $R(t/s,S) = 93.84\%$. Here observe the ***Weibull strength*** family is W($\beta_s = 1.336161$, $S_y = 800$ mpa).

Step 12. The basic Weibull tan(θ_i) values for each one of the Y_i elements are given in Table 2.

Step 13. The expected pair of principal stresses σ_{1i} and σ_{2i} values for each one of the Y_i elements are given in Table 2.

Step 14. The reliability $R(t_i/s)$ values for each one of the Y_i elements are given in Table 2.

Table 2: Weibull analysis for mechanical field.

				Weibull Stress		Weibull-q Data							Weibull Stress		Weibull-q Data		
(2.7)	(2.9)	(4.6)	(4.5)	(4.7)		(2.12)	(2.9)	(4.7)	(2.7)	(2.9)	(4.6)	(4.5)	(4.7)		(2.12)	(2.9)	(4.7)
N	Y_i	$tan(\theta_i)$	θ_i	σ_{1i}	σ_{2i}	σ_{sqi}	$ln(\sigma_{sqi})$	$R(t_i/s)$	n	Y_i	$tan(\theta_i)$	θ_i	σ_{1i}	σ_{2i}	σ_{sqi}	$ln(\sigma_{sqi})$	$R(t_i/s)$
1	-3.403	0.0776	4.43	1332.50	8.03	12.0965	2.49291	0.9673	12	-0.234	0.8388	39.98	123.32	86.76	130.701	4.87292	0.4533
	-2.724	**0.1293**	**7.36**	**800.00**	**13.37**	**20.1482**	**3.00312**	**0.9365**	13	-0.105	0.9240	42.73	111.95	95.58	143.978	4.96967	0.4065
2	-2.491	0.1540	8.75	671.89	15.93	23.990	3.17764	0.9206	14	0.021	1.0166	45.47	101.75	105.16	158.411	5.06520	0.3598
3	-2.003	0.2221	12.52	465.67	22.98	34.613	3.54425	0.8738	15	0.149	1.1188	48.21	92.45	115.73	174.341	5.16101	0.3131
4	-1.661	0.2871	16.02	360.25	29.70	44.742	3.80093	0.8271	16	0.279	1.2339	50.97	83.83	127.63	192.265	5.25888	0.2664
5	-1.394	0.3509	19.33	294.74	36.30	54.686	4.00162	0.7804	17	0.416	1.3667	53.80	75.69	141.37	212.957	5.36109	0.2196
6	-1.172	0.4147	22.52	249.42	42.90	64.624	4.16859	0.7336	18	0.562	1.5256	56.75	67.80	157.81	237.730	5.47114	0.1729
7	-0.979	0.4793	25.60	215.82	49.58	74.684	4.31327	0.6869	19	0.727	1.7270	59.92	59.90	178.64	269.111	5.59513	0.1262
8	-0.807	0.5453	28.60	189.68	56.41	84.977	4.44238	0.6402	20	0.929	2.0094	63.54	51.48	207.86	313.120	5.74659	0.0794
9	-0.650	0.6136	31.53	168.59	63.47	95.607	4.56025	0.5935	21	1.229	2.5178	68.33	41.08	260.45	392.341	5.97213	0.0327
10	-0.504	0.6846	34.39	151.09	70.82	106.684	4.66987	0.5467							μ=140	μx=4.639	
11	-0.366	0.7594	37.21	136.21	78.55	118.332	4.77350	0.5000							σ=100.83	σx=0.882	

From Table 2, observe that because the average of the Weibull-q family is also $\mu = 140$ Mpa, and from its log-mean, η_s is also ($\eta_s = \exp\{\mu_x\} = 103.4403$ Mpa), then the addressed Weibull stress family completely represents the applied stresses.

On the other hand, notice from Table 2 that the σ_{1i} value which corresponds to *R(t/s)* = 0.9365 is $\sigma_{1i} = 800$ Mpa, and in using $Sy = 800$ Mpa in Equation (4.5), *R(t/s)* is also *R(t/s)* = 0.9365, then we conclude that for *R(t)* > 0.90, the *R(t/s)* values of Table 2 and those given from Equation (4.5), are similar. Therefore, the σ_{1i} column of Table 2 can be used as a guide to select the minimum yield strength scale S_y parameter which corresponds to any desired *R(t/s,S)* index.

Moreover, from the minimum applied stress σ_2 value and the *Weibull stress* and *Weibull strength* scale parameters, the minimum yield strength value which we must select from the material engineering handbook, in order for the designed product to meet the desired reliability, is given as

$$S_{y\min} = \frac{\sigma_2 \eta_S}{\eta_s}, \quad S_{y\max} = \frac{\eta_s \eta_S}{\sigma_2} \tag{4.9}$$

For example, if the minimal applied stress is $\sigma_2 = 45.66$ mpa, the *Weibull stress* parameter is $\eta_s = 103.44$ mpa and the *Weibull strength* parameter is $S_y = 800$ mpa, then from Equation (4.9) the minimum material's strength value to be selected from the material engineering handbook is $S_{ymin} = 353.14$ mpa. Similarly, the corresponding expected maximum value is $S_{ymax} = 1812.36$ mpa. As a summary of this section we have that:

(1) The Weibull-q family W($\beta_s = 1.336161$, $\eta_{sq} = 155.8244$ mpa) allows us to validate that the Weibull stress family W($\beta_s = 1.336161$, $\eta_{sq} = 103.44$ mpa), completely represents the quadratic form *Qs* elements.

(2) The expected σ_{1i} elements given by the W($\beta_s = 1.336161$, $\eta_s = 103.44$ mpa) family can be used as the minimum Weibull strength eta value to formulate the corresponding minimum Weibull strength family W($\beta_s = 1.336161$, $\eta_s = 800$ mpa).

(3) The reliability *R(t/s)* indices given by the W($\beta_s = 1.336161$, $\eta_s = 103.44$ mpa) family and that given by the stress/strength function *R(t/s,S)* defined in Equation (4.5) are both similar for higher reliability percentiles (say, higher than 0.90).

Now let present the analysis for the quality field.

5. Quality Field Analysis

In this section, the analysis to determine the *Weibull stress* and the *Weibull-q stress* parameters as well as the numerical application to the quality field is presented. In the quality field, the analysis of a process is generally performed in two stages. In the first, the process' output is determined in such a way that the process' output fulfills both the performance and the quality requirements. In this first stage, generally the three Taguchi's phases are applied. The Taguchi's phases (see [6] Chapter 14), are: (1) *The system design phase*. This consists of determining the first functional design (or prototype), and determining the process' factors and functional relationship between the addressed factors (ideal function) and the desired quality and functional requirement to be met. (2) *The parameter design phase*. This consists of determining the set of the significant factors, and the factor's levels at which it the process is expected to present the desired output. (3) *The tolerance design phase*. This consists of determining the tolerance to those process' factors which must be controlled to reach the desired process' output. In the second stage, the performance through the time of the process is determined. This is made by analyzing the effect that the environmental factors have on the process' output. Therefore, because the environmental factors' behavior is random, a probability density function (*pdf*) is used to model the desired process' outputs in this second stage. Thus, to determine the parameters of the used *pdf*, a response surface polynomial, such as the one given in Equation (2.1), is fitted from an experiment design data. Then, from its quadratic form (*Qq*) elements, the corresponding *pdf* parameters are determined. Here, the Weibull *pdf* is used to perform the analysis, and the steps to fit the Weibull parameters from the Qq matrix elements are as follows.

5.1 Steps to determine the quality weibull families

Step 1. From the analyzed product or process, determine the performance and quality (or functional) characteristic of the process (or product) to be measured, as well as the set of significant factors.

Step 2. By using the corresponding experiment design data (here a Taguchi orthogonal array is used), determine the levels of the factors which fulfill the performance and quality requirements. Here, a capability index $cp = 2$, or six sigma behavior, and ability index $cpk = 1.67$ are used [16]. They are estimated as

$$cp = (USL - LSL)/6\sigma \tag{5.1}$$

$$cpk = \min\left((USL-\mu)/3\sigma \quad or \quad (\mu - LSL)/3\sigma\right) \tag{5.2}$$

Where USL is the upper specification limit, LSL is the lower specification limit, μ is the process' mean, and σ is the process' standard deviation.

Step 3. Determine the set of environmental factors which lower the process performance (output and quality).

Step 4. By applying the response surface methodology [4], determine the optimal second order polynomial model which relates the environmental factors of step 3 and the quality characteristic of step 1. Here, notice the response surface analysis is performed only to the addressed optimal (or robust) levels of step 2.

Step 5. By using the quadratic and interaction effects of the fitted response surface polynomial of step 4, form the Qq matrix as

$$Q_q = \begin{bmatrix} b_{11} & b_{12}/2 \\ b_{21}/2 & b_{22} \end{bmatrix} \tag{5.3}$$

Step 6. By using the b_{11}, b_{22} and $b_{12}/2$ elements from step 5 in Equation (3.6), estimate the *Weibull stress* quality η_s parameter.

Step 7. By using the b_{11} and b_{22} elements from step 5 in Equation (3.5), determine the arithmetic mean μ.

Step 8. By using μ from step 7 and η_s from step 6 in Equation (3.4), determine the maximum λ_1 and the minimum λ_2 eigenvalues.

Step 9. Determine the desired reliability *R(t)* index to perform the analysis, and by using it in Equation (2.7), determine the corresponding sample size n value.

Step 10. Following steps 1 to 3 of Section 2.1, determine the corresponding Y_i elements and its mean μ_y and standard deviation σ_y.

Step 11. By using the λ_1 and λ_2 values from step 8, and μ_y from step 10 in Equation (3.7), determine the quality Weibull β parameter.

Step 12. By using η_s from step 6 and β from step 11, form the quality *Weibull stress* family $W(\beta, \eta_s)$.

Step 13. By using the β value and the Y_i elements of step 10, determine the basic Weibull values

$$\tan(\theta_i) = \exp\{Y_i/\beta\} \tag{5.4}$$

Step 14. By using the basic Weibull values from step 13 and the η_s value from step 6, determine the expected eigenvalues λ_{1i} and λ_{2i} as

$$\lambda_{1i} = \eta_s / t_{oi} \quad \text{and} \quad \lambda_{2i} = \eta_s * t_{oi} \tag{5.5}$$

Step 15. By using β from step 11, μ_y from step 10 and η_s from step 6 in Equation (2.15), determine the Weibull-q parameter η_{sq}. The estimated β value in step 10 and the η_{sq} value of this step are the *Weibull-q* distribution $W(\beta, \eta_{sq})$.

Now let us present the numerical application.

5.2 *Quality improvement application*

In this section, we use data published in [6] (pg.452). Data is as follows. "The weather strip in an automobile is made of rubber. In the rubber industry, an extruder is used to mold the raw rubber compound into the desired shapes. Variation in output from the extruders directly affects the dimensions of the weather strip as the flow of the rubber increases or decreases." The Taguchi L8(2^7) experiment design, given in Table 3, was conducted in order to find the appropriate control factor levels for smooth rubber extruder output. The analysis is as follows.

Step 1. The seven significant factors with the experimented data are given in Table 3. The required weather strip's dimension (mm) is 350 ± 35 mm. Therefore, the upper allowed limit is *USL* = 385 mm and the lower allowed limit is *LSL*=315 mm.

Table 3: Experiment taguchi data for the quality field.

	Factors							Output for 30 Seconds									
N	A	B	C	D	E	F	G	1	2	3	4	5	6	7	8	9	10
1	1	1	1	1	1	1	1	268.4	262.9	268.0	262.2	265.1	259.1	261.5	267.4	264.7	270.2
2	1	1	1	2	2	2	2	302.9	295.3	298.6	302.7	314.4	305.5	295.2	286.3	302.0	299.2
3	1	2	2	1	1	2	2	332.7	336.5	332.8	342.3	332.2	334.6	334.8	335.5	338.2	326.8
4	1	2	2	2	2	1	1	221.7	215.9	219.7	221.2	221.5	230.1	228.3	228.3	214.6	213.2
5	2	1	2	1	2	1	2	316.6	326.5	320.4	327.0	311.4	310.8	314.4	319.3	310.0	314.5
6	2	1	2	2	1	2	1	211.3	222.0	218.2	218.6	218.6	216.5	214.8	217.4	210.8	223.9
7	2	2	1	1	2	2	1	210.7	210.0	211.6	211.7	210.1	206.5	203.4	207.2	208.0	219.3
8	2	2	1	2	1	1	2	287.5	299.2	310.6	289.9	290.0	294.5	294.2	297.4	293.7	325.6

Step 2. By using in Minitab the signal to noise ratio nominal best given by

$$S / N = 10\log(10)\left(\hat{\mu}^2 / \hat{\sigma}^2\right) \tag{5.6}$$

The signal to noise response Table and the mean response Table are

Table 4: S/N ratio nominal the best.

Level	A	B	C	D	E	F	G
1	**34.84**	**34.57**	32.95	**35.99**	**34.59**	32.85	**34.31**
2	32.70	32.96	**34.59**	31.55	32.95	**34.69**	33.23
Delta	2.14	1.61	1.63	4.44	1.64	1.84	1.08
Rank	**2**	6	5	**1**	4	**3**	7

Table 5: Response nominal the best.

Level	A	B	C	D	E	F	G
1	**280.3**	**274.9**	268.3	**281.6**	**278.8**	**275.4**	228.4
2	260.6	266.0	**272.6**	259.3	262.1	265.5	**312.5**
Delta	19.7	8.8	4.3	22.4	16.6	10.0	84.2
Rank	**3**	6	7	**2**	4	5	**1**

From Table 4 and Table 5, the factor levels which are closer to the weather strip requirement of 350 ± 35 mm are: *Setting 1*, (A1 B1 C2 D1 E1 F1 G2). And *Setting 2*, (A1 B1 C1 D1 E1 F1 G2). Therefore, by using the Taguchi polynomial model given by

$$T = \sum_{i=1}^{k} \hat{\mu}_i - (k-1)\mu \tag{5.7}$$

Where $\hat{\mu}_i$ is the mean of the corresponding factor's levels, and μ is the overall mean, the predicted mean and standard deviation of the *Setting 1* (A1 B1 C2 D1 E1 F1 G2) are μ = 353.415 and σ = 4.77323 mm, respectively. And to the *Setting 2*, (A1 B1 C1 D1 E1 F1 G2), they are μ = 349.135 and σ = 6.37765 mm. Thus, because from *Setting 2*, μ = 349.135 is closer to the nominal value of 350, and since from Equation (5.1) and Equation (5.2), its corresponding capability indices are cp = 1.83 and cpk = 1.98, which are close to six sigma performance (cp = 2, cpk = 1.67), Setting 2 (A1 B1 C1 D1 E1 F1 G2) is implemented.

Step 3. Suppose we found that two environmental noise factors (*Z1* and *Z2*) affect the selected Setting 2 process output.

Step 4. The central composite design and the corresponding experimented data for the environmental factors are given in Table 6. By using Minitab, the Anova analysis is given in Table 7. The fitted second order polynomial model is $Dim = 349.20 + 15\,Z_1 + 10\,Z_2 + 100Z_1 * Z_1 + 70Z_2 * Z_2 + 80Z_1 * Z_2$.

Table 6: Environment data.

No	Z1	Z2	Dim
1	-1	-1	574.20
2	1	-1	444.20
3	-1	1	434.20
4	1	1	624.20
5	-1.4142	0.0000	527.99
6	1.4142	0.0000	570.41
7	0.0000	-1.4142	475.06
8	0.0000	1.4142	503.34
9	0	0	353.00
10	0	0	356.00
11	0	0	344.00
12	0	0	341.00
13	0	0	352.00

Table 7: Anova analysis for setting 2.

Source	DF	Adj SS	Adj MS	F-Value	P-Value
Model	5	120723	24144.60	1038.16	0.000
Linear	2	2600	1300.00	55.90	0.000
Z1	1	1800	1800.00	77.40	0.000
Z2	1	800	800.00	34.40	0.001
Square	2	92523	46261.50	1989.13	0.000
Z1*Z1	1	69565	69565.00	2991.13	0.000
Z2*Z2	1	34087	34087.00	1465.66	0.000
2-Way	1	25600	25600.00	1100.74	0.000
Z1*Z2	1	25600	25600.00	1100.74	0.000
Error	7	163	23.30		
Lack-of-Fit	3	0	0.00	0.00	1.000
Pure Error	4	163	40.75		
Total	12	120886			

Step 5. From the fitted polynomial, the quadratic Qq matrix is Qq = [100 40; 40 70].

Step 6. From the determinant of Qq, the *Weibull stress* parameter is η_s = 73.4847 mm.

Step 7. From the Qq elements, the arithmetic mean is μ = 85 mm.

Step 8. From Equation (3.4), the eigenvalues of Qq are λ_1 = 127.72 mm and λ_2 = 42.28 mm.

Step 9. Suppose the desired reliability index is $R(t)$ = 0.9535. Thus, from Equation (2.7), n = 21.

Step 10. The Y_i elements, its mean μ_y and standard deviation σ_y are given in Table 8. Here, remember that, for n = 21, μ_y = –0.545624 and μ_y =1.175117 are both constant.

Step 11. From Equation (3.7), the Weibull shape parameter is β = 1.984692.

Table 8: Weibull analysis for the Quality field.

Equations														
(2.7)	(2.9)	(5.4)	(5.4)	(5.5)		(4.8)	(2.7)	(2.9)	(5.4)	(5.4)	(5.5)		(4.8)
N	Y_i	$tan(\Theta_i)$	Θ_i	λ_{1i}	λ_{2i}	$R(t_i)$	n	Y_i	$tan(\Theta_i)$	Θ_i	λ_{1i}	λ_{2i}	$R(t_i)$
1	-3.403	-1.7148	10.20	**408.28**	13.23	**0.9673**	11	-0.366	-0.1846	39.73	88.39	61.09	0.5000
2	-2.491	-1.2554	15.90	**257.89**	20.94	**0.9206**	12	-0.234	-0.1179	41.62	82.69	65.31	0.4533
3	-2.003	-1.0094	20.02	201.65	26.78	0.8738	13	-0.105	-0.0530	43.48	77.49	69.69	0.4065
4	-1.661	-0.8372	23.40	169.75	31.81	0.8271	14	0.021	0.0110	45.31	72.68	74.30	0.3598
5	-1.394	-0.7025	26.35	148.36	36.40	0.7804	15	0.149	0.0753	47.15	68.15	79.23	0.3131
6	-1.172	-0.5905	28.98	132.64	40.71	0.7336	16	0.279	0.1410	49.02	63.82	84.61	0.2664
	-1.097	**0.5753**	**29.91**	**127.72**	**42.28**	**0.7162**	17	0.415	0.2095	50.96	59.59	90.62	0.2196
7	-0.979	-0.4934	31.40	120.37	44.86	0.6869	18	0.562	0.2834	53.01	55.35	97.56	0.1729
8	-0.807	-0.4068	33.65	110.38	48.92	0.6402	19	0.727	0.3666	55.27	50.93	106.03	0.1262
9	-0.650	-0.3277	35.77	101.99	52.95	0.5935	20	0.929	0.4682	57.94	46.01	117.37	0.0794
10	-0.504	-0.2542	37.79	94.75	56.99	0.5467	21	1.229	0.6195	61.71	39.55	136.54	0.0327

Step 12. From steps 6 and 11, the *Weibull stress* family is W(β = 1.984692, η_s = 73.4847 mm).

Step 13. The basic Weibull elements are given in Table 8.

Step 14. The expected eigenvalues λ_{1i} and λ_{2i} elements are given in Table 8.

Step 15. From Equation (2.10), η_{sq} = 96.7367 mm. Therefore, the *Weibull-q* distribution is W(β = 1.984692, η_{sq} = 96.7367 mm).

On the other hand, notice that, as in Table 2, from Table 8 we have that a product with strength of 257.89 presents a reliability of *R(t)* = 0.9206. If the product has a strength of 408.28, then it will present a reliability of *R(t)* = 0.9673. Finally, it is important to mention that for several control factor settings, the Weibull parameters can be directly determined from the Taguchi analysis, as in [17]. Now, let us present the principal components field analysis.

6. Principal Components Field Analysis

The principal component analysis consists of determining the significant variables of the analyzed process. The selection is based on the magnitude of the eigenvalues of a variance and covariance matrix *Qc* [18]. Therefore, its diagonal elements contain the variance of the significant variables, and the elements out of the diagonal represent the covariance between the corresponding pair of significant variables. The *Qc* matrix is

$$Q_c = \begin{bmatrix} \sigma_1^2 & \sigma_1\sigma_2 \\ \sigma_2\sigma_1 & \sigma_2^2 \end{bmatrix} \tag{6.1}$$

On the other hand, in the case of the normal multivariate *T^2 Hotelling chart*, and in the case of the non-normal *R-chart*, all the analyzed output variables (Y_1, Y_2, ..., Y_k) must be correlated with each other. Hence, in the multivariate control field, the *Qc* matrix always exists. Thus, the decision-making process has to be performed based on the eigenvalues of the *Qc* matrix. However, first it is important to mention that, in the multivariate control process field, the *Qc* matrix is determined in such a way that it represents the process and customer requirements. Also, it is determined in the phase 1 of the multivariate control process [7]. In practice, this phase 1 is performed by using only conformant products. Therefore, the *Qc* matrix always represents the allowed variance and covariance expected behavior. For details of phase 1, see [7] Section 2. However, because the output process' variables are random, the Weibull distribution is used here to determine the random behavior of the eigenvalues of the *Qc* matrix. Now, let us give the steps to determine the *Weibull stress* and the *Weibull-q* families from the *Qc* matrix.

6.1 Steps to determine the corresponding weibull parameters

Step 1. For the process to be monitored, determine the set of quality and/or functional output variables to be controlled. The selected variables must be correlated each other. If they are not correlated, then use a univariate control chart to monitor each output variable in separated form.

Step 2. From at least 100 products, which fulfill with the required quality and/or functional requirements, collect the corresponding functional and quality measurements of the set of response variables of step 1.

Step 3. From the set of collected data of step 2, determine the variance and covariance matrix *Qc* defined in Equation (6.1), and the mean vector μ as

$$\mu = [\mu_1, \mu_2, \ldots] \tag{6.2}$$

Step 4. From the *Qc* matrix of step 3, determine the corresponding eigenvalues $\lambda_1, \lambda_2, \ldots, \lambda_k$. (here Mathlab was used).

Step 5. By using the maximum and the minimum eigenvalues of step 4, determine the corresponding *Weibull stress* parameter η_s as

$$\eta_s = \sqrt{\lambda_{\max}\lambda_{\min}} \tag{6.3}$$

Step 6. Determine the desired *R(t)* index, and by using it in Equation (2.7) determine the corresponding *n* value.

Step 7. Following steps 1 to 3 of Section 2.1, determine the Y_i elements and its mean μ_y and standard deviation σ_y. Here, remember that for $n = 21$, $\mu_y = -0.545624$ and $\mu_y = 1.175117$ are both constant.

Step 8. By using λ_{max} and λ_{min} from step 4, and from step 6 in Equation (6.4), determine the Weibull shape β_c parameter as

$$\beta_c = \frac{-4\mu_Y}{0.9973 * \ln(\lambda_{\max} / \lambda_{\min})} \tag{6.4}$$

Step 9. By using η_s from step 5 and β_c from step 8, form the principal components *Weibull stress* family $W(\beta_c, \eta_s)$.

Step 10. By using β_c the parameter of step 8, and the Y_i elements of step 7, determine the logarithm of the basic Weibull values as

$$\ln[\tan(\theta_i)] = \{Y_i / \beta_c\} \tag{6.5}$$

Step 11. From the logarithm basic Weibull values of step 10, determine the corresponding basic Weibull values as

$$\tan(\theta_i) = \exp\{Y_i / \beta_c\} \tag{6.6}$$

Step 12. By using the basic Weibull values of step 11 and the Weibull stress η_s parameter from step 5, determine the expected pair of eigenvalues λ_{max} and λ_{min} as

$$\lambda_{\max} = \eta_s / \tan(\theta_i) \quad \text{and} \quad \lambda_{\min} = \eta_s * \tan(\theta_i) \tag{6.7}$$

Step 13. By using η_s from step 5 and β_c from step 8 in Equation (2.15), determine the parameter. Then, form the corresponding *Weibull-q* distribution $W(\beta_c, \eta_s)$.
Now let us present the application.

6.2 Principal components application

Step 1. In the analysis, data given in [19] is used. Data represents a set of three correlated output process variables (Y_1, Y_2, ..., Y_k). Data is monitored by using the non-normal multivariate R-chart of Liu [20].

Step 2. Collected conformant data of phase I is given in Table 9.

Step 3. From Table 9 and Equations (6.1) and (6.2), the mean vector μ and Qc variance and covariance matrix are

$$\mu = [28.350186 \;\; 27.012524 \;\; 33.145115] \;\; Qc = \begin{bmatrix} 218.854738 & 36.7103257 & -12.891229 \\ & 270.749317 & -10.780954 \\ Sym & & 338.200441 \end{bmatrix}$$

Step 4. From Matlab the eigenvalues Qc are $\lambda_1 = 342.3$, $\lambda_2 = 286.2$ and $\lambda_3 = 199.3$.

Step 5. By using $\lambda_1 = 342.3$ and $\lambda_3 = 199.3$ in Equation (6.3), the *Weibull scale* parameter is $\eta_s = 261.1903$.

Step 6. The desired reliability for the analysis is $R(t) = 0.9535$, hence, from Equation (2.7), $n = 21$.

Step 7. The Y_i, μ_y and σ_y values are given in Table 10.

Step 9. The principal component *Weibull stress* family is W($\beta_c = 4.128837$, $\eta_s = 261.1903$).

Step 10. The logarithm of the basic tangent Weibull values is given in Table 10.

Step 8. From Equation (6.4), the Weibull shape parameter is $\beta_c = 4.128837$.

Step 11. The basic Weibull values are given in Table 10.

Step 12. The expected pair of λ_{max} and λ_{min} eigenvalues are given in Table 10.

Step 13. From Equation (2.15), the *Weibull time* scale parameter is $\eta_{sq} = 298.091$. Therefore, the *Weibull-q stress* distribution is W($\beta_c = 4.128837$, $\eta_{sq} = 298.091$).

As a summary of this section we have that, by using λ_{max} and λ_{min} eigenvalues, their random behavior can be determined using the Weibull distribution. Also, notice that the above Weibull analysis can be performed to any desired pair of eigenvalues of Table 10, or to any desired pair of eigenvalues of the analyzed Qc matrix. For example, following the steps above, the *Weibull stress* parameters to $\lambda_1 = 342.3$ and $\lambda_2 = 286.2$ are W($\beta_c = 12.476153$, $\eta_s = 312.9956$). And for $\lambda_2 = 286.2$ and $\lambda_3 = 199.3$, they are W($\beta_c = 6.171085$, $\eta_s = 238.8298$). However, notice that, by using λ_{max} and λ_{min}, we determine the maximum expected dispersion, as can be seen by observing that $\beta_c = 4.128837 < \beta_c = 6.171085 < \beta_c = 12.476153$). Finally notice from Table 10 that the higher eigenvalue of 595.6 has a cumulated probability of $F(t) = 1–0.9673 = 0.0327$ of being observed. Thus, the eigenvalues of the analyzed process should be monitored as in [21].

Now let us present the Vibration field analysis.

7. Vibration Field Analysis

In this section, the Weibull parameters which represent random vibration data are determined. Data corresponds to a sprung mass product attached to a car. The testing profile used is the one given in the test IV of the norm ISO16750-3. In the analysis, the following [22] three assumptions are made. (1) The first row in the testing's profile represents the most severe scenario in the test. (2) The whole test's profile is considered as a compact cycle which is repeated n times until failure occurs. (3) The generated damage in one cycle is cumulated from the first row of the testing's profile to the last.

Table 9: Data of a non-normal multivariate process.

No	Y_1	Y_2	Y_3	*No*	Y_1	Y_2	Y_3	*No*	Y_1	Y_2	Y_3	*No*	Y_1	Y_2	Y_3
1	43.8640	10.0494	33.1157	26	32.4801	17.1588	46.6231	51	43.7319	3.6033	21.1823	76	67.3737	66.0977	31.0451
2	26.2650	27.9573	61.9268	27	36.3008	19.6027	7.1347	52	40.9684	54.8082	21.5374	77	17.0418	16.4991	51.4294
3	10.6819	40.9290	52.3804	28	55.3878	5.7238	69.6335	53	32.7180	9.7179	41.4377	78	22.0765	48.3536	29.1045
4	21.2870	16.6687	8.1463	29	38.4896	37.4908	9.7581	54	33.7553	10.3044	34.2749	79	41.2681	16.7305	14.4449
5	19.8809	33.2707	30.6555	30	26.2484	43.7533	35.3334	55	25.4028	10.9298	23.6971	80	24.1916	2.4764	71.3878
6	14.9761	17.7642	38.6981	31	46.0915	32.5348	53.2886	56	52.6536	53.4061	13.7265	81	43.1579	29.8683	47.5383
7	11.5074	8.4100	24.7898	32	55.9161	40.9254	49.6362	57	17.0959	70.3467	14.8262	82	9.7300	19.2400	5.5050
8	42.7372	55.3353	21.3860	33	24.7075	25.8449	15.4058	58	24.2714	35.3896	47.1922	83	5.0324	2.6088	47.9323
9	13.9595	73.1556	76.0348	34	27.7981	42.7312	71.3045	59	12.9169	17.8995	30.6870	84	21.4039	2.1993	48.3945
10	21.7754	51.9827	27.6237	35	17.3796	12.2502	18.3954	60	18.9455	28.2305	34.9089	85	36.8481	19.6955	15.8172
11	20.0170	37.6447	74.3218	36	9.8233	32.1899	36.2559	61	49.5829	52.7012	24.4299	86	24.9993	29.8108	65.8987
12	30.6683	18.1314	15.8034	37	17.8899	5.5475	41.3209	62	30.6876	15.0500	30.4575	87	8.1575	35.3204	19.9059
13	1.9585	33.2475	32.3899	38	29.0605	21.1432	8.7047	63	5.6582	16.3089	17.6452	88	29.0943	20.4490	39.4786
14	34.5505	18.4945	59.5922	39	23.6911	61.2302	25.8783	64	27.2570	63.1206	23.3912	89	29.2487	6.4393	43.5499
15	19.3629	24.3424	65.7042	40	42.1966	28.0071	26.0839	65	23.9339	20.1540	30.9628	90	50.7453	43.8424	7.5473
16	17.9148	8.6084	24.1414	41	36.4545	36.2384	33.1884	66	25.0717	11.6396	43.1879	91	35.9559	31.6888	76.9695
17	14..3565	26.5591	13.1767	42	27.4647	25.2031	6.5683	67	26.3030	10.1967	27.9118	92	21.4835	30.4616	43.6526
18	30.8967	20.2957	12.2357	43	11.2996	21.2611	31.2990	68	33.1910	8.2672	31.3328	93	45.1344	30.2467	38.5232
19	43.0528	29.1823	14.7479	44	68.6678	12.3167	29.7939	69	24.2425	37.4854	13.0109	94	31.0069	23.1948	45.0889
20	32.8374	7.4563	32.3237	45	46.4753	15.2157	44.6967	70	35.0409	32.6236	10.7002	95	3.6266	30.9451	61.7411
21	12.8864	57.8727	5.7250	46	17.0357	22.3405	55.3367	71	46.4863	31.6622	43.4032	96	29.8396	22.3479	65.3039
22	68.6678	57.8671	12.7594	47	30.7515	25.3002	41.9358	72	42.1082	23.1356	13.8176	97	4.3065	17.5413	24.7333
23	25.1271	2.8157	28.7178	48	5.1723	19.2992	36.3460	73	50.9111	34.8608	46.2676	98	15.9923	6.9033	9.7309
24	24.4596	20.2257	10.9644	49	27.6552	36.1258	43.3015	74	42.9994	22.2937	28.2687	99	10.4957	28.7806	33.2900
25	4.8921	16.2747	14.3619	50	11.1145	48.9268	43.8018	75	25.1423	16.5281	11.4357	100	41.5974	25.9443	24.0576

Table 10: Weibull analysis for the principal components field.

Equations													
(2.7)	(2.9)	(6.5)	(6.6)	(6.7)		(4.8)	(2.7)	(2.9)	(6.5)	(6.6)	(6.7)		(4.8)
N	Y_i	$ln(tang(\Theta_i))$	$tang(\Theta_{ij})$	λ_{max}	λ_{min}	$R(t_i)$	n	Y_i	$ln(tang(\Theta_i))$	$tang(\Theta_{ij})$	λ_{max}	λ_{min}	$R(t_i)$
1	-3.403	-0.8243	0.4385	595.60	114.54	0.9673	11	-0.366	-0.0887	0.9150	285.43	239.00	0.5000
2	-2.491	-0.6034	0.5469	477.57	142.84	0.9206	12	-0.234	-0.0567	0.9448	276.42	246.79	0.4533
3	-2.003	-0.4852	0.6155	424.31	160.77	0.8738	13	-0.105	-0.0255	0.9748	267.93	254.61	0.4065
4	-1.661	-0.4024	0.6686	390.60	174.65	0.8271	14	0.021	0.0053	1.0053	259.80	262.58	0.3598
5	-1.394	-0.3377	0.7133	366.12	186.33	0.7804	15	0.149	0.0362	1.0368	251.90	270.82	0.3131
6	-1.172	-0.2838	0.7528	346.92	196.64	0.7336	16	0.279	0.0677	1.0701	244.07	279.50	0.2664
	-1.116	**-0.2704**	**0.7630**	**342.30**	**199.30**	**0.7208**	17	0.415	0.1007	1.1059	236.15	288.87	0.2196
7	-0.979	-0.2372	0.7888	331.11	206.03	0.6869	18	0.562	0.1362	1.1459	227.92	299.31	0.1729
8	-0.807	-0.1955	0.8223	317.60	214.79	0.6402	19	0.727	0.1762	1.1927	218.98	311.52	0.1262
9	-0.650	-0.1575	0.8542	305.75	223.11	0.5935	20	0.929	0.2250	1.2524	208.54	327.12	0.0794
10	-0.504	-0.1221	0.8849	295.13	231.14	0.5467	21	1.229	0.2978	1.3469	193.91	351.80	0.0327

Based on these assumptions, the steps to determine the corresponding vibration *Weibull stress* and *Weibull testing time* families are as follows.

7.1 Steps to determine the weibull families

Step 1. For the analyzed product, determine its location in the car and the norm to be used. And based on the location, determine the testing's type to be applied. Here, the ISO16750-3 norm [23] is used.

Step 2. From the selected testing's type, determine the testing's profile parameters; frequency (Hz), energy ((rms^2=m/s^2)^2/Hz), and testing time t (hrs).

Step 3. Determine the desired reliability $R(t)$, and from Equation (2.7) determine the corresponding sample size n to be tested.

Step 4. Following steps 1 to 3 of Section 2.1, determine the Y_i elements, its mean μ_y and its standard deviation σ_y. Here, remember that for $n = 21$, $\mu_y = -0.545624$ and $\mu_y = 1.175117$ are both constant.

Step 5. Take the product of the applied frequency and energy from the first row of the testing's profile of step 2 as the minimum eigenvalue

$$\lambda_{\min} = f_1 * G_1 \tag{7.1}$$

Step 6. Take the total cumulated energy as the maximum eigenvalue, given by

$$\lambda_{\max} = \sum_{i=1}^{k} A_i \tag{7.2}$$

Where A_i represents the area of the *th*-row of the testing's profile, given as

$$A_i = 10\log(2)\frac{APSD_i}{10\log(2)+m}\left[f_i - f_{i-1}\left(\frac{f_{i-1}}{f_i}\right)^{m/10\log(2)}\right] \tag{7.3}$$

where $APSD_i$ is the applied energy and f_i is the frequency of the *i-th* row of the used testing's profile, *f(i-1)* is the frequency of the *(th-1)*-row of the testing's profile, and m is the slope given as

$$m = dB/octaves \tag{7.4}$$

where

$$dB = 10\log(APSD_i / APSD_{i-1}) \tag{7.5}$$

$$octaves = \log(f_i / f_{i-1}) / \log(2) \tag{7.6}$$

Step 7. By using the μ_y value from step 4, and the addressed λ_{min} and λ_{max} values from steps 5 and 6, determine the corresponding Weibull vibration β_v parameter as

$$\beta_v = \frac{-4\mu_Y}{0.9947 * \ln(\lambda_{\max} / \lambda_{\min})} \tag{7.7}$$

Step 8. By using the testing's time of step 2, the $R(t)$ index from step 3, and the β_v value from step 7, determine the corresponding *Weibull-q* scale parameter as

$$\eta_{tq} = \frac{t_i}{\exp\left\{\dfrac{\ln(-\ln(R(t)))}{\beta_v}\right\}} \tag{7.8}$$

Note 8. The *Weibull testing time* family is W(β_v, η_{tq}).

Step 9. By using the square root of the A_i value of step 6, the reliability index from step 3, and the β_v value from step 7, determine the corresponding *Weibull stress* vibration scale parameter as

$$\eta_s = \frac{\sqrt{A_i}}{\exp\left\{\frac{\ln(-\ln(R(t)))}{\beta_v}\right\}} \tag{7.9}$$

Note 9. The addressed *Weibull stress* family is W(β_v, η_{tq}).

Step 10. By using the β_v parameter of step 7, and the Y_i elements of step 4, determine the logarithm of the basic Weibull values as

$$\ln[\tan(\theta_i)] = \{Y_i / \beta_v\} \tag{7.10}$$

Step 11. From the logarithm of the basic Weibull values of step 10, determine the corresponding basic Weibull values

$$\tan(\theta_i) = \exp\{Y_i / \beta_v\} \tag{7.11}$$

Step 12. By using the basic Weibull values of step 11 and the η_{tq} parameter from step 8, determine the expected testing times as

$$t_i = \eta_{tq} * \tan(\theta_i) \tag{7.12}$$

Step 13. By using the basic Weibull values of step 11 and the η_s parameter from step 8, determine the expected vibration levels as

$$S_{\text{i}} = \eta_s * \tan(\theta_i) \tag{7.13}$$

Now let us present the application.

7.2 Vibration application

In the application, the vibration test IV given in the ISO16750-3 norm is used. The ISO16750-3 norm applies to electric and electronic systems/components for road vehicles, and because in the vehicle the vibration stress can occur together with extremely low or high temperatures, the vibration test is generally performed with a superimposed temperature cycle test. For details of temperature cycle test see the ISO16750-4 norm and the appendices F and G of the guide of the norm GMW3172 [24]. The numerical analysis is as follows.

Step 1. The analyzed product is a sprung mass mounted in a passenger car. Therefore, the testing's type to be applied is the random vibration test IV given in the ISO16750-3 norm Section 4.1.2.4.

Step 2. For test IV, the product must be tested by 8 hrs in each one of the X, Y and Z directions. Thus, the total experimental testing time is t = 24 hrs. The testing frequencies and their corresponding applied energy are given in Table 11. And the corresponding testing's profile is plotted in Figure 5.

Step 3. The desired reliability to be demonstrated is $R(t)$ = 0.97. Therefore, from Equation (2.7) n = 32.8308 ≈ 33 parts must be tested.

Table 11: Random vibration profile.

Freq(Hz)	ASD(G2/Hz)	dB	Oct	dB/Oct	Area	Grms
10.00	30.0000	*	*	*	**300.00**	17.32
400.00	0.2000	-21.76	5.32	-4.09	614.00	24.78
1000.00	0.2000	0.00	1.32	0.00	**734.00**	27.09

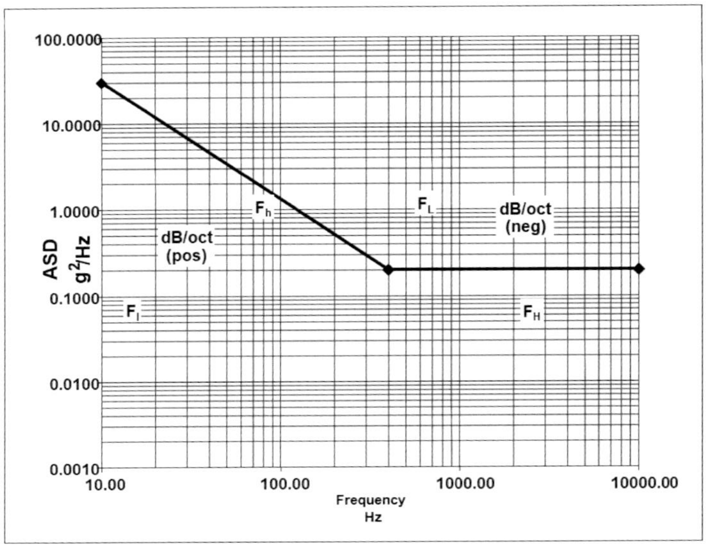

Figure 5: Testing Vibration Profile.

Step 4. The Y_i, μ_y and σ_y values are given in Table 12.

Step 5. From the first row of Table 11, the minimum eigenvalue is λ_{min} = 300 rms^2.

Step 6. From Equation (7.3) the maximum eigenvalue is λ_{max} = 734 rms^2.

Step 7. By using the μ_y value from step 4, and the addressed λ_{min} and λ_{max} values from steps 5 and 6 in Equation (7.7), the Weibull vibration parameter is $\beta_v = 2.5$.

Step 8. From Equation (7.8), η_{vt} = 96.9893 hrs. Therefore, the *Weibull testing time* family is W(β_v= 2.5, η_{vt} = 96.9893 hrs).

Step 9. From Equation (7.9), η_{vs} = 11.1538 Grms. Therefore, the *Weibull stress* vibration family is W(β_v =2.5, η_{vs} = 11.1538 Grms).

Step 10. The logarithm of the basic Weibull data is given in Table 12.

Step 11. The basic Weibull data is given in Table 12.

Step 12. The expected testing times are given in Table 12.

Step 13. The expected vibration levels are given in Table 12.

From Table 12, the row for t = 24 hrs and vibration level of 2.76 Grms, on which the analysis was performed, was added. The row between rows 5 and 6 was added to show that the above analysis can be used as an accelerated life time analysis [24]. For example, from this row, we have that *R(t)* = 0.97 can also be demonstrated by testing 6 parts at constant vibration level of 2.76 Grms, each for 47.366 hrs (we must test each part in the X, Y and Z axes for 47.366/3 = 15.7885 hrs each). This is true due to the fact that the total Weibull testing time (T_a) is the same for any n_i and t_i elements of Table 12. The total testing time is given as

$$T_a = n_i * t_i^{\beta v} \tag{7.14}$$

And since T_a is the same for any row of Table 12, then the *Weibull testing time* scale parameter is also the same for any n_i and t_i elements of Table 12. And since it is

Table 12: Weilbull testing time and in vabration analysis.

	Equations									Equations							
	(2.8)	(2.9)	(4.8)	(2.7)	(7.10)	(7.11)	(7.12)	(7.13)		(2.8)	(2.9)	(4.8)	(2.7)	(7.10)	(7.11)	(7.12)	(7.13)
n	$F(t_i)$	Y_i	$R(t_i)$	n_i	$ln(tan(\Theta_i))$	$tan(\Theta_i)$	t_i	S_i	n	$F(t_i)$	Y_i	$R(t_i)$	n_i	$ln(tan(\Theta_i))$	$tan(\Theta_i)$	t_i	S_i
1	0.021	-3.855	0.979	47.213	-1.542	0.214	20.754	2.387	17	0.500	-0.367	0.500	1.443	-0.147	0.864	83.763	9.633
	0.030	**-3.491**	**0.970**	**32.831**	**-1.397**	**0.247**	**24.000**	**2.760**	18	0.530	-0.281	0.470	1.325	-0.112	0.894	86.672	9.967
2	0.051	-2.952	0.949	19.143	-1.181	0.307	29.780	3.425	19	0.560	-0.198	0.440	1.218	-0.079	0.924	89.619	10.306
3	0.081	-2.473	0.919	11.863	-0.989	0.372	36.061	4.147	20	0.590	-0.115	0.410	1.122	-0.046	0.955	92.620	10.651
4	0.111	-2.142	0.889	8.517	-0.857	0.425	41.172	4.735	21	0.620	-0.034	0.380	1.034	-0.013	0.987	95.694	11.005
5	0.141	-1.886	0.859	6.594	-0.754	0.470	45.611	5.245	22	0.650	0.048	0.350	0.953	0.019	1.019	98.862	11.369
	0.154	-1.792	0.846	6.000	-0.717	0.488	47.366	5.447	23	0.680	0.130	0.320	0878	0.052	1.053	102.148	11.747
6	0.171	-1.676	0.829	5.344	-0.670	0.512	49.611	5.705	24	0.710	0212	0.290	0.809	0.085	1.089	105.582	12.142
7	0.201	-1.497	0.799	4.466	-0.599	0.550	53.302	6.130	25	0.740	0.297	0.260	0.743	0.119	1.126	109.205	12.559
8	0.231	-1.339	0.769	3.816	-0.536	0.585	56.766	6.528	26	0.769	0.383	0.231	0.682	0.153	1.166	113.067	13.003
9	0.260	-1.198	0.740	3.314	-0.479	0.619	60.060	6.907	27	0.799	0.474	0.201	0.622	0.190	1209	117.239	13.482
10	0.290	-1.070	0.710	2.915	-0.428	0.652	63.224	7.271	28	0.829	0.570	0.171	0.566	0.228	1.256	121.822	14.009
11	0.320	-0.951	0.680	2.589	-0.381	0.683	66.289	7.623	29	0.859	0.673	0.141	0.510	0.269	1.309	126.974	14.602
12	0.350	-0.841	0.650	2.319	-0.336	0.714	69.281	7.967	30	0.889	0.789	0.111	0.454	0.315	1.371	132.957	15.290
13	0.380	-0.737	0.620	2.090	-0.295	0.745	72.218	8.305	31	0.919	0.922	0.081	0.398	0.369	1.446	140.268	16.131
14	0.410	-0.639	0.590	1.894	-0.256	0.775	75.120	8.639	32	0.949	1.091	0.051	0.336	0.436	1.547	150.068	17.258
15	0.440	-0.545	0.560	1.724	-0.218	0.804	78.002	8.970	33	0.979	1.352	0.021	0.259	0.541	1.717	166.569	19.155
16	0.470	-0.454	0.530	1.575	-0.182	0.834	80.878	9.301									

$$\eta_{vt} = \sqrt[\beta v]{T_a} = n_i^{\beta v} * t_i \tag{7.15}$$

Then the reliability function defined in Equation (4.8) in terms of Equation (7.14) is also given as

$$R(t) = \exp\left\{-\left(\frac{t}{\sqrt[\beta v]{Ta}}\right)^{\beta v}\right\} \tag{7.16}$$

On the other hand, because the relation between the applied vibrations S_i and the testing times t_i for any two rows of Table 12 always holds, from the relation

$$\left(\frac{S_{accel}}{S_{norm}}\right) = \left(\frac{t_{norm}}{t_{accel}}\right) \tag{7.17}$$

we can use the S_i column of Table 12 as the accelerated vibration level for any desired sample size value. For example, we can demonstrate $R(t) = 0.97$ by testing 6 parts for 24 hrs each at constant vibration level of 5.447 Grms. (we must test each part in the X, Y and Z axes by 8 hrs each). The accelerated test parameters and its corresponding testing's profile are given in Table 13 and in Figure 6.

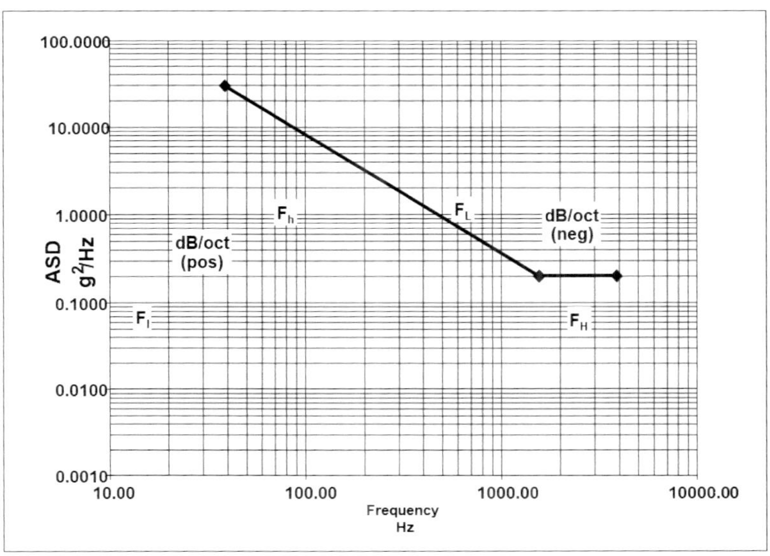

Figure 6: Accelerated Testing Profile.

Table 13: Accelerated random profile.

Freq(Hz)	ASD(G2/Hz)	dB	Oct	dB/Oct	Area	Grms
38.95	30.0000	*	*	*	**1168.49**	34.18
1557.99	0.2000	-21.76	5.32	-4.09	2391.50	48.90
3894.97	0.2000	0.00	1.32	0.00	**2858.89**	53.47

For deeper Weibull/vibration analysis see [24]. Now let us present the relations between the Weibull parameters and the cycling formulas which can be used to perform the corresponding fatigue analysis.

8. Weibull Fatigue Formulation

In this section, the objective is to present the formulas which can be used to perform a fatigue analysis [25]. The analysis is made by considering that the applied stress has the general cyclical behavior given in Figure 7. The given formulas are derived from the Weibull/stress theory given in [9]. Hence, readers interested in deeper analysis are invited to consult [9]. In the analysis, the stress data given in Section 4.2 is used to present the numerical analysis. The steps to determine the *Weibull stress* and *Weibull time* parameters for fatigue analysis are as follows.

8.1 Steps to determine the weibull stress and time fatigue analysis

Step 1. From the applied normal σ_x and σ_y and shear τ_{xy} stresses values, determine the arithmetic mean stress as

$$\mu = (\sigma_x + \sigma_x)/2 \tag{8.1}$$

Step 2. By using the applied normal σ_x and σ_y and shear τ_{xy} stresses from step 1, determine the fatigue *Weibull stress* parameter as

$$\eta_f = \sqrt{\sigma_x \sigma_y - \tau_{xy}^2} \tag{8.2}$$

Step 3. By using the mean μ and the η_f values from steps 1 and 2, determine the maximum (σ_1) and the minimum (σ_2) principal stress values as

$$\sigma_1, \sigma_2 = \mu \pm \sqrt{\mu^2 - \eta_f^2} \tag{8.3}$$

Step 4. By using the principal stress values from step 3, determine the alternating stress as

$$S_a = (\sigma_1 - \sigma_2)/2 \tag{8.4}$$

Step 5. Following steps 1 to 3 of Section 2.1, determine the Y_i elements and its mean μ_y and standard deviation σ_y.

Step 6. By using the μ_y value from step 5 and the principal stress values from step 3 in Equation (3.7) determine the corresponding Weibull shape parameter β.

Note 10: The η_f parameter from step 2 and the β parameter from this step are the *Weibull stress* family W(β, η_{tf}).

Step 7. By using μ_y from step 5 and the β value from step 6 in Equation (2.10), determine the *Weibull time* scale parameter η_{tf}. Thus, the *Weibull time* family is W(β, η_{tf}).

Step 8. By using the β parameter from step 6 and the Y_i elements from step 5, determine the logarithm of the basic Weibull elements as

$$\ln[\tan(\theta_i)] = \{Y_i / \beta\} \tag{8.5}$$

Step 9. From the logarithm of the basic Weibull values of step 8, determine the corresponding basic Weibull values as

$$\tan(\theta_i) = \exp\{Y_i / \beta\} \tag{8.6}$$

Step 10. By using the basic Weibull values of step 9 and η_f the value from step 2, determine the expected maximum and minimum stresses values as

$$S_{1i} = \eta_f / \tan(\theta_i),\ S_{2i} = \eta_f * \tan(\theta_i) \tag{8.7}$$

Step 11. Determine the basic Weibull value which corresponds to the principal stress values of step 3 as

$$\tan(\theta_{\lambda 1,\lambda 2}) = \sqrt{\lambda_2 - \lambda_1} \tag{8.8}$$

Step 12. By using the expected stress values of step 10, determine the corresponding Weibull angle as

$$\theta_i = \tan^{-1}\left(\sqrt{S_{2i} / S_{1i}}\right) \tag{8.9}$$

Step 13. Determine the reliability index which corresponds to the principal stress values as

$$R(t) = \exp\left\{-(\sqrt{\lambda_2 - \lambda_1})^{\beta s}\right\} \tag{8.10}$$

Now let us present the application.

8.2 Weibull/Fatigue application

The application is based on the stress data of Section 4.2.

Step 1. The applied stresses values are $\sigma_x = 90$ mpa, $\sigma_y = 190$ mpa and $\tau_{xy} = 80$ mpa. Therefore, from Equation (8.1), $\mu = 140$ mpa.

Step 2. From Equation (8.2), the *Weibull stress* parameter is $\eta_f = 103.44$ mpa.

Step 3. From Equation (8.3), the principal stresses are $\sigma_1 = 234.34$ mpa and $\sigma_2 = 45.66$ mpa.

Step 4. From Equation (8.4), the alternating stress is $S_a = 94.34$ mpa.

Step 5. From Equation (2.9), the Y_i elements as well as its mean μ_y and its standard deviation σ_y are both given in Table 14.

Step 6. From Equation (3.7) the Weibull shape parameter is $\beta = 1.3316$. Therefore, the Weibull stress family is W($\beta = 1.3316$, $\eta_f = 103.44$ mpa).

Step7. From Equation (2.10), η_{tf} = 155.8244 mpa. Therefore, the Weibull time family is W($\beta = 1.3316$, $\eta_f = 155.8244$ mpa).

Step 8. From Equation (8.5), the logarithm of the basic Weibull elements are given in Table 14.

Step 9. From Equation (8.6), the basic Weibull elements are given in Table 14.

Step 10. From Equation (8.7), the expected stress values are given in Table 14.

Step 11. From Equation (8.8), the basic Weibull value which corresponds to the principal stresses of step 3 is $\theta_{\lambda 1,\lambda 2} = 0.441413$.

Step 12. From Equation (8.9), the corresponding Weibull angles are given in Table 14.

Step13. From Equation (8.10), the reliability which corresponds to the principal stresses of step 3 is *R(t)*=0.7142.

Table 14: Weibull fatigue analysis for mechanical data of Section 4.2.

Equations							
(2.7)	(2.9)	(8.5)	(8.6)	(8.7)		(8.9)	
N	Y_i	$ln(tan(\theta_i))$	$tan(\theta_i)$	σ_{1i}	σ_{2i}	$Ang(\theta_i)$	$R(t_i)$
1	-3.4035	-2.5558	0.0776	1332.5045	8.0300	4.439	0.9673
2	-2.4917	-1.8711	0.1540	671.8874	15.9252	8.752	0.9206
3	-2.0035	-1.5045	0.2221	465.6722	22.9775	12.524	0.8738
4	-1.6616	-1.2478	0.2871	360.2496	29.7015	16.021	0.8271
5	-1.3944	-1.0471	0.3509	294.7447	36.3025	19.339	0.7804
6	-1.1721	-0.8801	0.4147	249.4209	42.8992	22.525	0.7336
	-1.0890	**-0.8178**	**0.4414**	**234.3400**	**45.6600**	**23.817**	**0.7142**
7	-0.9794	-0.7355	0.4793	215.8224	49.5776	25.608	0.6869
8	-0.8074	-0.6063	0.5453	189.6810	56.4103	28.605	0.6402
9	-0.6505	-0.4885	0.6136	168.5917	63.4668	31.531	0.5935
10	-0.5045	-0.3789	0.6846	151.0868	70.8200	34.397	0.5467
11	-0.3665	-0.2752	0.7594	136.2141	78.5526	37.213	0.5000
12	-0.2341	-0.1758	0.8388	123.3234	86.7635	39.989	0.4533
13	-0.1053	-0.0791	0.9240	111.9509	95.5773	42.737	0.4065
	0.0000	**0.0000**	**1.0000**	**103.4406**	**103.4406**	**45.000**	**0.3679**
14	0.0219	0.0165	1.0166	101.7512	105.1581	45.472	0.3598
15	0.1495	0.1123	1.1188	92.4541	115.7327	48.210	0.3131
16	0.2798	0.2101	1.2339	83.8350	127.6312	50.976	0.2664
17	0.4160	0.3124	1.3667	75.6891	141.3673	53.806	0.2196
18	0.5625	0.4224	1.5256	67.8020	157.8119	56.756	0.1729
19	0.7276	0.5464	1.7270	59.8955	178.6440	59.928	0.1262
20	0.9293	0.6979	2.0094	51.4772	207.8582	63.543	0.0794
	1.0890	**0.8178**	**2.2655**	**45.66**	**234.34**	**66.183**	**0.0512**
21	1.2297	0.9234	2.5178	41.0830	260.4473	68.339	0.0327
μy = -0.545624		σy = 1.175117					

From Table 14, we observe that

(1) the reliability of the addressed principal stresses σ_1 = 234.34 mpa and σ_1 = 45.66 mpa is $R(t)$ = 0.7142. However, notice that this reliability corresponds to a component that presents a Weibull strength parameter with η_s = 234.34 mpa. Therefore, if the Weibull component's strength is of η_s = 1332.5045 mpa, then its reliability is $R(t)$ = 0.9673.

(2) The cyclical stress behavior is shown in Figure7, in Table 14 it is given between rows 6 and 21. Thus, as it is shown from row 1 to row 6 of Table 14, it is expected that higher stresses values will occur. For example, a stress of 1332.50 mpa will occur with cumulated probability of $F(t)$ = 1–0.9673 = 0.0327.

(3) From the columns σ_{1i} and σ_{2i} we have that the addressed Weibull stress distribution, W(β = 1.3316, η_f= 103.44 mpa) completely models the principal stresses random behavior. Also, since from [9] the Weibull distribution can be represented by a circle centered in the arithmetic mean μ, the Weibull distribution can effectively be used to model fatigue data, as in Figure 7, and to cumulate the generated damage as in [26].

(4) As can be seen from Table 14, since the *Weibull stress* scale η_f parameter always occurs at an angle of 45°, and due to the ratio between η_f and σ_1 and σ_2 being the same for both principal stresses as

$$R = \frac{\sigma_1}{\eta_f} = \frac{\eta_f}{\sigma_2} \tag{8.11}$$

the Weibull analysis given in Table 14 can be used to perform the corresponding modified Goodman diagram to determine the threshold between finite and infinite life.

(5) From Equation (8.11), by using the λ_i column of Table 14 as the Weibull strength scale parameter η_s, we can determine the minimum material's strength (say the yields S_y value) which we must select in order for the designed component to present the desired reliability. Based on the selected η_s value, the minimum and maximum strength S_y values to be selected from an engineering handbook are given as

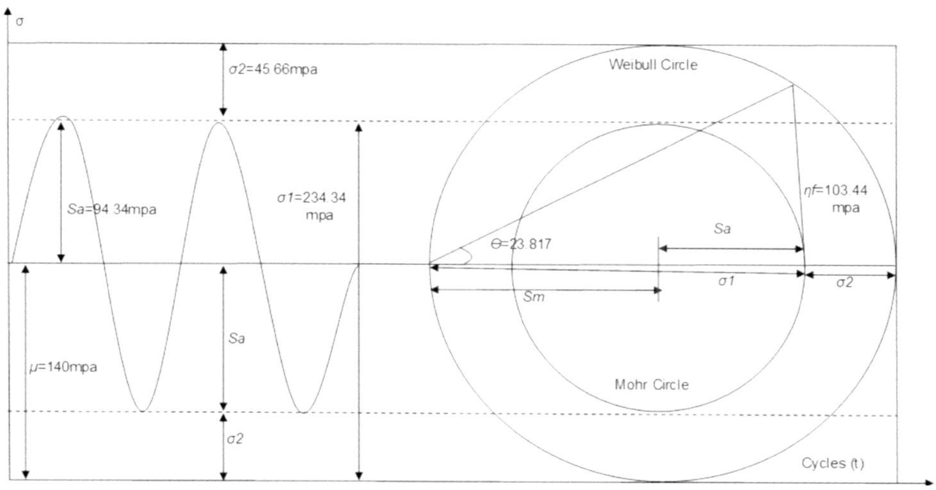

Figure 7: Weilbull/Mohr circle representation.

$$S_{y\min} = \frac{\sigma_2 \eta_S}{\eta_f}, \quad S_{y\max} = \frac{\eta_f \eta_S}{\sigma_2} \tag{8.12}$$

For example, suppose we want to design an element with a minimum reliability of $R(t)$ = 0.9673, thus, from Table 14, the corresponding value is η_s = 1332.5045 mpa, and since Table 14 was constructed with η_f = 103.44 mpa, then, by using these values with the minimal stress of σ_2 = 45.66 mpa in Equation (8.12), the minimum material's strength to be selected from a engineering handbook is S_y = 588.1843 mpa.

9. Conclusions

(1) When a quadratic form represents an optimum (maximum or minimum), its random behavior can be modeled by using the two parameter Weibull distribution.

(2) In the proposed Weibull analysis, the main problem consists of determining the maximum and the minimum stress (λ_1 and λ_2) values that generate the failure. However, once they are determined, both the Weibull stress and the Weibull time families are determined. Therefore, the stress values used in Equation (3.7) *must be those stresses values that generate the failure*. Here, notice that the constant 0.9947 value used in Equation (3.7) was determined only for the given application. The general method to determine it for any λ_1 and λ_2 values is given in [9] Section 4.1.

(3) The columns σ_{1i} and σ_{2i} are the maximum and minimum expected stress values which generate the failure, and thus, σ_{1i} represents the minimum Weibull strength value that the product must present to withstand the applied stress. Therefore, column σ_{1i} can be used as a guide to select the minimal strength material as it is given in Equations (4.9) and (8.12).

(4) From Table 12, the columns $R(t_i)$, n_i, t_i, and S_i can be used to design any desired accelerated testing scenario. For example, suppose we want to demonstrate $R(t_i)$ = 0.9490, then from Table 12 we have that by fixing the testing time t = 24 hrs, we can test n = 19.143 (testing 18 parts by 24 hrs each one and one part by 1.143*24 hrs) at a constant stress of S_i = 3.425 Grms.

(5) In any Weibull analysis, the n value addressed in Equation (2.7) is the key variable in the analysis. This because for the used β value it always let us to determine the basic Weibull elements as $\tan(\theta_i) = 1/n_i^{1/\beta}$. This fact can be seen by combining Equation (2.7) and Equation (4.8) or directly from Equations (43) and (53) in [9].

(6) From the applications, we have that although they appear to be very different, because all of them use a quadratic form in their analysis, they can all be analyzed by using the Weibull distribution. Generalizing, we believe that the Weibull distribution can always be used to model the random behavior of any quadratic form when the cumulated damage process can be modeled by using an additive damage model, such as that given in [27]. When the damage is not additive, then the log-normal distribution, which is based on the Brownian motion [28], could be used.

(7) Finally, it is important to mention that by using the maximum and minimum applied stresses, the given theory could be used in the contact ball bearing analysis [29] to determine the corresponding Weibull shape parameter and to determine the corresponding Weibull stress scale parameter from the equivalent stress and/or from the corresponding dynamic load as it is given in [30]. Similarly, the Weibull time scale parameter can be determined from the desired L10 life proposed by [31].

References

[1] Weibull, W. 1939. A statistical theory of the strength of materials. Proceedings, R Swedish Inst. Eng. Res. 151: 45.

[2] Rinne, H. 2009. The Weibull distribution a handbook. CRC PRESS. ISBN-13:978-1-42008743-7; http: //dx.doi.org/10.1201/9781420087444.

[3] Montgomery, D.C. 2004. Design and Analysis of Experiments. Limusa Wiley, New York, USA. ISBN. 968-18-6156-6.

[4] Box, G.E.P. and Draper, N.R. 1987. Empirical Model-Biulding and Response Surfaces, Wiley, New York USA. ISBN-13. 978-0471810339.

[5] Steven, R. Schmid, Bernard J. Hamrock and Bo O. Jacobson. 2014. Fundamentals of Machine Elements SI Version Third Edition. Taylor and Francis Group. Boca Raton Fl. ISBN-13: 978-1-4822-4750-3 (eBook - PDF).

[6] Kay Yang and Basem El-Haik. 2003. Design for Six Sigma; A Roadmap for Product Development. McGraw-Hill. ISBN-0-07-141208-5.

[7] Piña-Monarrez, M.R. 2013. Practical Decomposition Method for T^2 Hotelling Chart. International Journal of Industrial Engineering Theory Applications and Practice. 20(5-6): 401–411.

[8] Howard Anton, Irl Bivens, Stephen Davis. Calculus: Early Transcendentals Combined. Somerset, New Jersey, 8th Edition, 2005. ISSN-13:978-0471472445.

[9] Piña-Monarrez, M.R. 2017. Weibull Stress Distribution for Static Mechanical Stress and its Stress/strength Analysis. Qual Reliab Engng Int. 2018; 34: 229–244. DOI:10.1002/qre.2251.

[10] Mischke, C.R. 1979. A distribution-independent plotting rule for ordered failures. Journal of Mechanical Design; 104: 593–597. DOI: 10.1115/1.3256391.

[11] Piña-Monarrez, M.R., Ramos-López, M.L., Alvarado-Iniesta, A, Molina-Arredondo, R.D. 2016. Robust sample size for Weibull demonstration test plan. DYNA.; 83: 52–57.

[12] Piña-Monarrez, M.R. 2016. Conditional Weibull control charts using multiple linear regression. Qual Reliab Eng Int.; 33: 785–791. https://doi.org/10.1002/qre.2056.

[13] Taguchi, G., Subir, C. and Wu, Y. 2005. Taguchi's Quality Engineering Handbook. John Wiley and Soons. ASI Consulting Group, LLC. Livonia, Michigan. ISBN: 0-471-41334-8.

[14] Kececioglu, D.B. 2003. Robust Engineering Design-By-Reliability with Emphasis on Mechanical Components and Structural Reliability. Pennsylvania: DEStech Publications Inc. ISBN:1-932078-07-X.

[15] Budynas, N. 2006. Shigley's Mechanical Engineering Design. 8th ed. New York: McGraw-Hill.

[16] Piña-Monarrez, M.R., Ortiz-Yañez, J.F., Rodríguez-Borbón, M.I. 2015. Non-normal capability indices for the Weibull and lognormal distributions. Qual Reliab Eng Int.; 32: 1321–1329. https://doi.org/10.1002/qre.1832.

[17] Piña-Monarrez, M.R. and Ortiz-Yañez, J.F. 2015. Weibull and Lognormal Taguchi Analysis Using Multiple Linear Regression. Reliab Eng. Syst. Saf; 144: 244–53. doi:10.1016/j.ress.2015.08.004.

[18] Peña, D. 2002. Análisis de Datos Multivariantes, Mc Graw Hill. ISBN: 84-481-3610-1.

[19] Piña-Monarrez, M.R. 2018. Generalization of the Hotelling´s T^2 Decomposition Method to the R-Chart. International Journal of Industrial Engineering, 25(2): 200–214.

[20] Liu, R.Y. 1995. Control Charts for Multivariate Processes. Journal of the American Statistical Association. 90(432): 1380–1387.

[21] Piña-Monarrez, M.R. 2019. Probabilistic Response Surface Analysis by using the Weibull Distribution. Qual Reliab Eng Int. 2019; in Press.

[22] Piña-Monarrez, M.R. 2019. Weibull Analysis for Random Vibration Testing. Qual Reliab Eng Int. 2019; in Press.

[23] SS-ISO16750-3:(2013). Road vehicles – Environmental conditions and testing for electrical and electronic equipment – Part 3: Mechanical loads (ISO 16750-3:2012, IDT) A https://www.sis.se/api/document/preview/88929/.

[24] Larry Edson. 2008. The GMW3172 Users Guide. The Electrical Validation Engineers Handbook Series. Electrical Component Testing. https://ab-div-bdi-bl-blm.web.cern.ch/ab-div-bdi-bl-blm/RAMS/Handbook_testing.pdf.

[25] Enrique Castillo, Alfonso Fernández-Canteli, Roland Koller, María Luisa Ruiz-Ripoll and Alvaro García. 2009. A statistical fatigue model covering the tension and compression Wöhler fields. Probabilistic Engineering Mechanics 24, 199–209. doi:10.1016/j.probengmech.2008.06.003.
[26] Yun Li Lee, Jwo Pan, Richard Hathaway and Mark Barkey. 2005. Fatigue testing and analysis, Theory and practice. Elsevier Butter Worth Heineman, New York. ISBN:0-7506-7719-8.
[27] Nakagawa, T. 2007. Shock and Damage Models in Reliability Theory, vol. 54. Springer-Verlag: London. DOI:10.1007/978-1-84628-442-7.
[28] Marathe, R.R. and Ryan, S.M. 2005. On the validity of the geometric Brownian motion assumption. The Engineering Economist, 50: 159–192. doi:10.1080/00137910590949904.
[29] Erwin, V. Zaretsky. 2013. Rolling Bearing Life Prediction, Theory and Application (NA SA/TP—2013-215305) National Aeronautics and Space Administration Glenn Research Center, Cleveland, Ohio 44135. Available electronically at http://www.sti.nasa.gov.
[30] Palmgren, A. 1924. The Service Life of Ball Bearings. Z. Ver. Deut. Ingr. (NASA TT F–13460), 68(14): 339–341.
[31] Lundberg, G. and Palmgren, A. 1947. Dynamic Capacity of Rolling Bearings. Acta Polytech. Mech. Eng. Ser., 1(3).

Index

A

Algorithms for Warehouses 167–169
Ambient Intelligence 47, 48
applied artificial intelligence 161
Aquaculture industry 186

B

Big data applied to the Automotive Industry 232
breast cancer 1–3

C

Caloric Burning 10–12, 15, 16, 18
Capability indices 250
Children color blindness 81
classification 216, 218, 221, 222, 226, 227
Clinical Dashboard 47, 48
Cognitive Architecture 89, 90, 95, 96, 98–104
Cognitive Innovation Model for Smart Cities 89
Color blindness 75–81, 87
conservation 28–33
crowd simulation 147, 154

D

data analytics 48, 108, 114
data indexing 112
deep learning 1–4
Diabetes 47–52, 66, 69, 73, 74
Diabetes Complications 47
distribution 23, 26, 27, 29–31, 33

E

E-governance 162
Electronic colorblindness 75
Evaluation of business leaders 207–209

F

feature extraction 252–254
floods 135, 136, 144
Fuzzy Logic Type 2 for decision makings 229

H

Human Avalanche 117, 123, 133
humanitarian logistics 135, 145–147

I

ICT for industrial 203
Industrial IoT 229
Industry 4.0 203, 235, 236

K

KMoS-RE 90, 96–101, 105
Knowledge Management 90, 96, 98, 105
Koi Fish 187, 198, 199, 201

M

Mechanical design 242
menge simulation 138, 139, 144, 146, 147, 149, 151, 154
Mental workload 34, 39
Metabolic Equivalent 11, 12, 15
micro-enterprises 25, 26, 31
Mobile APP 47, 49, 51, 52, 57, 59, 66, 69, 73, 74
mobile cooler 28–30
Monitoring 47–49, 69, 74
multi agent tool 117
Multicriteria Analysis 47, 49
myoelectric signals 216, 227

O

Order Picking Heuristics 176–180

P

pattern recognition and deep learning
perishable foods 22–24, 31, 32
Principal components 251–253, 255
processing 216, 218, 219, 221–223, 227

Q

Quadratic form 235–237, 239, 240, 242, 247, 264
Quality improvement 249

R

Random vibration 236, 253, 257
Reliability 235, 236, 238, 239, 243–245, 247, 248, 250, 251, 253, 256, 257, 259, 261–263
rugby football 117, 119

S

Serious Game 10–12, 14–20
shelf life 23, 24

simulation 117, 118, 122, 123, 126–133
Smart Cities 1, 89, 90, 111, 155, 156, 158–160, 162
Smart Manufacturing 229, 235
social data mining and multivariable analysis 19
Statistical multivariate control process 252
strategic design 38
stress in bus drivers 39, 40

T

Taguchi method 243
Technological competencies in the industry 203, 205, 206

U

urban computing 107, 108, 110–112, 114, 115

W

Weibull fatigue analysis 262